# NETWORKS ON CHIPS

**The Morgan Kaufmann Series in Systems on Silicon**
*Series Editor: Wayne Wolf, Princeton University*

The rapid growth of silicon technology and the demands of applications are increasingly forcing electronics designers to take a systems-oriented approach to design. This has lead to new challenges in design methodology, design automation, manufacture and test. The main challenges are to enhance designer productivity and to achieve correctness on the first pass. *The Morgan Kaufmann Series in Systems on Silicon* presents high-quality, peer-reviewed books authored by leading experts in the field who are uniquely qualified to address these issues.

*The Designer's Guide to VHDL, Second Edition*
Peter J. Ashenden

*The System Designer's Guide to VHDL-AMS*
Peter J. Ashenden, Gregory D. Peterson and Darrell A. Teegarden

*Readings in Hardware/Software Co-Design*
Edited by Giovanni De Micheli, Rolf Ernst and Wayne Wolf

*Modeling Embedded Systems and SoCs*
Axel Jantsch

*ASIC and FPGA Verification: A Guide to Component Modeling*
Richard Munden

*Multiprocessor Systems-on-Chips*
Edited by Ahmed Amine Jerraya and Wayne Wolf

*Comprehensive Functional Verification*
Bruce Wile, John Goss and Wolfgang Roesner

*Customizable Embedded Processors: Design Technologies and Applications*
Edited by Paolo Ienne and Rainer Leupers

*Networks on Chips: Technology and Tools*
Giovanni De Micheli and Luca Benini

*Designing SOCs with Configured Cores: Unleashing the Tensilica Diamond Cores*
Steve Leibson

*VLSI Test Principles and Architectures: Design for Testability*
Edited by Laung-Terng Wang, Cheng-Wen Wu, and Xiaoqing Wen

**Contact Information**

Charles B. Glaser
Senior Acquisitions Editor
Elsevier
(Morgan Kaufmann; Academic Press; Newnes)
(781) 313-4732
*c.glaser@elsevier.com*
*http://www.books.elsevier.com*

Wayne Wolf
Professor
Electrical Engineering, Princeton University
(609) 258 1424
*wolf@princeton.edu*
*http://www.ee.princeton.edu/~wolf/*

# NETWORKS ON CHIPS:
## TECHNOLOGY AND TOOLS

Luca Benini and Giovanni De Micheli

AMSTERDAM • BOSTON • HEIDELBERG • LONDON
NEW YORK • OXFORD • PARIS • SAN DIEGO
SAN FRANCISCO • SINGAPORE • SYDNEY • TOKYO
Morgan Kaufmann is an imprint of Elsevier

| | |
|---|---|
| Publishing Director | Denise Penrose |
| Senior Acquisitions Editor | Charles B. Glaser |
| Publishing Services Manager | George Morrison |
| Project Manager | Brandy Lilly |
| Assistant Editor | Michele Cronin |
| Composition | Charon Tec |
| Interior printer | The Maple-Vail Book Manufacturing Group |
| Cover printer | Phoenix Color |

Morgan Kaufmann Publishers is an imprint of Elsevier.
500 Sansome Street, Suite 400, San Francisco, CA 94111

This book is printed on acid-free paper.

**Library of Congress Cataloging-in-Publication Data**
Benini, Luca, 1967–
  Networks on chips : technology and tools / Luca Benini and Giovanni De Micheli.
    p. cm.—(The Morgan Kaufmann series in systems on silicon)
  Includes bibliographical references and index.
  ISBN-13: 978-0-12-370521-1 (casebound : alk. paper)
  ISBN-10: 0-12-370521-5 (casebound : alk. paper)
  1. Systems on a chip. 2. Computer networks—Equipment and supplies.
I. De Micheli, Giovanni. II. Title.
  TK7895.E42B45 2006
  621.3815—dc22                                                    2006018657

ISBN-13: 978-0-12-370521-1
ISBN-10: 0-12-370521-5

For information on all Morgan Kaufmann publications,
visit our Web site at www.mkp.com or www.books.elsevier.com

Typeset by Charon Tec Ltd, Chennai, India
www.charontec.com

Printed and bound by CPI Group (UK) Ltd, Croydon, CR0 4YY

Transferred to Digital Print 2011

# CONTENTS

# ABOUT THE AUTHORS

**Luca Benini** is Professor at the Department of Electrical Engineering and Computer Science (DEIS) of the University of Bologna. He also holds a visiting faculty position at the Ecole Polytechnique Federale de Lausanne. He held position at Stanford University and the Hewlett-Packard Laboratories. He received a Ph.D. degree in electrical engineering from Stanford University in 1997.

Dr. Benini's research interests are in the design and computer-aided design of integrated circuits, architectures and system, with special emphasis on low-power applications. He has published more than 270 papers in peer-reviewed international journals and conferences, three books and several book chapters.

He has been Program Chair and Vice-chair of Design Automation and Test in Europe Conference. He has been a member of the technical program committee and organizing committee of a number of technical conferences, including the *Design Automation Conference, International Symposium on Low Power Design, the Symposium on Hardware–Software Codesign*.

He is Associate Editor of the *IEEE Transactions on Computer Aided Design of Circuits and Systems and the ACM Journal on Emerging Technologies in Computing Systems*. He is a Senior Member of the IEEE.

**Giovanni De Micheli** (Ph.D. U. Berkely 1983) is Professor and Director of the Integrated Systems Centre at EPF Lausanne, Switzerland, and President of the Scientific Committee of CSEM, Neuchatel, Switzerland. Previously, he was Professor of Electrical Engineering at Stanford University.

His research interests include several aspects of design technologies for integrated circuits and systems, with particular emphasis on systems and networks on chips, design technologies and low-power design. He is author of Synthesis and Optimization of Digital Circuits, McGraw-Hill, 1994, co-author and/or co-editor of six other books and of over 300 technical articles. He is, or has been, member of the technical advisory board of several companies, including Magma Design Automation, Coware, IROC Technologies, Ambit Design Systems and STMicroelectronics.

Dr. De Micheli is the recipient of the 2003 IEEE Emanuel Piore Award for contributions to computer-aided synthesis of digital systems. He is a Fellow of ACM and IEEE. He received the Golden Jubilee Medal for outstanding contributions to the IEEE CAS Society in 2000. He received the 1987 D. Pederson Award for the best paper on the IEEE Transactions on CAD/ICAS and two Best Paper Awards at the Design Automation Conference, in 1983 and in 1993, and a Best Paper Award at the Date Conference in 2005.

He was President of the IEEE CAS Society (2003) and President Elect of the IEEE Council on EDA (2005–2006). He was Editor in Chief of the IEEE Transactions on CAD/ICAS in 1987–2001, Program Chair (1996–1997) and General Chair (2000) of the Design Automation Conference (DAC), Program (1988) and General Chair (1989) of the International Conference on Computer Design (ICCD), Program Chair of pHealth (2006) and VLSI System on Chip Conference (2006).

# LIST OF CONTRIBUTORS

**Davide Bertozzi**
*Engineering Department, University of Ferrara,*
*Ferrara, Italy*

**Israel Cidon**
*Technion – Israel Institute of Technology*
*Technion City, Haifa, Israel*

**Kees Goossens**
*Philips Research Laboratories*
*Eindhoven, The Netherlands*

**Hoi-Jun Yoo, Kangmin Lee, Se-Joong Lee and Kwanho Kim**
*KAIST – Korea Advanced Institute of Science and Technology*
*Daejeon, Republic of Korea*

**Srinivasan Murali**
*Stanford University*
*Palo Alto, California, USA*

# CHAPTER 1

# NETWORKS ON CHIP

The reason for the growing interest in *networks on chips* (NoCs) can be explained by looking at the evolution of integrated circuit technology and at the ever-increasing requirements on electronic systems. The integrated microprocessor has been a landmark in the evolution of computing technology. Whereas it took monstrous efforts to be completed, it appears now as a simple object to us. Indeed, the microprocessor involved the connection of a computational engine to a layered memory system, and this was achieved using busses. In the last decade, the frontiers of integrated circuit design opened widely. On one side, complex *application-specific integrated circuits* (ASICs) were designed to address-specific applications, for example mobile telephony. These systems required multiprocessing over heterogeneous functional units, thus requiring efficient on-chip communication. On another side, multiprocessing platforms were developed to address high-performance computation, such as image rendering. Examples are Sony's emotion engine [25] and IBM's cell chip [26], where on-chip communication efficiency is key to the overall system performance.

At the same time, the shrinking of processing technology in the *deep submicron* (DSM) domain exacerbated the unbalance between gate delays and wire delays on chip. Accurate physical design became the bottleneck for *design closure*, a word in jargon to indicate the ability to conclude successfully a tape out. Thus, the on-chip interconnection is now the dominant factor in determining performance. Architecting the interconnect level at a higher abstraction level is a key factor for system design.

We have to understand the introduction of NoCs in *systems-on-chip* (SoCs) design as a gradual process, namely as an evolution of bus interconnect technology. For example, there is not a strict distinction between multi-layer busses and crossbar NoCs. We have also to credit C. Seitz and W. Dally for stressing the need of network interconnect for high-performance multiprocessing, and for realizing the first prototypes of networked integrated multiprocessors [9]. But overall, NoC has become a broad topic of research and development in the new millennium, when designers were confronted with technological limitations, rising hardware design costs and increasingly higher system complexity.

■ **FIGURE 1.1**

Traffic pattern in a large-scale system. Limited parallelism is often a cause of congestion.

## 1.1   WHY ON-CHIP NETWORKING?

Systems on silicon have a complexity comparable to skyscrapers or aircraft carriers, when measured in terms of number of basic elements. Differently from other complex systems, they can be cloned in a straightforward way but they have to be designed in correctly, as repairs are nearly impossible. SoCs require design methodologies that have commonalities with other types of large-scale system design (Fig. 1.1). In particular, when looking at on-chip interconnect design methods, it is useful to compare the on-chip interconnect to the worldwide interconnect provided by the Internet. The latter is capable of taming the system complexity and of providing reliable service in presence of local malfunctions. Thus, networking technology has been able to provide us with *quality of service* (QoS), despite the heterogeneity and variability of the Internet nodes and links. It is then obvious that networking technology can be instrumental for the bettering of *very-large-scale integration* (VLSI) circuit/system design technology.

On the other hand, the challenges in marrying network and VLSI technologies are in leveraging the essential features of networking that are crucial to obtaining fast and reliable on-chip communication. Some novices think that on-chip networking equates to porting the *Transmission*

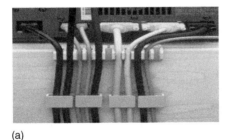

(a)                                    (b)

■ **FIGURE 1.2**

Distributed systems communicate via a limited number of cables. Conversely, VLSI chips use up to 10 levels of wires for communicating.

*Control Protocol/Internet Protocol* (TCP/IP) to silicon or achieving an on-chip Internet. This is not feasible, due to the high latency related to the complexity of TCP/IP. On-chip communication must be fast, and thus networking techniques must be simple and effective. Bandwidth, latency and energy consumption for communication must be traded off in the search for the best solution.

On the bright side, VLSI chips have wide availability of wires on many layers, which can be used to carry data and control information. Wide data busses realize the parallel transport of information. Moreover, data and control do not need to be transported by the same means, as in networked computers (Fig. 1.2). Local proximity of computational and storage unit on chip makes transport extremely fast. Overall, the wire-oriented nature of VLSI chips makes on-chip networking both an opportunity and a challenge.

In summary, the main motivation for using on-chip networking is to achieve performance using a system perspective of communication. This reason is corroborated by the fact that simple on-chip communication solutions do not scale up when the number of processing and storage arrays on chip increases. For example, on-chip busses can serve a limited number of units, and beyond that, performance degrades due to the bus parasitic capacitance and the complexity of arbitration. Indeed, it is the trend to larger-scale on-chip multiprocessing that demands on-chip networking solutions.

## 1.2  TECHNOLOGY TRENDS

In the current projections [37] of future silicon technologies, the operating frequency and transistor density will continue to grow, making energy dissipation and heat extraction a major concern. At the same time, on-chip supply voltages will continue to decrease, with adverse impact on signal integrity. The voltage reduction, even though beneficial, will not suffice to mitigate the energy consumption problem, where a major contribution is due to leakage. Thus, SoCs will incorporate *dynamic power management* (DPM) techniques in various forms to satisfy energy consumption bounds [4].

Global wires, connecting different functional units, are likely to have propagation delays largely exceeding the clock period [18]. Whereas signal pipelining on interconnections will become common practice, correct design will require knowing the signal delay with reasonable accuracy. Indeed, a negative side effect of technology downsizing will be the spreading of physical parameters (e.g., variance of wire delay per unit length) and its relative importance as compared to the timing reference signals (e.g., clock period).

The spreading of physical parameters will make it harder to achieve high-performing chips that safely meet all timing constraints. Worst-case timing methodologies, that require clocking period larger than the worst-case propagation delay, may underuse the potentials of the technology, especially when the worst-case propagation delays are rare events. Moreover, it is likely that varying on-chip temperature profiles (due to varying loads and DPM) will increase the spread of wiring delays [2]. Thus, it will be mandatory to go beyond worst-case design methodology, and use fault-tolerant schemes that can recover from timing errors [11, 30, 35].

Most large SoCs are designed using different voltage *islands* [23], which are regions with specific voltage and operation frequencies, which in turn may depend on the workload and dynamic voltage and frequency scaling. Synchronization among these islands may become extremely hard to achieve, due to timing skews and spreads. Global wires will span multiple clock domains, and synchronization failures in communicating between different clock domains will be rare but unavoidable events [12].

### 1.2.1 Signal Integrity

With forthcoming technologies, it will be harder to guarantee error-free information transfer (at the electrical level) on wires because of several reasons:

- Reduced *signal swings* with a corresponding reduction of voltage noise margins.

- *Crosstalk* is bound to increase, and the complexity of avoiding crosstalk by identifying all potential on-chip noise sources will make it unlikely to succeed fully.

- *Electromagnetic interference* (EMI) by external sources will become more of a threat because of the smaller voltage swings and smaller dynamic storage capacitances.

■ The probability of occasional *synchronization failures* and/or metastability will rise. These erroneous conditions are possible during system operation because of transmission speed changes, local clock frequency changes, timing noise (jitter), etc.

■ *Soft errors* due to collision of thermal neutrons (produced by the decay of cosmic ray showers) and/or alpha particles (emitted by impurities in the package). Soft errors can create spurious pulses, which can affect signals on chip and/or discharge dynamic storage capacitances.

Moreover, SoCs may be willfully operated in error-prone operating conditions because of the need of extending battery lifetime by lowering energy consumption via supply voltage over-reduction. Thus, specific run-time policies may trade-off signal integrity for energy consumption reduction, thus exacerbating the problems due to the fabrication technology.

### 1.2.2 Reliability

System-level *reliability* is the probability that the system will operate correctly at time, *t*, as a function of time. The expected value of the reliability function is the *mean time to failure* (MTTF). Increasing MTTF well beyond the expected useful life of a product is an important design criterion. Highly reliable systems have been object of study for many years. Beyond traditional applications, such as aircraft control, defense applications and reliable computing, there are many new fields requiring high-reliable SoCs, ranging from medical applications to automotive control and more generally to embedded systems that are critical for human operation and life.

The increased demand of high-reliable SoCs is counterbalanced by the increased failure rates of devices and interconnects. Due to technology downscaling, failures in the interconnect due to electromigration are more likely to happen (Fig. 1.3). Similarly, device failure due to dielectric breakdown is more likely because of higher electric fields and carrier speed (Fig. 1.4). Temperature cycles on chip induce mechanical stress, that has counter-productive effects [28].

For these reasons, SoCs need to be designed with specific resilience toward hard (i.e., permanent) and soft (i.e., transient) malfunctions. System-level solutions for hard errors involve redundancy, and thus require the on-line connection of a stand-by unit and disconnection of the faulty unit. Solutions for soft errors include design techniques for error containment, error detection and correction via encoding. Moreover, when soft errors induce timing errors, system based on double-latch clocking can be used for detection and correction. NoCs can provide resilient solutions toward hard errors (by supporting seamless connection/disconnection of units) and soft errors (by layered error correction).

■ **FIGURE 1.3**

Failure on a wire due to electromigration.

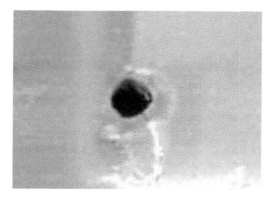

■ **FIGURE 1.4**

Failure on a transistor due to oxide breakdown.

### 1.2.3  Non-determinism in SoC Modeling and Design

As SoC complexity scales, it will be more difficult, if not impossible, to capture their functionality with fully deterministic models of operation. In other words, system models may have multiple implementations. Property abstraction, which is key to managing complexity in modeling and design, will hide implementation details and designers will have to relinquish control of such details.

Whereas abstract modeling and automated synthesis enables complex system design, such an approach increases the variability of the physical and electrical parameters. In summary, to ensure correct and safe realizations, the system architecture and design style have to be resilient against errors generated by various sources, including:

■  process technology (parameter spreading, defect density, failure rates);

- environment (temperature variation, EMI, radiation);

- operation mode (very-low-voltage operation);

- design style (abstraction and synthesis from non-deterministic models).

### 1.2.4 Variability, Design Methodologies and NoCs

Dealing with *variability* is an important matter affecting many aspects of SoC design. We consider here a few aspects related to on-chip communication design.

The first important issue deals with malfunction containment. Traditionally, malfunctions have been avoided by putting stringent rules on physical design and by applying stringent tests on signal integrity before tape out. Rules are such that variations of process parameters can be tolerated, and integrity analysis can detect potential problems such as crosstalk. This approach is conservative in nature, and leads to perfecting the physical layout of circuits. On the other hand, the downscaling of technologies has unveiled many potential problems and as a result the physical design tools have grown in complexity, cost and time to achieve design closure. At some point, correct-by-construction design at the physical level will no longer be possible. Similarly, the increasingly larger amount of connections on chip will make signal integrity analysis unlikely to detect all potential crosstalk errors.

Future trends will soften requirements at the physical and electrical level, and require higher-level mechanisms for error correction. Thus, electrical errors will be considered inevitable. Nevertheless, their effect can be contained by techniques that correct them at the logic and functional levels. In other words, the error detection/correction paradigm applied to networking will become a standard tool in on-chip communication design.

Timing errors are an important side effect of variability. Timing errors can be originated by a wide variety of causes, including but not limited to: incorrect wiring delay estimate, overaggressive clocking, crosstalk and soft (radiation-induced) errors. Timing errors can be detected by double latches, gated by different clocking signals, and by comparing the latched data. When the data differs, it means that most likely the signal settled after the first latch was gated, that is, that a timing error was on the verge of being propagated. (Unfortunately, errors can happen also in the latch themselves.)

Asynchronous design methodologies can make the circuit resilient to delay variations. For example, speed-independent and delay-insensitive circuit families can operate correctly in presence of delay variations in gates and interconnects. Unfortunately, design complexity often make the application of an integral asynchronous design methodology impractical. A viable compromise is the use of *globally asynchronous locally*

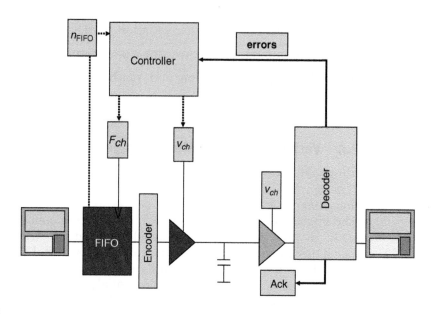

**▪ FIGURE 1.5**

The voltage swing on communication busses is reduced, even though signal integrity is partially compromised [35]. Encoding techniques are used to detect corrupted data which is retransmitted. The retransmission rate is an input to a closed-loop *dynamic voltage scaling* (DVS) control scheme, which sets the voltage swing at a trade-off point between energy saving and latency penalty (due to data retransmission).

*synchronous* (GALS) circuits, that use asynchronous handshaking protocols to link various synchronous domains possibly clocked at various frequencies.

NoCs are well poised to deal with variability because networking technology is layered and error detection, containment and correction can be done at various layers, according to the nature of the possible malfunction. There are several paradigms that deal with variability for NoCs. *Self-calibrating* circuits are circuits that adapt on-line to the operating conditions. There are several embodyments of self-calibrating circuits, as shown in Figs 1.5–1.7.

## 1.3  SoC OBJECTIVES AND NoC NEEDS

There are several hardware types of SoC designs that can be defined according to the required functionality and market. In general, SoCs can be classified in terms of their versatility (i.e., support for programming) and application domains. A simple taxonomy is described next:

- *General-purpose on-chip multiprocessors* are high-performance chips that benefit from spatial locality to achieve high performance.

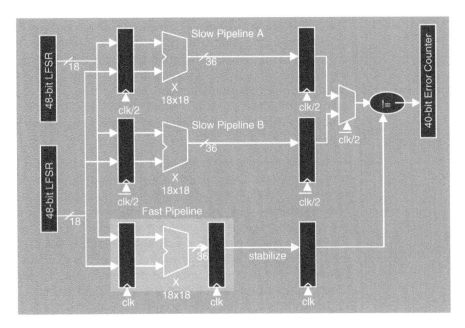

■ **FIGURE 1.6**

*Razor* [11] is another realization of self-calibrating circuits, where a processor's supply is lowered till errors occur. The correct operation of the processor is preserved by an error detection and pipeline adjustment technique. As a result, the processor settles on-line to an operating voltage which minimizes the energy consumption even in the presence of variation of technological parameters.

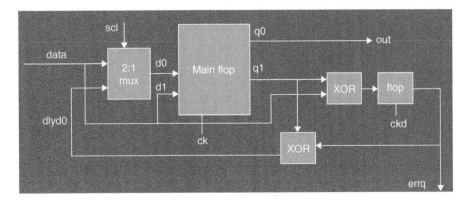

■ **FIGURE 1.7**

*T-error* is a timing methodology for NoCs where data is pipelined through double latches, where the former used an aggressive period and the latter a safe one. For most patterns, T-error will forward data from the first latch. When the slowest patterns are transmitted that fail the deadline at the first latch, correct but slower operation is performed by the second latch [30].

They are designed to support various applications, and thus the processor core usage and traffic patterns may vary widely. They are the evolution of on-board multiprocessors, and they are typified by having a homogeneous set of processing and storage arrays. For these reasons, on-chip network design can benefit from the experience on many architectures and techniques developed for on-board multiprocessors, with the appropriate adjustments to operate on a silicon substrate.

- *Application-specific SoCs* are hardware chips dedicated to an application. In some cases, as for all mobile applications, energy consumption is a major concern. Most application-specific SoCs are programmable, but their application domain is limited and the software characteristics are known *a priori*. Thus, some knowledge of the traffic pattern is available when the NoC is designed. In many cases, these systems contain fairly heterogeneous computing elements, such as processors, controllers, *digital signal processors* (DSPs) and a number of domain-specific hardware accelerators. This heterogeneity may lead to specific traffic patterns and requirements, thus requiring NoCs with specialized architectures and protocols.

- *SoC platforms* are application-specific SoCs dedicated to a family of applications in a specific domain. Examples are SoCs for GSM telephony support and platforms for automotive control. A platform is more versatile in nature, as it can be used in different (embedded) systems by different manufacturers. Thus, versatility and programmability are preferred to customization, yielding SoCs that can be produced in high volumes, and thus offset the *non-recurrent engineering* (NRE) costs. Whereas the processing and storage unit may differ in nature and performance, the traffic patterns are harder to guess *a priori* as the application software may vary widely.

- *Field-programmable gate arrays* (FPGAs) are hardware systems where the functionality is determined after manufacturing by connecting and configuring components. Components vary in size and in functionality and are connected by reprogrammable networks. These networks are simple and provide bit-level connectivity with little or no control. Nevertheless we expect FPGAs to grow substantially over the coming years and require effective NoC communication.

## 1.3.1  Some Design Examples

One of the first multiprocessor designed around an NoC is the RAW architecture [32]. This is a fully programmable SoC consisting of an array of identical computational tiles with local storage. Full programmability means that the compiler can program both the function of each tile and the interconnections among them. The name RAW stems from the fact that the "raw" hardware is fully exposed to the compiler. To accomplish programmable communication, each tile has a router. The compiler programs the routers on all tiles to issue a sequence of commands that determine exactly which set of wires connect at every cycle. Moreover, the compiler pipelines the long wires to support high clock frequency. The RAW architecture implementation is described in detail in Chapter 9.

The cell processor [26] was developed by Sony, Toshiba and IBM to build a general-purpose processor for a computer, even though it is primarily targeted for Sony's Playstation 3. Its architecture resembles multiprocessor vector supercomputers, targeting high-performance distributed computing. The architecture comprises one 64-bit *power processor element* (PPE), eight *synergistic processor elements* (SPEs), memory and interconnection. The PPE is a dual issue, dual threaded in-order RISC processor, with 512K cache. Each SPE is a self-contained in-order vector processor which acts as an independent processor. Each contains a $128 \times 128$-bit register, four (single-precision) floating point units and four integer units. The *element interconnection bus* (EIB) connects the PPE, the eight SPEs and the memory interface controller (Fig. 1.8). The EIB has independent networks for commands (requests for data from other sources) and for the data being moved. Commands are filtered through address concentrators which handle collision detection and prevention, and ensure that all units have equal access to the command bus. There are multiple address concentrators, all of which forward data to a single-serial command reflection point. Data transfer is elaborate. There are four "rings," each of which is a chain connecting all data ports. Data can move down a ring only in one direction. For instance, a connection that allows data to move from

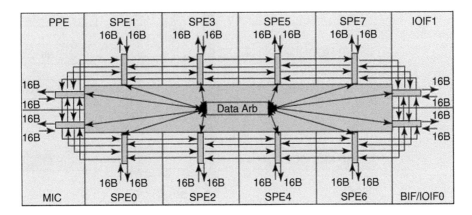

**■ FIGURE 1.8**

The EIB in cell.

the PPE to SPE1 cannot be used to move data from SPE1 back to the PPE. Two rings go clockwise and two counterclockwise, and all four rings have the components attached in the same order. Each ring can move 16 bytes at a time from any position on the ring to any other position. In fact, each ring can transmit three concurrent transfers, but those transfers cannot overlap.

The Nexperia architecture, developed by Philips, is a platform for handling digital video and audio in consumer electronics (Fig. 1.9). It uses one or more 32-bit MIPS CPUs for control processing, and one or more 32-bit Trimedia processors for streaming data. Moreover, the platform can house a flexible range of programmable modules, such as an MPEG decoder, a UART, etc. To connect the CPUs and other modules with each other and with the main external memory, a high-speed memory access network, and two *device control and status* (DCS) networks are used. These DCS networks enable each processor to control and observe on chip the status of the other modules. One of the advantages of the platform is the variable number of CPUs used, thus making Nexperia fit well various applications. A specific implementation of Nexperia, the PNX8550 system chip, houses 10 million gates in 62 cores, out of which five are hard (including the MIPS and Trimedia CPUs) and the others are soft cores [15].

The Xilinx Spartan-II FPGA chips are rectangular arrays of *configurable logic blocks* (CLBs). Each block can be programmed to perform a specific

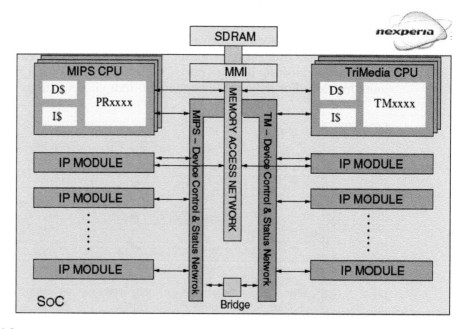

■ **FIGURE 1.9**

The Nexperia architecture.

logic function. CLBs are connected via a hierarchy of routing channels. A more complex and interesting family of products is the Xilinx Virtex-II and Virtex-II Pro. These FPGAs have various complex elements, such as CLBs, RAMs, processor cores, multipliers and clock managers. Programmable interconnection is achieved by routing switches. Each programmable element is connected to a switch matrix, allowing multiple connections to the general routing matrix. All programmable elements, including the routing resources, are controlled by values stored in static memory cells. Thus, Virtex-II can be also seen as NoC over a heterogeneous fabric of components.

The complexity of the chip designs described above has prompted the development of infrastructure to support communication. For example, STMicroelectronics has developed the STBus kit that can provide various functions including full (and partial) crossbar connection. A similar framework is provided by the *advanced microcontroller bus architecture* (AMBA) multi-layer bus system (see Chapter 8).

## 1.3.2  Distinguishing Characteristics of NoCs

SoCs differ from wide-area networks because of local proximity and because they exhibit much less non-determinism. Indeed, despite the undesirable variability features of **DSM CMOS** technologies, it is still possible to predict many physical and electrical parameters with reasonable accuracy.

On the other hand, on-chip networks have a few distinctive characteristics, namely *low communication latency*, *energy consumption constraints* and *design-time specialization*. Latency of communication on chip needs to be small, that is, in the order of few clock periods. The shortest latency implementations can be achieved by fully hard-wired implementations, which defeat the flexibility required by on-chip networks. Clearly, smart protocols for communication may add to the latency of the signals. Thus, to be competitive in performance, NoCs require streamlined protocols.

Energy consumption in NoCs is often a major concern, because whereas computation and storage energy greatly benefits from device scaling (smaller gates and smaller memory cells), the energy for global communication does not scale down. On the contrary, projections based on current delay optimization techniques for global wires [18, 29, 31] show that global communication on chip will require increasingly higher energy consumption. Hence, communication-energy minimization will be a growing concern in future technologies. Furthermore, network traffic control and monitoring can help in better managing the power consumed by networked computational resources. For instance, clock speed and voltage of end nodes can be varied according to available network bandwidth.

Design-time specialization is another facet of NoC design, and it is relevant to application-specific and platform SoCs. Whereas macroscopic networks emphasize general-purpose communication and modularity, in

NoCs these constraints are less restrictive because most on-chip solutions are proprietary. Thus, NoC implementation may separate data from control, use arbitrary bus width and control flow schemes. Such a flexibility needs to be mitigated at the NoC boundary, that is, where the communication infrastructure connects to end nodes (e.g., processors). Existing standards like the *Open Core Protocol* (OCP) are extremely useful in defining the interface between processor/storage arrays and NoCs. Interestingly enough, the flexibility in tailoring the NoC to the specific application can be used effectively to design low-energy communication schemes.

## 1.4   ONCE OVER LIGHTLY

The topic of NoC design is vast and complex. For this reason, we want to cover once over lightly the issues related to NoC design. This will give the reader a short plan of the upcoming chapters, where NoC protocols are described bottom-up, starting from the physical up to the software layer. Moreover, since some design considerations are out of order with respect to the bottom-up description, this section will help the reader making the connection among issues at various abstraction layers and described in different chapters. This section will also help to put in perspectives the various topics of this book and to summarize them.

### 1.4.1   NoC Architectures

There is a large literature on architectures for local- and wide-area networks and more specifically for single-chip multiprocessors [1, 5, 6, 13, 20, 33]. These architectures can be classified by their topology, structure and parameters. The most common on-chip communication architecture is the shared-medium architecture, as exemplified by the shared bus. Unfortunately, bus performance and energy consumption are penalized when the number of end nodes scales up. Point-to-point architectures such as *mesh*, *torus* and *hypercube* have been shown to scale up despite a higher complexity in their design. There are a few *ad hoc* NoC architectures, such as *Octagon* [20], which have been designed for silicon implementation and have no counterpart in macroscopic networks.

Some properties of NoC architectures depend on the topology, while some others depend on specific choices of protocols. For example, *deadlock*, *livelock* and *starvation* depend also on the forwarding scheme and routing algorithms chosen for the specific architecture. NoC architectures are described in Chapter 2. NoC protocols are typically organized in layers, in a fashion that resembles the OSI protocol stack (Fig. 1.10). They are described in Chapters 3–7, and we summarize next the major issues at the physical, data-link, network and transport, and software layers.

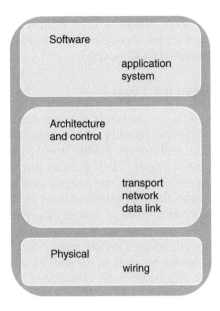

Micro-network stack.

### 1.4.2  Physical Layer

Global wires are the physical implementation of the communication channels. Physical layer signaling techniques for lossy transmission lines have
been studied for a long time by high-speed board designers and microwave
engineers [3, 12]. Traditional rail-to-rail voltage signaling with capacitive
termination, as used today for on-chip communication, is definitely not
well suited for high-speed, low-energy communication on future global
interconnect [12]. Reduced swing, current-mode transmission, as used in
some processor-memory systems, can significantly reduce communication
power dissipation while preserving speed of data communication.

Nevertheless, as the technology trends lead us to use smaller voltage
swings and capacitances, the error probabilities will increase. Thus, the
trend toward faster and lower-power communication may decrease reliability as an unfortunate side effect. Reliability bounds as voltages scale
can be derived from theoretical (entropic) considerations [17] and can be
measured also by experiments on real circuits.

NoCs support a design paradigm shift. Former SoC design styles consider wiring-related effects as undesirable parasitics, and try to reduce or
cancel them by specific and detailed physical design techniques. Recent
and future NoC design will de-emphasize physical design, by allowing circuits to produce errors. Errors will be contained, detected and corrected
at the data-link level, by using, for example, *error correcting codes* (ECCs).
This is possible in NoCs due to the layering of protocols. Thus, emphasis

on physical layer design will be mainly on signal drivers and receivers, as well as design technologies for restoring and pipelining signals. These issues are covered in Chapter 3.

### 1.4.3   Data-Link Layer

The *data-link layer* abstracts the physical layer as an unreliable digital link, where the probability of bit errors is small but not negligible (and increasing as technology scales down). Furthermore, reliability can be traded off for energy [17]. The main purpose of data-link protocols is to increase the reliability of the link up to a minimum required level, under the assumption that the physical layer by itself is not sufficiently reliable.

An additional source of errors is contention in shared-medium networks. Contention resolution is fundamentally a non-deterministic process because it requires synchronization of a distributed system, and for this reason it can be seen as an additional noise source. In general, non-determinism can be virtually eliminated at the price of some performance penalty. For instance, centralized bus arbitration in a synchronous bus eliminates contention-induced errors, at the price of a substantial performance penalty caused by the slow bus clock and by bus request/release cycles.

Future high-performance shared-medium on-chip micro-networks may evolve in the same direction as high-speed local-area networks, where contention for a shared communication channel can cause errors because two or more transmitters are allowed to concurrently send data on a shared medium. In this case, provisions must be made for dealing with contention-induced errors.

An effective way to deal with errors in communication is to *packetize* data. If data is sent on an unreliable channel in packets, error containment and recovery are easier, because the effect of errors is contained by packet boundaries, and error recovery can be carried out on a packet-by-packet basis. At the data-link layer, error correction can be achieved by using standard ECCs that add redundancy to the transferred information. Error correction can be complemented by several packet-based error detection and recovery protocols. Several parameters in these protocols (e.g., packet size, number of outstanding packets, etc.) can be adjusted depending on the goal to achieve maximum performance at a specified residual error probability and/or within given energy consumption bounds. Data-link issues, including arbitration strategies, will be covered in Chapter 4.

### 1.4.4   Network and Transport Layers

At the *network layer*, packetized data transmission can be customized by the choice of *switching* and *routing* algorithms. The former establishes the type of connection while the latter determines the path followed by a message through the network to its final destination. Popular packet switching techniques include *store-and-forward* (SAF), *virtual cut-through* (VCT) and *wormhole*. SAF forwarding inspects each packet's content before forwarding it to the next stage. While SAF enables more elaborated routing algorithms (e.g., content-aware packet routing), it introduces extra packet delay at every router stage. Furthermore, SAF also requires a substantial amount of buffer spaces because the switches need to store multiple complete packets at the same time. The VCT scheme can forward a packet to the next stage before its entirety is received by the current switch. Therefore, VCT switching reduces the delay, as compared to SAF, but it still requires storage because when the next stage switch is not available, the entire packet still needs to be stored in the buffers.

With wormhole switching, each packet is further segmented into *flits* (flow control unit). The header flit reserves the routing channel of each switch, the body flits will then follow the reserved channel and the tail flit will later release the channel reservation. One major advantage of wormhole switching is that it does not require the complete packet to be stored in the switch while waiting for the header flit to route to the next stages. Thus, wormhole switching not only reduces the SAF delay at each switch, but it also requires much smaller buffer spaces. On the other hand, with wormhole, one packet may occupy several intermediate switches at the same time. Thus, it may block the transmission of other packets. *Deadlock* and *livelock* are the potential problems in wormhole schemes [8, 13]. A number of recently proposed NoC prototype implementations are indeed based on wormhole packet switching [1, 7, 10, 16].

Switching is tightly coupled to routing. Routing algorithms establish the path followed by a message through the network to its final destination. The classification, evaluation and comparison of on-chip routing schemes [13] involve the analysis of several trade-offs, such as predictability versus average performance, router complexity and speed versus achievable channel utilization, and robustness versus aggressiveness. A coarse distinction can be made between *deterministic* and *adaptive* routing algorithms. Deterministic approaches always supply the same path between a given source–destination pair, and they are the best choice for uniform or regular traffic patterns. In contrast, adaptive approaches use information on network traffic and channel conditions to avoid congested regions of the network. They are preferable in presence of irregular traffic or in networks with unreliable nodes and links. Among other routing schemes, probabilistic broadcast algorithms [14] have been proposed for NoCs.

At the *transport layer*, algorithms deal with the decomposition of messages into packets at the source and their assembly at destination. Packetization granularity is a critical design decision because the behavior of most network control algorithms is very sensitive to packet size. Packet size can be application specific in SoCs, as opposed to general networks. In general, flow control and negotiation can be based on either deterministic or statistical procedures. Deterministic approaches ensure that traffic meets specifications, and provide hard bounds on delays or message losses. The main disadvantage of deterministic techniques is that they are based on worst cases, and they generally lead to significant under-utilization of network resources. Statistical techniques are more efficient in terms of utilization, but they cannot provide worst-case guarantees. Similarly, from an energy viewpoint, we expect deterministic schemes to be more inefficient than statistical schemes because of their implicit worst-case assumptions. Network and transport issues are described in Chapter 5.

### 1.4.5  Software Layers

Current and future SoCs will be highly programmable, and therefore their power consumption will critically depend on software aspects. Software layers comprise system and application software. The system software provides us with an abstraction of the underlying hardware platform, which can be leveraged by the application developer to safely and effectively exploit the hardware's capabilities. The *hardware abstraction layer* (HAL) is described in Chapter 6, and it is tightly coupled to the design of wrappers for processor cores, that act as *network interfaces* between processing cores and the NoC.

Current SoC software development platforms are mostly geared toward single microcontroller with multiple coprocessors architectures. Most of the system software runs on the control processor, which orchestrates the system activity and farms off computationally intensive tasks to domain-specific coprocessors. Microcontroller–coprocessor communication is usually not data-intensive (e.g., synchronization and reconfiguration information), and most high-bandwidth data communication (e.g., coprocessor–coprocessor and coprocessor–IO) is performed via shared memories and *direct memory access* (DMA) transfers. The orchestration activities in the microcontroller are performed via run-time services provided by single-processor RTOSes (e.g., VxWorks, Micro-OS, Embedded Linuxes, etc.), which differentiate from standard operating systems in their enhanced modularity, reduced memory footprint, and support for real-time scheduling and bounded time interrupt service times.

Application programming is mostly based on manual partitioning and distribution of the most computationally intensive kernels to data coprocessors (e.g., VLIW multimedia engines, DSPs, etc.). After partitioning, different code generation and optimization tool chains are used for each target coprocessor and the control processor. Hand-optimization at the assembly level is still quite common for highly irregular signal processors, while advanced optimizing compilers are often used for VLIW engines and fine-grained reconfigurable fabrics. Explicit communication via shared memory is usually supported via storage classes declarations (e.g., non-cacheable memory pages) and DMA transfers from and to shared memories are usually set up via specialized system calls which access the memory-mapped control registers of the DMA engines.

Even though the standard single-processor software design flows have been adapted to deal with architectures with some degree of parallelism, NoCs are communication-dominated architectures. Thus, they require much more fundamental work on software abstraction and computer-aided software development frameworks. On one hand, more aggressive and effective techniques for automatic parallelism discovery and exploitations are needed; on the other hand however, programming languages and environments should enable explicit description of parallel computation without obfuscating functional specification behind the complexity of low-level parallelism management tasks. In our view, software issues are among the most critical and less understood in NoC. We believe that the full potential of on-chip networks can be effectively exploited only if adequate software abstractions and programming aids are developed to support them. Software issues for NoCs are described in Chapter 7.

### 1.4.6  NoC Design Tools and Design Examples

Designing NoCs requires specialized environments and tools. On one hand, *analysis* tools are important to evaluate the performance of an NoC of interest, as well as to trade-off alternative implementations and tune parameters. On the other hand, *synthesis* of NoCs is important, in view of the possibility of using *ad hoc* architectures and of customizing NoCs for given applications. Synthesis tools aim at taking abstract, high-level views of network topologies and protocols, and at generating an implementation instance. This instance can be a high-level functional implementation of an NoC; yet it may be used as input to other, lower level and more standard, synthesis tools. It is important to note that the relation between the functional view and the physical view is important in both analysis and synthesis. Indeed the SoC floorplan determines wiring lengths and delays that represent constraints for the NoC realization. Analysis and synthesis tools are described in Chapter 8, while Chapter 9 is devoted to the discussion of some recent NoC realizations in silicon, to their advantages and limitations.

## 1.5  PERSPECTIVES

NoCs were conceived with the goal of boosting system performance by taking a system view of on-chip communication and by rationalizing communication resources. At the same time, NoCs promise to solve some of the problems of DSM technology, by providing a means to deal with signal delay variability and with unreliability of the physical interconnect. NoCs are also well poised to realize the communication infrastructure for SoCs with tens or hundreds of computational cores, where parallel processing creates a significant traffic that needs to be managed with run-time techniques. NoCs can be thought as the evolution of high-performance busses, which today are available in multi-layer, multi-level instantiations. Yet NoCs require a new vision of SoC design, which incorporates fault tolerance and layering.

A few advanced chips use NoCs as communication substrate, and it is conjectured that NoCs will be the backbone of all SoCs of significant complexity designed with the 65 nm technology node and below. NoCs will be an integral part of SoC platforms dedicated to specific application domains, and programming platforms with NoCs will be simpler due to regularity and predictability. Moreover, FPGAs represent a large and ever-increasing sector of the semiconductor market. FPGAs represent another embodyment of NoCs. Within advanced FPGA architectures, both hardware computing elements and their interconnects are programmable, thus achieving an unprecedented flexibility.

Some design and implementation issues are still in search of efficient solutions. Problems get harder as we move up from the physical layer because most hardware design problems are well understood while system software issues, from handling massive parallelism to generating predictable code are still being investigated. It is the purpose of this book to shed some light on this important and timely technology, and to gather together results and experiences of several researchers.

## REFERENCES

[1] A. Adriahantenaina, H. Charlery, A. Greiner, L. Mortiezand and C. Zeferino, "SPIN: A Scalable, Packet Switched, On-Chip Micro-network,"*DATE – Design, Automation and Test in Europe Conference and Exhibition*, 2003, pp. 70–73 .

[2] A.H. Ajami, K. Banerjee and M. Pedram, "Modeling and Analysis of Nonuniform Substrate Temperature Effects on Global ULSI Interconnects," *IEEE Transactions on CAD*, Vol. 24, No. 6, June 2005, pp. 849–861.

[3] H. Bakoglu, *Circuits, Interconnections, and Packaging for VLSI*, Addison-Wesley, Upper Saddle River, NJ, 1990.

[4] L. Benini, A. Bogliolo and G. De Micheli, "A Survey of Design Techniques for System-Level Dynamic Power Management," *IEEE Transactions on Very Large-Scale Integration Systems*, Vol. 8, No. 3, June 2000, pp. 299–316.

[5] W.O. Cesario, D. Lyonnard, G. Nicolescu, Y. Paviot, S. Yoo, L. Gauthier, M. Diaz-Nava and A.A. Jerraya, "Multiprocessor SoC Platforms: A Component-Based Design Approach," *IEEE Design and Test of Computers*, Vol. 19, No. 6, November–December 2002, pp. 52–63.

[6] W. Dally and B. Towles, *Principles and Practices of Interconnection Networks*, Morgan Kaufmann, San Francisco, CA, 2004.

[7] W. Dally and B. Towles, "Route Packets, Not Wires: On-Chip Interconnection Networks," *Proceedings of the 38th Design Automation Conference.* 2001.

[8]   W.J. Dally and H. Aoki, "Deadlock-Free Adaptive Routing in Multicomputer Networks Using Virtual Channels," *IEEE Transactions on Parallel and Distributed Systems*, Vol. 4, No. 4, April 1993, pp. 466–475.

[9]   W. Dally and C. Seitz, "The Torus Routing Chip," *Distributed Processing*, Vol. 1, 1996, pp. 187–196.

[10]  M. Dall'Osso, G. Biccari, L. Giovannini, D. Bertozzi and L. Benini, "Xpipes: A Latency Insensitive Parameterized Network-on-Chip Architecture for Multiprocessor SoCs," *International Conference on Computer Design*, 2003, pp. 536–539.

[11]  D. Ernst, S. Das, S. Lee, D. Blaauw, T. Austin, T. Mudge, N. S. Kim and K. Flautner, "Razor: Circuit-Level Correction of Timing Errors for Low-Power Operation," *IEEE Micro*, Vol. 24, No. 6, November–December 2004, pp. 10–20.

[12]  W. Dally and J. Poulton, *Digital Systems Engineering*, Cambridge University Press, Cambridge, MA, 1998.

[13]  J. Duato, S. Yalamanchili and L. Ni, *Interconnection Networks: An Engineering Approach*, Morgan Kaufmann, San Francisco, CA, 2003.

[14]  T. Dumitra, S. Kerner and R. Marculescu, "Towards On-Chip Fault-Tolerant Communication," *ASPDAC – Proceedings of the Asian-South Pacific Design Automation Conference*, 2003, pp. 225–232.

[15]  S. Goel, K. Chiu, E. Marinissen, T. Nguyen and S. Oostdijk, "Test Infrastructure Design for the Nexperia Home Platform PNX8550 System Chip," *DATE – Proceedings of the Design Automation and Test Europe Conference*, 2004.

[16]  K. Goossens, J. van Meerbergen, A. Peeters and P. Wielage, "Networks on Silicon: Combining Best Efforts and Guaranteed Services," *Design Automation and Test in Europe Conference*, 2002, pp. 423–427.

[17]  R. Hegde and N. Shanbhag, "Toward Achieving Energy Efficiency in Presence of Deep Submicron Noise," *IEEE Transactions on VLSI Systems*, Vol. 8, No. 4, August 2000, pp. 379–391.

[18]  R. Ho, K. Mai and M. Horowitz, "The Future of Wires," *Proceedings of the IEEE*, January 2001.

[19]  J. Hu and R. Marculescu, "Energy-Aware Mapping for Tile-Based NOC Architectures Under Performance Constraints," *Asian-Pacific Design Automation Conference*, 2003.

[20]  F. Karim, A. Nguyen and S. Dey, "On-Chip Communication Architecture for OC-768 Network Processors," *Proceedings of the 38th Design Automation Conference*, 2001.

[21]  B. Khailany, et al., "Imagine: Media Processing with Streams," *IEEE Micro*, Vol. 21, No. 2, 2001, pp. 35–46.

[22]  S. Kumar, et al., "A Network on Chip Architecture and Design Methodology," *VLSI on Annual Symposium, IEEE Computer Society ISVLSI 2002*.

[23]  D. Lackey, P. Zuchowski, T. Bednar, D. Stout, S. Gould and J. Cohn, "Managing Power and Performance for Systems on Chip Design Using Voltage Islands," *ICCAD – International Conference on Computer Aided Design*, 2002, pp. 195–202.

[24]  P. Lieverse, P. van der Wolf, K. Vissers and E. Deprettere, "A Methodology for Architecture Exploration of Heterogeneous Signal Processing Systems," *Journal of VLSI Signal Processing for Signal, Image and Video Technology*, Vol. 29, No. 3, 2001, pp. 197–207.

[25]  M. Oka and M. Suzuoki, "Designing and Programming the Emotion Engine," *IEEE Micro*, Vol. 19, No. 6, November–December 1999, pp. 20–28.

[26]  D. Pham, et al., "Overview of the Architecture, Circuit Design, and Physical Implementation of a First-Generation Cell Processor," *IEEE Journal of Solid-State Circuits*, Vol. 41, No. 1, January 2006, pp. 179–196.

[27]  A. Pinto, L. Carloni and A. Sangiovanni-Vincentelli, "Constraint-Driven Communication Synthesis," *Design Automation Conference*, 2002, pp. 195–202.

[28]  K. Skadron, et al., "Temperature-Aware Computer Systems: Opportunities and Challenges," *IEEE Micro*, Vol. 23, No. 6, November–December 2003, pp. 52–61.

[29]  D. Sylvester and K. Keutzer, "A Global Wiring Paradigm for Deep Submicron Design," *IEEE Transactions on CAD/ICAS*, Vol. 19, No. 2, February 2000, pp. 242–252.

[30]  R. Tamhankar, S. Murali and G. De Micheli, "Performance Driven Reliable Link for Networks on Chip," *ASPDAC – Proceedings of the Asian Pacific Conference on Design Automation*, Shahghai, 2005, pp. 749–754.

[31]  T. Theis, "The Future of Interconnection Technology," *IBM Journal of Research and Development*, Vol. 44, No. 3, May 2000, pp. 379–390.

[32]  E. Waingold, et al., "Baring It All to Software: Raw Machines," *IEEE Computer*, Vol. 30, No. 9, September 1997, pp. 86–93.

[33]  J. Walrand and P. Varaiya, *High-Performance Communication Networks*, Morgan Kaufmann, San Francisco, CA, 2000.

[34]  M. Wolfe, *High Performance Compilers for Parallel Computing*, Addison-Wesley, Upper Saddle River, NJ, 1995.

[35]  F. Worm, P. Ienne, P. Thiran and G. De Micheli, "An Adaptive Low-Power Transmission Scheme for On-Chip Networks," *ISSS, Proceedings of the International Symposium on System Synthesis*, Kyoto, October 2002, pp. 92–100.

[36]  H. Zhang, V. George and J. Rabaey, "Low-Swing On-Chip Signaling Techniques: Effectiveness and Robustness," *IEEE Transactions on VLSI Systems*, Vol. 8, No. 3, June 2000, pp. 264–272.

[37]  http://public.itrs.net/

# CHAPTER 2

# NETWORK ARCHITECTURE: PRINCIPLES AND EXAMPLES

*Networks on chip* (NoCs) are designed using principles that were investigated for multiprocessor computers as well as for local and wide area networks. Networks are characterized by architectures and protocols. The former embody the structural relations among the constituents of the network, while the latter specify the ways in which the network operates under various conditions. In this chapter we review the principles of network architecture, starting from a general taxonomy and revisit of the most commonly used topologies. Next, we will describe some novel architectures that have been proposed specifically for on-chip realizations. Finally, we will consider the impact of the physical substrate limitations and evaluate the merits of classic and novel architectures for NoCs.

## 2.1  NETWORK ARCHITECTURE

The network architecture specifies the *topology* and physical organization of the interconnection network. The physical organization on chip and the proximity among nodes have a strong impact on the effectiveness of the architectures, as the signal propagation delay is often higher than, or comparable to, the processors' cycle time. A network architecture operates according to a set of protocols, which principally determine the *switching*, *routing* and *control flow*.

The choice of an architecture and protocol is usually done to meet some general goals, which include, but are not limited to, performance (latency and throughput), energy consumption, reliability, scalability and implementation cost. Note that energy consumption is an important metric for on-chip applications, while it is less relevant for other networks. Implementation cost relates to resource usage and, in on-chip networks, it depends ultimately on silicon area usage.

The elements of a network architecture are the processing and storage units, which we call *nodes*, the *switches* and the physical *links*. Nodes

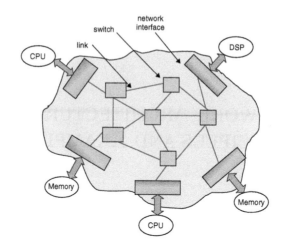

Conceptual realization of an NoC.

are computational elements and (or combined with) storage arrays. As far as we are concerned here, we do not distinguish among the nature of the nodes. Nodes can incorporate switches, often called *routers*. Physical links are wires which can be interrupted by *repeaters* that have the task to amplify the signal and to steepen its edges. Repeaters can incorporate clocked registers; in this case the wires are *pipelined* (Fig. 2.1).

Network architectures can be classified into four groups according to their topology [15]:

1. *Shared-medium networks*: The transmission medium (link) is shared by all nodes, and only one node at a time can send information.

2. *Direct networks*: Each node has a router and point-to-point links to other nodes.

3. *Indirect networks*: Each node is connected to a switch and switches have point-to-point links to other switches.

4. *Hybrid networks*: A mixture of the previous approaches.

It is interesting to note that on-chip communication has been traditionally done by shared medium, bus-based networks. Recent extensions to direct and indirect solutions have been driven by both performance and energy consumption reasons. In the following subsections we will give a short overview of the key characteristics of the four groups listed above. The reader interested in a more complete treatment of these concepts

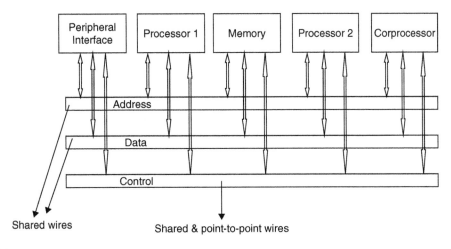

Shared-medium backplane bus.

should refer the detailed discussion provided by recent textbooks on traditional multiprocessor networks (14, 15).

## 2.1.1  Shared-Medium Networks

Most *systems on chip* (SoCs) use interconnection architectures which fall within the *shared-medium* class [4, 10, 26, 29, 31]. These are the simplest interconnect structures, in which the transmission medium is shared by all communication devices. Only one device, called *initiator* (or *master*), can drive the network at a time and connect to one, or more, *targets* (or *slaves*). Every device connected to the network has a network interface, with requester, driver and receiver circuits. The network is usually passive and it does not generate control or data messages. A critical issue in the design of shared-medium networks is the *arbitration strategy* that assigns the mastership of the medium and resolves access conflicts. A distinctive property of these networks is the support for broadcast of information, which is very advantageous when communication is highly asymmetric, that is, the flow of information originates from few transmitters to many receivers.

Within current technologies, the most common embodiment of the on chip, shared-medium structure is the *backplane bus* (Fig. 2.2). This is a very convenient, low-overhead interconnection structure for a small number of active processors (i.e., bus masters) and a large number of passive modules (i.e., bus slaves) that only respond to requests from bus masters. The information units on a bus belong to three classes, namely: *data*, *address* and *control*. Data, address and control information can either be time multiplexed on the bus or they can travel over dedicated busses/wires, spanning the tradeoff between performance and hardware cost (area). Indeed, a distinguishing characteristics of on-chip network implementation is the availability of physical wires, and thus parallel transmission of data, addresses and controls is often preferred for performance reasons.

A critical design choice for on-chip busses is synchronous versus asynchronous operation. In synchronous operation, all bus interfaces are synchronized with a common clock, while in asynchronous busses all devices operate with their own clock and use a handshaking protocol to synchronize with each other. The tradeoff involved in the choice of synchronization approach is complex and depends on a number of constraints, such as testability, ease of debugging and simulation, and the presence of legacy components. Currently, all commercial on-chip busses are synchronous (even though commercial asynchronous bus technology is starting to emerge [34]), but the bus clock is slower than the clock of fast masters. Hence, simplicity and ease of testing/debugging is prioritized over sheer performance.

Bus arbitration mechanisms are required when several processors attempt to use the bus simultaneously. Arbitration in current on-chip busses is performed in a centralized fashion by a bus arbiter module. A processor wishing to communicate must first gain bus mastership from the arbiter. This process implies a control transaction and communication performance loss, hence arbitration should be as fast as possible and as rare as possible. Together with arbitration, the response time of slow bus slaves may cause serious performance losses, because the bus remains idle while the master waits for the slave to respond. To minimize the waste of bandwidth, *split transaction protocols* have been devised for *high-performance busses*. In these protocols, bus mastership is released just after a request has completed, and the slave has to gain access to the bus to respond, possibly several bus cycles later. Thus the bus can support multiple outstanding transactions. Needless to say, bus masters and bus interfaces for split transaction busses are much more complex than those of simple single-transaction busses. Even though early on-chip busses supported only single transactions, split transaction organizations have been used in high-performance chips [1] and are now becoming commonplace [28, 35].

There are many industrial players supporting and promoting bus standards, such as large semiconductor firms (e.g., *coreConnect* by IBM and *STBUS* by ST Microelectronics), core vendors (e.g., *advanced microcontroller bus architecture* (AMBA) by ARM) and interconnect intellectual property vendors (e.g., *CoreFrame* by PalmChip, *WishBone* by Silicore and *SiliconBackPlane* by Sonics).

**Example 2.1.** The AMBA 2.0 bus standard has been designed for the ARM processor family [4]. It is fairly simple and widely used today. Within

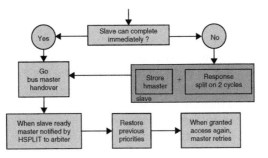

■ **FIGURE 2.3**

Support for split transactions in AMBA.

AMBA, the AHB protocol connects high-performance modules, such as ARM cores and on-chip RAM. The bus supports separate address and data transfers. A bus transaction is initiated by a bus master, which requests access from a central arbiter. The arbiter decides priorities when there are conflicting requests. The arbiter is implementation specific, but it must adhere the ASB protocol. Namely, the master issues a request to the arbiter. When the bus is available, the arbiter issues a grant to the master. Arbitration address and data transfer are pipelined to increase the bus effective bandwidth, and burst transactions are allowed to amortize the performance penalty of arbitration. However, multiple outstanding transactions are supported only to a very limited degree: the bus protocol allows a split transaction, where a burst can be temporarily suspended (by a slave request). New bus transactions can be initiated during a split burst (Fig. 2.3).

The most advanced version of the AMBA bus protocol (AMBA *advanced extensible interface* (AXI)) fully supports multiple outstanding transactions and out of order transaction completion. These features are supported by providing the designer with a higher-level view of the protocol. Namely AXI uses transaction-level (or session layer) models, as compared to AHB which uses transport layer models.

Shared-medium network organizations are well understood and widely used, but their scalability is seriously limited. The bus-based organization is still convenient for current SoCs that integrate less than five processors and rarely more than ten bus masters. Indeed bus-based systems are often not scalable, as the bus becomes the performance bottleneck when more processors are added. Another major critical limitation of shared-medium networks is their energy inefficiency, not only because the switched capacitance on the bus gets larger with the number of nodes, but also because they create functional congestion which implies energy waste at the system level. Thus, energy considerations severely limit the use of large-scale bus-based systems.

### 2.1.2   Direct Networks

The *direct* or *point-to-point* network is a network architecture that overcomes the scalability problems of shared-medium networks. In this architecture, each node is directly connected with a subset of other nodes in the network, called *neighboring* nodes. Nodes are on-chip computational units which can include one (or more) level of memory. In direct networks, nodes contain a network interface block, often called a *router*, which is directly connected with the routers of the neighboring nodes through *channels*. Channels can be of the following types: *input, output or bidirectional*, and usually each node has a fixed number of channels. Differently from shared-medium architectures, as the number of nodes in the system increases, the total communication bandwidth also increases. Direct interconnect networks are therefore very popular for building large-scale systems.

Direct networks have been extensively studied, and their properties can be defined on the basis of a graph model. The *network graph $G(N, C)$* is a pair where the vertex set $N$ represents the nodes, and the edge set $C$ represent the communication channels. Directed and undirected edges can be used to represent unidirectional and bidirectional channels, respectively. The basic network properties can be defined using the graph model. The *node degree* is the number of channels connecting a node with its neighbors; the *network diameter* is the maximum distance between two nodes in a network. A network is *regular* when all nodes have the same degree and *symmetrical* when the network looks alike from every node. A more loosely defined property is *path diversity*, which a network has when (most) node pairs have multiple (minimal) paths between them. Path diversity adds to robustness of the network. An important characteristic of a network is the *bisection width*, that is the minimum number of edges that must be cut when the graph is divided into two equal sets of nodes [15]. The collective bandwidth of the links associates with these edges is termed *bisection bandwidth*, and is a common measure of performance of a topology.

**Example 2.2.** Figure 2.4 shows a crossbar with four nodes connected to each other. The network graph is isomorphic to the network, and it is the complete graph $K_4$. The diameter is 1, as each node can be reached in one hop from each neighbor. The degree of each node is 3, and thus the network is regular. The network is symmetric and it shows path diversity, because each node can be reached from each other node through two alternative paths, when the corresponding edge (link) fails. Note that the alternative paths are non-minimal. The bisection width of $k_4$ is 4.

The fundamental tradeoff in direct network design is between connectivity and cost. Higher connectivity relates to higher performance and is therefore desirable, but its corresponding area and energy consumption costs for physical connections and routers provide serious limitations. A fully connected topology, where each node is connected directly to each

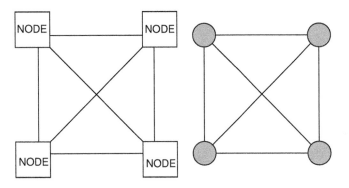

■ **FIGURE 2.4**

A crossbar topology and related graph.

other node, is represented by a complete graph. It has a prohibitive cost even for networks of moderate size. Therefore many topologies have been proposed where messages may traverse several intermediate nodes before reaching their destinations.

Most practical implementations of direct networks adopt an *orthogonal topology*, where nodes can be arranged in an *n*-dimensional orthogonal space, in such a way that every link produces a displacement in a single dimension. Control of message paths (i.e., routing) in these networks is simple and it can be implemented efficiently in hardware. Examples of popular, orthogonal networks are the *n-dimensional mesh*, the *torus* and the *hypercube*.

**Example 2.3.** A simple and common architecture is the *n*-dimensional mesh, as shown in Fig. 2.5(a). Meshes are neither regular nor symmetrical, as there is discontinuity at the external boundary nodes.

The *k*-ary *n*-cube, also called torus, is an *n*-dimensional grid with *k* nodes in each dimension. The torus is a symmetric network. An example is given in Fig. 2.5(b).

A hypercube, or an *n*-dimensional cube, has $k = 2$ (Fig. 2.5(c)).

A major advantage of orthogonal topologies is that routing can be made very simple, and thus implemented in hardware. Indeed, nodes can be labeled using their coordinates in the *n*-dimensional space. The distance between two nodes can be computed by the sum of the dimension offsets, because each link traverses a dimension and every node has a link in any dimension.

There are other direct network topologies. An interesting topology is the tree. Trees are acyclic graphs. A *k*-ary tree has all nodes, but the leaves, with the same number *k* of descendants. A balanced tree is such that the distance from the root to all leaves is the same. A drawback of a tree topology is that there is only a path between each pair of nodes. Another drawback

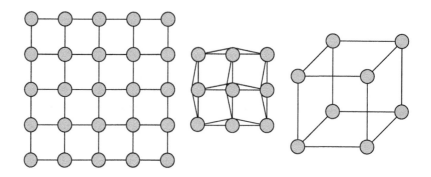

■ **FIGURE 2.5**

(a) A two-dimensional mesh; (b) a 3-ary 2-cube (or torus), and (c) an $n = $ two-dimensional hypercube.

is that the root and its neighbors have higher traffic. This disadvantage can be alleviated by allocating higher bandwidth to the channels close to the root, even though this solution does not help when messages are pipelined. The *fat tree* topology obviates some of the problem of the general tree. Since it is an indirect network, it is described in the following section.

### 2.1.3    Indirect Networks

*Indirect* or *switch-based* networks are an alternative to direct networks for scalable interconnection design. In these networks, a connection between nodes has to go through a set of *switches*. The network adapter associated with each node connects to a port of a switch. Switches do not perform information processing. Their only purpose is to provide a programmable connection between their ports, or, in other words, to set up a communication path.

Indirect networks can be abstracted as graphs $G(N, C)$, where $N$ is the set of switches and $C$ is the set of channels between switches. Similarly to direct networks, it is possible to define diameter, regularity and symmetry of the network, bisection (band)width as well as path diversity. Many topologies have been proposed for indirect networks. The simplest indirect network is the *crossbar*, a structure where each processing element can be connected through any other by traversing just a single switch without contention. Similarly to the fully-connected direct network, the crossbar does not scale well, as its area costs increases with $N^2$.

**Example 2.4.** Several bus standards, such as AMBA and STBUS, support now parallel transactions between more than one initiators and targets. This feature is provided by overlaying multiple bus structures. When all initiators have a corresponding bus, the structure is equivalent to a full crossbar. In the case of AMBA, this configuration is referred to as AMBA

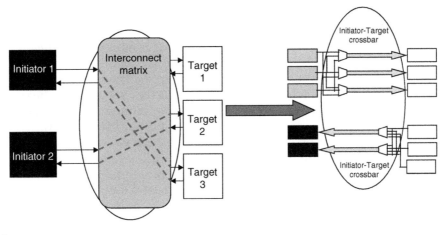

■ **FIGURE 2.6**

A multilayer bus.

multilayer. Please note that these bus-based configurations cannot be considered as shared-medium architectures, because the buses are effectively not shared. Thus these are example of indirect on-chip networks (Fig. 2.6).

In many cases, partial cross-bars are effective implementation, because either some initiator–target paths are unused or they are seldom used. In the latter case, arbitration just adds a small performance loss. On the other hand, partial crossbars may be significantly smaller and less energy consuming as compared to full cross bars. Automatic methods for selecting effective partial crossbars have been proposed [23].

As crossbars are complex, it is possible to realize them by cascading intermediate smaller crossbars. This is an example of realizing a *multistage interconnection network* (MIN). In these types of network, the area complexity is traded for increasing the number of hops from source to destination, and thus the latency. In general, multistage networks can be classified into three groups:

1. *Blocking networks*: Connections between inputs and outputs may not always be possible because of shared resources. Thus, information may be temporarily blocked, or dropped, if contention happens. An example is the *Butterfly* network.

2. *Non-blocking networks*: Expensive but performing networks, that are equivalent to a full crossbar. An example is the *Clos* network.

3. *Rearrangeable networks*: Paths may have to be rearranged to provide a connection, and require an appropriate controller. An example is the *Benes* network.

Moreover, networks can be classified as unidirectional and bidirectional. In a bidirectional network, paths can be established by sending the data forward and backward, using *turnaround routing*.

A *fat tree* is an indirect network whose topology is a tree and where network nodes can be connected only to the leaves of the tree. Links among adjacent switches can be increased as they get closer to the root. Additionally, switches can also be duplicated to reduce contention. It is possible to see a fat tree as a folded butterfly network [15] (Fig. 2.8).

Indirect on-chip networks are at the basis of the communication fabric of most of today's *field-programmable gate arrays* (FPGAs). FPGAs can be seen as the archetype of future programmable SoCs: they contain a large number of computing elements, and they can support complex, programmable communication patterns among them. Moreover, as in the case of networks, FPGA can be dynamically reconfigured.

## 2.1.4   Hybrid Networks

Hybrid networks combine mechanisms and structures from shared medium, direct and indirect networks. Objectives include increasing bandwidth, with respect to shared-medium networks and reducing the distance between nodes, as compared to direct and indirect networks.

Hybrid networks can be abstracted as hypergraphs $H(N, C)$, where $N$ is the set of nodes and $C$ is the set of channels. In a hypergraph an edge can be incident to two or more nodes. Thus the semantic of the model is as follows. An edge connecting two nodes represents a point-to-point connection while an edge connecting more than two nodes represents a shared medium.

Hierarchical busses, or bridged busses, are examples of hybrid networks. Usually, a higher bus speed (bandwidth) is available at the bus at the top of the hierarchy. Hierarchical busses enable clustering highly interacting nodes on dedicated busses and provide lower bandwidth (and/or higher latency) inter-cluster communication. Compared to homogeneous, high-bandwidth architectures, they can provide comparable performance using a fraction of the communication resources, and at a fraction of the energy consumption. Thus, energy efficiency is a strong driver toward hybrid architectures (Fig. 2.7).

**Example 2.5.** The AMBA 2.0 standard [4] specifies three protocols: the AHB, the *advanced system bus* (ABS) and the *advanced peripheral bus* (APB). The latter is designed for lower-performance transactions related to peripherals. An AMBA-based chip can instantiate multiple bus regions which cluster components with different communication bandwidth requirements. Connections between two clusters operating with different protocols (and/or different clock speed) are supported

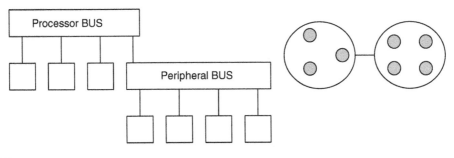

■ **FIGURE 2.7**

A hierarchical bus and its representation as a hypergraph, where edges are represented as segments and enclosures.

through bus bridges. Bridges perform protocol matching and provide the buffering and synchronization required to communicate across different clock domains. Note that bridges are also beneficial from an energy viewpoint, because the high transition rate portion of the bus is decoupled from the low speed, low transition rate, peripheral section.

## 2.2  NETWORK ARCHITECTURES FOR ON-CHIP REALIZATION

We consider here some classic and some novel architecture and their use for on-chip networks. The choice of topology is very important for NoCs, because it affects several figures of merit of the corresponding SoC, such as area and power dissipation, performance. We evaluate these factors next.

First, the static (silicon area) cost of an NoC depends directly on the topology. The area of routers and *node interfaces* (NIs) (see Section 2.4.3), which can be computed by gate-level synthesis and layout tools, sums up to the area cost of the links [8, 16]. In particular, the size of buffers in routers and NIs is a major contributor to the area cost of the NoC. The number of global wires (links times their width) gives a first approximation of the link area.

Second, the dynamic (power dissipation) cost of an NoC depends on the number of NoC components that are active, independently of data in the network (static power, clock power, etc.), and the power dissipation due to data in the NoC itself. The latter depends on the NI dissipation, the number of routers data has to traverse to arrive at the destination, and the dissipation in the links that are used. Topologies with a short routes and short links score well.

Third, the performance of an NoC depends on the interplay of many factors, which includes the topology. Topologies with a high-performance measure (e.g., large bisection bandwidth, small diameter, etc.) tend to have a high area cost (number of routers and links). Cost and performance must be traded off against one another.

Finally, different types of NoC objectives, such as being application specific, reconfigurable or general purpose, match better with different topologies. If much of the application is known in advance, the NoC topology can be tailored and optimized, for example, to avoid hot spots (local overloading of the NoC). For general purpose SoCs, on the other hand, the NoC must be generic and be able to handle different uses.

NoCs differ from general networks because router placement is limited to the two-dimensional plane of the integrated circuit. Moreover, links between routers can travel only in X or Y direction, in a limited number of planes (the number of metal layers of the IC process). As a result, many NoCs have topologies that can be easily mapped to a plane, such as low-dimensional (1–3) meshes and tori.

Early NoCs used mesh and torus topologies for the sake of simplicity. The RAW multiprocessor [3] consisted of a set of homogeneous tiles comprising processing elements, storage and routers. The tiles abutted to form a two dimensional mesh. Thus each tile has a direct connection to its neighbors. Similarly Dally and Towles [13] described the advantages of using regular NoCs. The *Nostrum* project [20] used mesh topology to show the feasibility of on-chip networked communication.

The area of meshes grows linearly with the number of nodes. Silicon realizations of mesh-based networks have a relatively large average distance between nodes, affecting power dissipation negatively. Moreover, their limited bisection bandwidth reduces performance under high load. Care has to be taken to avoid accumulation of traffic in the center of the mesh (creating a *hot spot*) [24].

Among the torus-based configurations, the $k$-ary 1-cube (one-dimensional torus or ring) is the simplest and is used in Proteo [5, 27]. The area and power dissipation costs (related to the number of routers and NIs, and the average distance) of the ring grow linearly with the number of IP cores. Performance decreases as the size of the NoC grows because the bisection bandwidth is very limited. The 3-ary 2-cube (two-dimensional) torus adds wrap-around links to the two dimensional mesh. To reduce the length of these links, the torus can be folded. Dally and Towles [13] advocated this topology. The area of tori is roughly the same as for meshes (when considering the wrap-around link areas), but the power dissipation and performance are better because the average distance is smaller than in meshes.

Meshes and tori can be extended with bypass links to increase the performance (i.e., increasing bisection bandwidth and reducing average distance) for a higher area cost. The resulting *express cubes* [12] are essentially used in FPGA architectures.

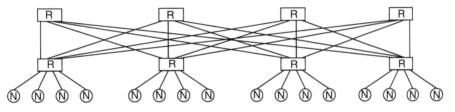

■ **FIGURE 2.8**

SPIN network topology. R: router; N: node.

As an example of an NoC with a *fat tree* topology, shown in Fig. 2.8, *SPIN* was one of the first realizations of an NoC [6, 18]. It is an indirect NoC with a *fat tree* topology [18]. The early realizations of SPIN had a fixed limitation of 256 nodes, so that they could be addressed with 1 byte. The fat tree architecture was chosen to provide a simple and effective routing scheme, which is deterministic on fat trees.

Note that the bisection bandwidth of a fat tree is very low, due to concentration of all traffic at the root of the tree. To solve this problem the root can be duplicated. The resulting *fat trees* [22] (or folded butterfly) have a large bisection bandwidth (and hence performance), but higher area cost. For larger number of nodes, the layout of the fat tree is more difficult in comparison with meshes or tori.

A simple but effective form of NoC can be achieved by connecting computing and/or storage nodes to different levels of a tree topology, and by restricting routing (usually circuit switching) to converge to the root. This structure is useful to implement a memory hierarchy, for example, with an external memory at the root, on-chip memories (e.g., caches) at lower levels of the tree, and computing nodes at the leaves. This topology scales well for a large number of master nodes, but not as well for many slaves (at or near the root). Examples are Sonics's SonicsMX [33] and Philips's PMAN [17].

*Octagon* is an example of a direct NoC [19]. It has been designed by ST Microelectronics for network processors. In an octagon network, eight processors are connected by an octagonal ring and with three diameters. Messages between any two processors require at most two hops, as shown in Fig. 2.9. If one node processor is used as the bridge node, more octagons can be tiled together. Recently, the octagon architecture has been generalized to a polygon with diameters. Such an extension is called *Spidergon* because of its spider-like connection [9].

Hierarchical hybrid networks have also proven to be useful NoC topologies, because they privilege high bandwidth within clusters of highly dependent nodes. Lee and coworkers [21] proposed a hierarchical hybrid network which is composed of several clusters of tightly connected nodes. Intra-cluster local links provide high-bandwidth connection on relatively short wires (Fig. 2.10).

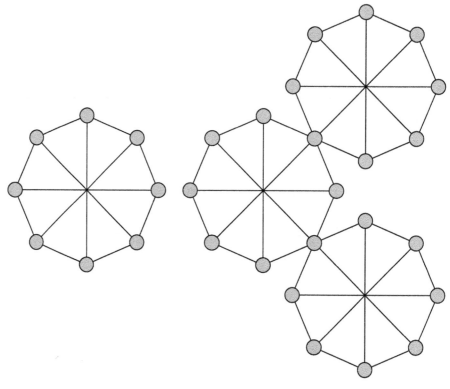

Octagon network topology and its modular extension.

## 2.3  AD HOC NETWORK ARCHITECTURES

Network architectures need to be tuned to the type of nodes and switches used as components as well as to the applications they support. In some cases, as for single chip parallel computing and for FPGAs with homogeneous fabric, the constituents of the network are regular. Thus the network benefits from having a regular and symmetric architecture. Indeed, most work on single-chip multiprocessors is based on well-known network topologies, which were studied for cabinet-based multiprocessors. On the other hand, new SoCs with dedicated functions, such as gaming consoles, communication terminals, embedded control systems, have often several heterogeneous nodes, such as processors, controllers, DSPs, various types of memory arrays and application-specific logic blocks (e.g., MPEG4 engines). Moreover, such nodes communicate with each other using a traffic which can be predicted to have different requirements from link to link.

In this perspective, application-specific NoCs are extremely beneficial. Such networks use *ad hoc* topologies and protocols. It is important to

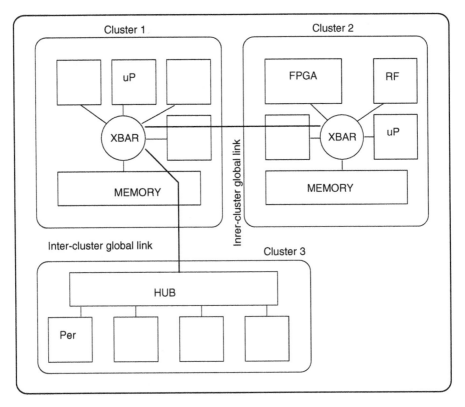

**■ FIGURE 2.10**

Cluster-based hierarchical network.

notice that especially in this wide class of SoCs, NoCs can outperform other styles of on-chip communication, such as busses. The design of *ad hoc* network architecture can be done by using tools (see Chapter 8 ) that can help designers mapping applications to network architectures and to instantiate the network to a level of detail so that it can be extensively simulated and synthesized. Moreover, optimization of network parameters, such as link bandwidth and latency, switch configuration, buffering and channel interface, as well as NI design is key to an effective design. Indeed, *ad hoc* on-chip networks are the interconnection platform for heterogeneous SoCs as well as for the new generation of FPGAs.

## 2.4   COMPONENT DESIGN FOR NoCs

We present now the essential features of NoC components and the most important consideration for integrated implementation.

### 2.4.1  Switch Design

A switch consists of a set of input buffers, an interconnection matrix, a set of output buffers and some control circuitry. Buffers at the inputs and outputs are used to enqueue data transmitted along the channels. Buffer allow for local storage of data that cannot be immediately routed. Unfortunately buffers have a high cost in terms of energy consumption, and thus many NoC implementation strive at having limited local storage. This constraint, in conjunction with the objective of achieving high performance, has implications on the type of architectures and routing protocols which are efficient on chip. For example, mesh architectures with deflection routing [20] avoid local storage by always forward routing or misrouting the data, with a corresponding added simplicity and reduced energy consumption.

The interconnection matrix can be realized by a single crossbar or by cascading various stages. From a logic standpoint the interconnection matrix is realized by multiplexing the input data to the output ports. From a physical design standpoint, the crossbar is often designed with a regular layout, to minimize the length of local interconnection and achieve minimal transmission delay. The control circuitry can serve ancillary tasks, such as arbitration, when contention is possible, and possibly error detection and correction. It also implements some functions of the control flow protocol.

*Virtual channels* are often used within switches, because of the following reasons. Buffers are usually organized as FIFO queues, and thus when a message occupies a buffer for a channel, the physical channel cannot be used by other messages, even when the original message is blocked. *Virtual channels* are used to avoid this inconvenience. Two or more virtual channels are multiplexed on the physical channel. Moreover, virtual channels are useful to avoid deadlock situations in wormhole-switched networks. Finally, with virtual channels messages can make progress rather than being blocked, thus improving message latency and network throughput. The drawback of using virtual channels is a more complex control protocol, as data corresponding to different messages which is multiplexed on the physical channel must be eventually separated.

### 2.4.2  Link Design

The physical connection between two nodes and/or switches can be realized by wires. Due to the relatively small wire pitch in silicon technologies, and to the availability of several layers for wiring, physical wires are a cheap commodity for networks. Thus, in NoCs, data and control are typically physically separated, with the result of a cleaner and more performing design. Nevertheless, signal delay and signal dispersion on chip wires are usually a problem. Indeed, delay grows quadratically with the wire length, because of the combined resistive and capacitive effects.

Global wires are typically segmented by repeaters, which are buffers that restore the voltage level on wires and provide local current sourcing and sinking capability. Sometimes repeaters include regenerative circuits, because of their sensing capability and fast output transitions.

When wires have significant length, repeaters may incorporate clocked registers, thus providing a pipelined transmission of data so that transmitted data arrives at a rate that matches consumption at the processing nodes. Typically global wires include 1–5 pipestages. Interestingly enough, pipelined wires provide local storage in a form of distributed buffering.

It is important to note that the use of specific repeaters may render the wiring channel unidirectional, and thus bidirectional links may need to be split over two sets of wires. It is also important to remember that the choice of pipelined wires has an impact on control flow, as local storage needs to be initialized and flushed at times.

### 2.4.3  NI Design

In direct networks, the switch is incorporated in the node and interfaced locally to the processing unit, while in indirect networks a switch is externally connected to the processing unit via an interface. The interface is just a wrapper that provides the computational unit a view of the network consistent with its I/O protocol. In other words, the NI is a protocol converter that maps the processing node I/O protocol into the protocol used within the NoC. A reason for the need of this conversion is that NoCs use streamlined protocols for the sake of efficiency.

The *open core protocol* (OCP) [32] is an open standard to ease the integration of processing cores within SoCs. The objectives of this standard are to enable processor core creation independently of network design, while sustaining parametrizeable design and high performance. Even though OCP is not meant to be an NoC protocol, it standardizes that design of NIs, as they do not need to be processing-core specific but just translators between OCP and the NoC protocol.

## 2.5  PROPERTIES OF NETWORK ARCHITECTURES

The properties and the corresponding effectiveness of an NoC in a specific environment depend on its architecture and protocols. Since switching, routing and control flow have not been presented yet, we limit ourselves to a few generic but important considerations that are related to the network architecture itself and to a generic abstract model of protocols. We assume that messages are transmitted in terms of packets. A detailed description of these issues is reported in Ref. [15].

The most important properties of a network is to be immune of the following causes of failure:

- *Deadlock*: A packet does not reach its destination, because it is blocked at some intermediate resource.

- *Livelock*: A packet does not reach its destinations, because it enters a cyclic path.

- *Starvation*: A packet does not reach its destination, because some resource does not grant access (while it grants access to other packets).

Note that in all cases one (or more) packet never reaches destination, and thus either some loss of information is tolerable or some recovery strategy is used. Deadlock, livelock and starvation are due to the sharing of a finite number of resources and to the policy of resource access.

Typically, the most critical resource is local storage. Buffer arrays are expensive in terms of area and their energy comsumption is significant. Hence, in the context of NoCs, buffer usage is limited.

Starvation can be avoided using a correct resource assignment scheme, such as a round-robin scheme that produces a fair allocation to requestors. Even though some packets may have higher priority, some bandwidth must be allocated to low-priority packets to avoid starvation, for example by reserving some virtual channels or buffers to low-priority packets.

Livelock can be avoided by using only minimal-length paths. While this choice seems to be obvious for performance reasons, sometimes non-minimal path are used for fault tolerance reasons. Alternatively, some routing algorithms (like deflection routing) use non-minimal paths to avoid local storage. In these conditions, livelock depends on the specifics of the routing algorithm and livelock can be probabilistically avoided [15].

Deadlock is the hardest problem to solve, and it has been attacked by several researchers. There are several ways to avoid deadlock, namely deadlock *prevention*, deadlock *avoidance* and deadlock *recovery*. Deadlock prevention is a conservative technique, where resources are granted to a packet so that a request never leads to a deadlock. This can be done by reserving the resources before a packet transmission. With deadlock avoidance, resources are allocated as packets advance through the network. In deadlock recovery, it is assumed that deadlock may happen sometimes (and hopefully rarely). A deadlock detection mechanism is in place, and in case a deadlock is detected some resources are deallocated and granted to other packets. Some packets may be aborted and retransmitted.

Overall, deadlock is due to cyclic dependencies among storage resources. Thus, its avoidance is related to choosing routing functions that avoid such dependencies. The drawback may be in terms of area and energetic cost of the network. Indeed, fully adaptive routing requires a large number of resources to avoid cyclic dependencies.

Even though these issues have been studied in great depths for macroscopic networks, additional challenges arise in the case of NoC. The critical issue in this context is that the interconnect fabric competes for silicon real estate with on-chip storage and functional units, hence severe constraints are enforced on the hardware cost of any solution to deadlock, livelock and starvation issues. Many effective, but expensive (in terms of buffer storage, logic) techniques that have been developed for macroscopic networks are therefore not usable directly in NoCs.

## 2.6  SUMMARY

NoCs differ from macroscopic networks because of the tight relation between design choices and parameters and the characteristics of the silicon substrate. They benefit from spatial proximity and from the wide availability of wires over different layers. The choice of a network architecture affects the area, performance and energy consumption of the NoC. Whereas maximizing performance is a general wish of system designers, coping with energy consumption and dissipation limitations is a specific constraining for NoCs. Application-specific NoCs may benefit also by tailoring the topology to the application of interest. The overall benefits and figures of merits of NoCs depend both on the architectures and on the related operating protocols. These will be presented and analyzed in the following chapters.

## REFERENCES

[1]  B. Ackland, et al., "A Single Chip, 1.6-Billion, 16-b MAC/s Multiprocessor DSP," *IEEE Journal of Solid-State Circuits*, Vol. 35, No. 3, March 2000.

[2]  A. Adriahantenana and A. Greiner, "Micro-network for SoC: Implementation of a 32-Bit SPIN Network," *Design Automation and Test in Europe Conference*, 2003, pp. 1128–1129.

[3]  A. Agrawal, "Raw Computation," *Scientific American*, Vol. 281, No. 2 August 1999, pp. 44–47.

[4]  P. Aldworth, "System-on-a-Chip Bus Architecture for Embedded Applications," *IEEE International Conference on Computer Design*, 1999, pp. 297–298.

[5]  M. Alho and J. Nurmi, "Implementation for Interface Router IP for Proteo Network on Chip," *Proceedings of the International Workshop on Design and Diagnostic of Electronic Circuits and Systems*, 2003.

[6]  A. Andriahantenaina and A. Grenier, "Micronetwork for SoC: Implementation of a 32-port SPIN Network," *DATE – Proceedings of Design Automation Test in Europe*, March 2003, pp. 1128–1129.

[7] L. Benini and G. De Micheli, "Networks on Chips: A New SoC Paradigm," *IEEE Computers*, January 2002, pp. 70–78.

[8] E. Bolotin, I. Cidon, R. Ginosaur and A. Kolodny, "QNoC: QoS Architecture and Design Process for Network on Chip," *Journal of System Architecture*, Vol. 50, No. 2–3, February 2004, 105–128.

[9] M. Coppola, R. Locatelli, G. Maruccia, L. Pieralisi and A. Scandurra, "Spidergon: A Novel On-Chip Communication network," *Proceedings of the 2004 International Symposium on System on Chip*, November 2004, p.15.

[10] B. Cordan, "An Efficient Bus Architecture for System-on-Chip Design," *IEEE Custom Integrated Circuits Conference*, 1999, pp. 623–626.

[11] M. Dall'Osso, G. Biccari, L. Giovannini, D. Bertozzi and L. Benini, "Xpipes: A Latency Insensitive Parameterized Network-on-Chip Architecture for Multi-Processor SoCs," *International Conference on Computer Design*, 2003, pp. 536–539.

[12] W. Dally, "Express Cubes: Improving the Performance of $K$-ary n-cube Interconnection Networks," *IEEE Transactions on Computers*, Vol. 40, No. 9, September 1991, pp. 1016–1023.

[13] W. Dally and B. Towles, "Route Packets, Not Wires: On-Chip Interconnection Networks," *DAC, Proceedings of the 38th Design Automation Conference*, pp. 684–689.

[14] W. Dally and B. Towles, *Principles and Practices of Interconnection Networks*, Morgan Kaufmann, San Francisco, CA, 2004.

[15] J. Duato, S. Yalamanchili and L. Ni, *Interconnection Networks: An Engineering Approach*, Morgan Kaufmann, San Francisco, CA, 2003.

[16] K. Goossens, J. Dielisssen, O. Gangwal, S. Pestana, A. Radulescu and E. Rijkema, "A Design Flow for Application Specific Networks on Chip with Guaranteed Performance to Accelerate SoC Design and Verification," *DATE – Proceedings of Design Automation Test in Europe*, March 2005, pp. 1182–1187.

[17] K. Goossens, O. Gangwal, J. Roever and A. Niranjan, "Interconnect and Memory Organization in SoCs for Advanced Set-Top Boxes and TV – Evolution, Analysis and Trends," in J. Nurmi, H. Tenhunen, J. Isoahao and A. Jantcsh (editors), *Interconnect-Centric Design for Advanced SoC and NoC*, Kluwer, Norwell, MA, 2004, Chapter 15, pp. 399–423.

[18] P. Guerrier and A. Grenier, "A Generic Architecture for On-Chip Packet-Switched Interconnections," *Design Automation and Test in Europe Conference*, 2000, pp. 250–256.

[19] F. Karim, A. Nguyen, S. Dey and R. Rao, "On-Chip Communication Architecture for OC-768 Network Processors," *DAC – Design Automation Conference*, 2001, pp. 678–683.

[20] S. Kumar, A. Jantsch, J. Soininen, M. Forsell, M. Millberg, J. Oberg, K. Tiensyrj and A. Hemani, "A Network on Chip Architecture and Design Methodology," *Proceedings of IEEE Computer Society Annual Symposium on VLSI*, April 2002, pp. 105–112.

[21] S.-Y. Lee, S.-J. Song, K. Lee, J.-H. Woo, S.-E. Kim, B.-G. Nam and H.-J. Yoo, "An 800 MHz Star-Connected On-chip Network for Application to Systems on a Chip," *IEEE Solid-State Circuits Conference*, 2003, pp. 468–469.

[22] C. Leiserson, "Fat-Trees: Universal Networks for Hardware-Efficient Supercomputing," *IEEE Transactions on Computers*, Vol. 34, No. 10, October 1985, pp. 892–901.

[23] S. Murali and G. De Micheli, "An Application-Specific Design Methodology for STbus Crossbar Generation," *DATE – Design, Automation and Test in Europe*, Vol. 2, 2005, pp. 1176–1181.

[24] M. Millberg, A. Nilsson, R. Thid and A. Jantsch, "The Nostrum Backbone – A Communication Protocol Stack for Networks on Chip," *Proceedings of the International Conference on VLSI Design*, 2004, pp. 693–696.

[25] C. Patel, S. Chai, S. Yalamanchili and D. Shimmel, "Power Constrained Design of Multiprocessor Interconnection Networks," *IEEE International Conference on Computer Design*, 1997, pp. 408–416.

[26] W. Remaklus, "On-chip Bus Structure for Custom Core Logic Design," *IEEE Wescon*, 1998, pp. 7–14.

[27] I. Saastamoinen, M. Alho and J. Nurmi, "Buffer Implementation for Proteo Network on Chip," *ISCAS – Proceedings of the International Symposium on Circuits and Systems*, 2003.

[28] G. Strano, S. Tiralongo and C. Pistritto, "OCP/STBUS Plug-in Methodology," *GSPX Conference*, 2004.

[29] W. Weber, "CPU Performance Comparison: Standard Computer Bus Versus SiliconBacplane," *www.sonicsinc.com*, 2000.

[30] D. Wingard, "MicroNetwork-based integration for SOCs," *Design Automation Conference*, 2001, pp. 673–677.

[31] S. Winegarden, "A Bus Architecture Centric Configurable Processor System," *IEEE Custom Integrated Circuits Conference*, 1999, pp. 627–630.

[32] www.ocpip.org/

[33] www.sonicsinc.com

[34] www.silistix.com

[35] www.arm.com

# CHAPTER 3

# PHYSICAL NETWORK LAYER*

Recent drive for "smaller, faster, cooler and cheaper" semiconductor technology achieves commercialization of 65 nm process and soon 45 nm process will be adopted in mass production by the industry [23]. Such *deep submicron* (DSM) technology brings up many new challenges and an increase of uncertainty (i.e., unpredictability) in process, design and market. The precise control of the fabrication process is almost impossible so that fluctuation of the dopant concentration, variation of the photolithographic resolution and thin film thicknesses are indispensable and become the causes of uncertainties in threshold, $g_m$ and $I_{DSAT}$ of MOSFET. The same problems can take place to interconnection lines implemented with 6–7 layers of metals, or polysilicon and diffusion layers. The process uncertainty leads to non-uniformity of the sheet resistance and increase of the coupling noises among interconnects. In addition, decreasing interconnect pitch and increasing aspect ratio with the technology advance accelerates these issues further. The interconnect is used to distribute clock and signals, and to provide power and ground to and among many functional blocks on a chip. Copper wires and *low k* material are widely used for the control of RC delays but still cannot meet the high-speed transmission needs with scaling of feature sizes. How to isolate those process dependent problems in interconnect from the performance of the *system on chips* (SoC) is the critical points in SoC design.

According to an ITRS prediction illustrated in Fig. 3.1, the gap between interconnection delay and the gate delay will increase to 9:1 with the 65 nm technology. In addition, total wire length on a chip is expected to amount to 1.78 km/cm$^2$ in the year of 2010 [23] (Fig. 3.2). Another observation is the increase of the power dissipation consumed in charging and discharging the interconnection wires on a chip. According to [8], the capacitance portion of the interconnect amounts for 30% of the total capacitance of a chip, and soon the interconnect will consume about 50 times more power than logic circuits [11]. This means that performance of the entire SoC, power consumption, speed and area, will be more affected by the interconnection rather than the gates on a chip in the near future (Fig. 3.3).

---

* This Chapter was provided by Hoi-Jun Yoo, Kangmin Lee, and Se-Joong Lee of Korea Advanced Institute of Science and Technology, Republic of Korea.

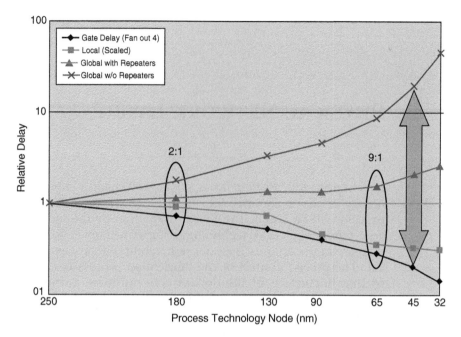

■ **FIGURE 3.1**

Relative delay comparison of wires versus process technology.

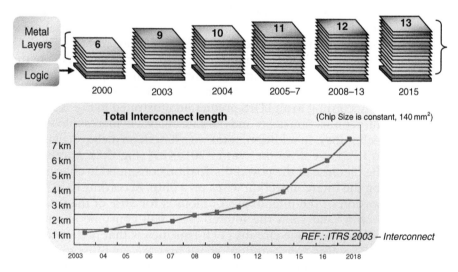

■ **FIGURE 3.2**

Trend of total interconnect length on a chip.

| Operation | Power (0.13 μm) | (0.05 μm) | |
|---|---|---|---|
| 32b ALU Operation | 5 pJ | 0.3 pJ | Operation |
| 32b Register Read | 10 pJ | 0.6 pJ | |
| Read 32b from 8 KB RAM | 50 pJ | 3 pJ | |
| Transfer 32b across chip (10 mm) | 100 pJ | 17 pJ | Global on-chip interconn. |
| Execute a uP instruction (SB-1) | 1.1 nJ | 130 pJ | |
| Transfer 32b off chip (2.5 G CML) | 1.3 nJ | 400 pJ | Off-chip interconnect |
| Transfer 32b off chip (200 M HSTL) | 1.9 nJ | 1.9 nJ | |

■ **FIGURE 3.3**

Power consumption on a 32 bits microprocessor chip [11].

Although a new design approach like *network on chips* (NoC) is required to get a more process independent interconnection architecture, still more emphasis should be put on the physical design of interconnection and more detail understanding of the physics in the interconnect of DSM technology is required than before. Four factors are referred to cause new challenges in DSM [41]. The first is the increase of the operation frequency so that the inductance characteristics are more conspicuous than before. The second is the signal reflection and transmission due to impedance mismatch. The third is the increase of fringe capacitance due to increased aspect ratio of the interconnection wire. The last is the increase of resistance with sheet resistance due to skin effect of high frequency and contact resistance.

The detail explanation on these general problems has been the topics of many books [2, 12, 25]. Here the scope is more narrowed to the NoC and only limited topics required for the design and real chip implementation of NoC will be covered. In Section 3.1 the physical issues of the wires will be explained. First the modeling of the wires including the transmission line model will be introduced for the clear understanding of the interconnect physics and then the performance metrics of the on-chip interconnect such as the signal integrity and noise immunity will be discussed. Circuit-level issues will be covered in Section 3.2 such as the high-speed logic, low-power interconnection and synchronization circuits for NoC. Section 3.3 will introduce the functional building blocks of NoC such as switches, buffer/memory and *serialization and deserialization* (SERDES).

## 3.1 INTERCONNECTION IN DSM SoC

### 3.1.1 Interconnection Models

This section will summarize circuit designer's viewpoints on the on-chip interconnect based on the basic physics. The effect of $L$ can be neglected if

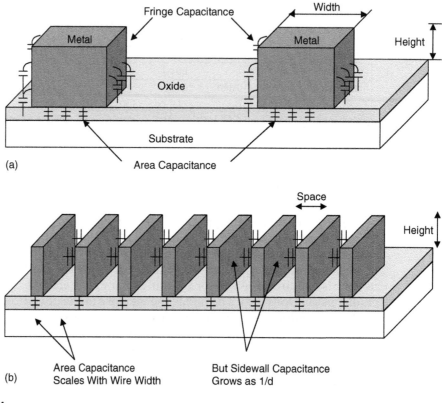

(a)

(b)

■ **FIGURE 3.4**

Parasitic capacitance of DSM interconnect.

$f \ll R/(2\pi L)$ which is true in most case of on-chip interconnect. On-chip interconnect can be modeled as an RC network, where $R$ is given as $\rho L/A$ and $C$ is given as $\varepsilon(LW)/t$ in Fig. 3.4(a). Even the resistance can be excluded and the interconnect is modeled as a lumped capacitor if the wire length is short ($<1000 \times$ critical dimension) and if the source and load impedances are significantly higher than the impedance of the wire $(L/C)^{1/2}$. The simple model was modified by [43] to include the effects of fringe capacitances as depicted in Fig. 3.4(b) and has been widely used in the very-large-scale integration (VLSI) design of many design projects. However, this model still needs refinement due to the advance of the process technology to DSM because more fringe effects should be taken into consideration [35].

More accurate behavior can be analyzed with transmission line model [6, 13, 14]. Assume that the interconnect is divided into many infinitesimal sections as depicted in Fig. 3.5, their parasitic circuit elements give a partial differential equation as follows:

$$\frac{\partial^2 V}{\partial x^2} = RC\frac{\partial V}{\partial t} + LC\frac{\partial^2 V}{\partial t^2}$$

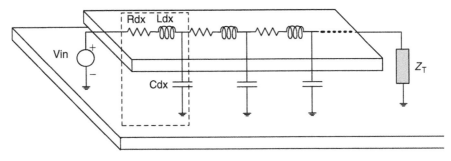

■ **FIGURE 3.5**

Distributed RLC model of the interconnect.

If $L = 0$ in the equation (3.1), it becomes a diffusion equation and the voltage wave diffuses along the interconnect getting wider pulse width and decreasing amplitude. Assume $R = 0$, the equation results in the traveling wave equation and the voltage wave travels along the interconnect line with velocity of $(LC)^{-1/2}$. If the interconnection length is larger than $0.1(t_r)(LC)^{-1/2}$, where $t_r$ is the shortest rise time in the system, wave nature of the signal should be taken into consideration [18]. If the wave sees any electrical discontinuity on the line, reflection wave occurs with the reflection coefficient given by $(Z_T - Z_0)/(Z_T + Z_0)$. Open circuit at the end terminal, $Z_T = \infty$, gives total reflection, matched termination, $Z_T = Z_0$ leads to no reflection and any mismatch results in a finite reflection.

Skin effect is conspicuous at high frequency too. At high frequency, the current cannot flow through the total cross-section of the interconnection but flow along only the surface of the wire so that the effective area for the current flow decreases and the effective resistance increases with frequency. The skin depth, $\delta$, the depth where the current has fallen off to $\exp(-1)$ of its normal value is given by:

$$\delta = \frac{1}{\sqrt{\pi f \mu \sigma}}$$

where $\sigma = 1/\rho$ is the conductivity of the material and $f$ is the frequency of the signal [12].

### 3.1.2 Signal Integrity and Noises on the Interconnect

The signal on the interconnection wire experiences degradation of its signal quality during its travel. The degradation comes from many different sources, intrinsic RLC circuit nature and extrinsic noises. For example, as explained in the previous section, RC delay and attenuation of the wave amplitude or dispersion of the wave, originate from the RC circuit nature of the interconnection wire. In addition, the ringing comes from the transmission line effect of the wire and can be estimated with LC circuit

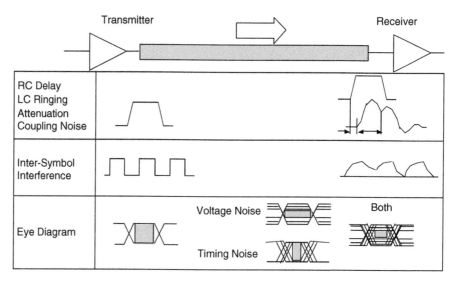

■ **FIGURE 3.6**

Signal integrity, ISI and eye diagrams.

modeling of the wire. The waves of sequential data overlap each other due to wave dispersion along the wire and each signal is corrupted by decaying tails created by previous signals [22]. The wave overlapping is called *inter-symbol interference* (ISI) and leads to higher error rate in the detection of signals transmitted through band-limited interconnection wire. The signal integrity can be visualized by the "eye diagram." The eye diagram is made by capturing repeatedly the sequential signals on an oscilloscope with regular time window and overlapping them on the same screen. If there is a voltage noise or insufficient voltage build-up, the height of the eye decreases. If there is a propagation delay or jitter, the width of the eye becomes narrow. In real signaling, both of the height and width get smaller leading to reduced opening of eye (Fig. 3.6).

The extrinsic sources of the degradation in signal quality are noises. Figure 3.7 shows the origins of the noises, offsets of transmitter and receivers, coupling noise and power supply noise [12]. The device offsets are more pronounced in DSM because of increased uncertainties in device parameters composing transmitter and receiver circuits. The crosstalk noises are perceived as the most significant noises in DSM. Through the coupling capacitance, switching on neighbor wires, aggressors, injects the charge into quite wires, victims. Crosstalk can affect delay and skew of the signals depending on whether the aggressor signals are switching in the same or the opposing direction to the victim signal, and it makes difficult to estimate the exact delay deviation. In DSM, the inductive coupling between on-chip wires becomes a noise source. Current spike in aggressors, when wide bus wires are switching simultaneously, induces

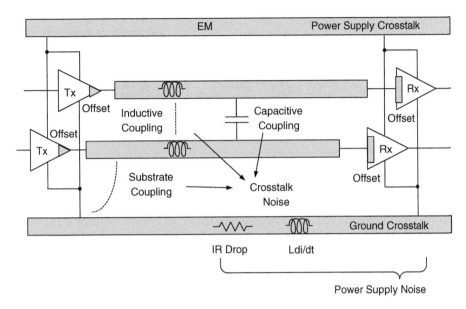

■ **FIGURE 3.7**

Origins of on-chip noises.

voltage noise on the victims via near field electromagnetic coupling [29]. When transmitters share the common power and ground lines, switching of a transmitter can affect the output signal of another transmitter through the power line or ground line. There are two components of power supply noise, low frequency and high frequency. The low frequency is the variation in the *direct current* (DC) power supply and ground levels by IR drop. The simultaneous switching of internal circuits produces the high-frequency "delta" noise. If the size of the output buffer of the transmitter is large, it draws a sudden large current and generates a bounce in the power line through IR drop. The glitch transfers to the output of the neighboring silent transmitter which is turned on earlier. Similarly, via the ground switching of circuits generates glitch noise on the quiet signal of neighboring circuits. Recently, in the mixed-mode IC the relatively quiet analog blocks are susceptible to the noise generated by digital blocks through the substrate. Guard rings to absorb the minority carriers through the substrate can protect the quiet wires in the analog blocks from digital noise.

### 3.1.3   Issues in Power and Clock Distribution

Recently low-power circuits, systems and even low-power NoC received high attention due to the increase of mobile applications such as cell phones. NoC should get its energy from the power grid and clock signals from clock trees. The power network should deliver power from the package to the NoC with little voltage drop by providing charge from

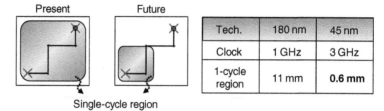

■ **FIGURE 3.8**

One clock cycle region with evolution of the clock frequency and technology.

the on-chip capacitances to control noise spikes which are generated by wide bus drivers. Large wires made of low-resistance material like copper should be used for power supply lines. Grid of the metal lines is used for the power supply distribution, which in return limits the useful distance for the layout of decoupling capacitors. Decoupling capacitors have to be located close enough to be reached in time for the efficient isolation of the noise from sensitive circuits. Cell type of decoupling capacitors is used in arrays for the easy design of large capacitance.

The background philosophy of the NoC is to make it easy to design and implement complicated SoC. Clock delivery, one of the hardest design issues in SoC, is no exception from this concept. As technology is evolved to DSM and the clock frequency increases, only a portion of the chip can be covered by a single clock cycle as illustrated in Fig. 3.8. For example, the distance that 1 GHz clock signal can travel in a one clock cycle time amounts to 11 mm large enough to cover most of chip area. However, 3 GHz clock signal can traverse only 0.6 mm in a clock cycle time far smaller than the chip size. It is clear that it will be more difficult to obtain global synchronization in DSM technology.

In addition, the processing cores are designed using different frequencies each other. The SoC designed by integration of multiple cores does not need to provide a singular clock signal to each core. Recent NoCs are supporting plesiochronous communication [32, 48] like the computer-to-computer Internet and no effort for the optimization of the clock delivery system will be necessary.

## 3.2  HIGH-PERFORMANCE SIGNALING

### 3.2.1  High-Speed Signaling

**Voltage-mode signaling**

The voltage-mode signaling is the conservative and safe approach: using simple circuits, appropriate performance/cost figure is obtained. As shown in Fig. 3.9(a), a driver and a receiver consist of simple CMOS inverters. The

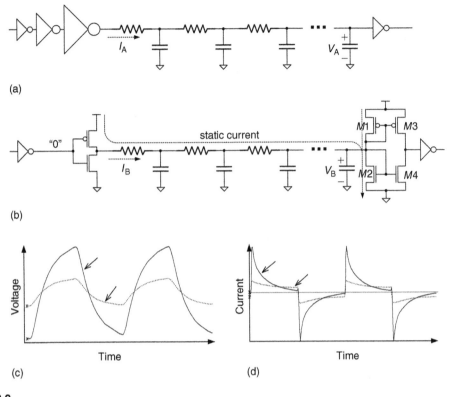

(a)

(b)

(c)    (d)

■ **FIGURE 3.9**

(a) Voltage-mode and (b) current-mode of the signaling circuits. (c) The voltage and (d) current waveforms of the circuits.

current from the driver charges the capacitive load of the link wire. As the current reaches at the input gate of the receiver, the current charges the gate capacitance of the receiver until the voltage of the input gate reaches to the $V_{DD}$. There is no static current dissipation thus the power consumption is proportional to the signal activity.

As the driver size increases in order to drive longer link wire, a large driver can cause the problems like peak-current and electro-migration. Moreover, the resistance of the metal wire becomes considerable, and the larger driver size does not reduce the delay time any more. Although it needs more efforts in placement and routing, one can use distributed repeaters along the link wire. In a multi-bit link, staggering the repeaters cancels down the capacitive noise effectively. Also, an active charge compensation scheme, where a reverse noise is injected to the victim line using anti-millers [16], can be applied.

**Current-mode signaling**

The current-mode signaling is one of the alternative selections, if only limited EDA tool support is available due to many analog circuits used in the chip design. Figure 3.9(b) shows an example circuit. With the same driver circuit as that of the voltage-mode scheme, the current from the driver charges the capacitive load of the link wire. As the current reaches at the input gate of the receiver, however, it just flows through the load resistance, $M1$ and $M2$. Before charging the input node of the receiver up to the $V_{DD}$, the small voltage change by the current at the load resistor can be detected by the receiver. The current sensing mechanism is faster because it can avoid the full-level charging and discharging of the capacitive load wire.

Current-mode signaling in **SRAM** devices has been studied in literatures [5, 24, 44]. It is because the large capacitive load of the bitline and the small cell transistors with the series pass transistor turn out very slow bitline voltage build-up and voltage-mode operation degrades the access speed seriously.

Similarly, in the works of Refs [3, 15, 26], global on-chip interconnects which have high capacitive load, have used current-mode signaling. One of the comprehensive studies on current-mode signaling in global interconnects is given in Ref. [3]. In the performance comparison between current-mode and voltage-mode signaling using $0.18\,\mu$m CMOS technology, the maximum data rate of a current-mode signaling outperforms that of a voltage-mode signaling by about twice if the same number of repeaters are used: for example, when they drive the interconnect line with the RC constant time of 1 ns, which is roughly equivalent to 4 mm line, the current-mode achieves about 8 Gb/s NRZ data rate using three repeaters while the voltage mode shows about 3 Gb/s performance with the same number of repeaters.

The major source of power dissipation in current mode is static current through the link wire. The static current does not scale according to data rate thus low-activity factor degrades the power efficiency significantly. For instance, when a current-mode link is designed to support 2-GHz frequency data, it consumes 25% less power than voltage-mode link, if the activity factor is 0.5. This implies that if the activity factor is less than about 0.4, the voltage-mode signaling is more power efficient.

**Wave-pipelining**

Wave-pipelining is an essential technique for global link of an *on-chip network* (OCN). One of the important motivation of the OCN is the fact that time of more than 10 clock cycles are necessary for a signal traversing across a chip in the near future SoCs. Intensive pipelining increases the throughput of the interconnect. However, it still suffers from considerable end-to-end latency. In order to minimize the end-to-end signal transmission latency while maximizing the throughput of the wire, wave-pipelining can be employed. The mesochronous interconnect schemes in Refs [32, 48, 50] basically contains the wave-pipelining technique for inter-clock domain communication. The elastic interconnect is one of the advanced techniques to hold and resume the signal propagation in a wave-pipelined link [37]. The *wave-front-train* (WAFT) scheme provides a new methodology for *on-chip serialization* (OCS) using the concept of the wave-pipelining [50].

For successful wave-pipelining, two issues should be tackled: synchronization at receiver terminal and skew across multiple link wires. In the wave-pipelining, series of data signals are pipelined without consideration of the receiver's clock information. Therefore, in order to find the exact latching time at a receiver, clock information should be transmitted together with the wave-pipelined data. Its synchronization issue will be described in Section 3.2.3. The skew across the multiple link wires, which is one of the signal integrity metric, occurs due to mismatches in electrical characteristics among the wires. The mismatches result from asymmetric effect of crosstalk, power supply noise and process variation. However, such factors are surfaced only after layout or fabrication, which means the factors are hard to be estimated at design time and designers need to put sufficient margins in the design of wave-pipelining for the safe operation.

### 3.2.2 Low-Power Interconnection Design

The power consumption of the on-chip interconnection can be simply described as:

$$P = \alpha \times C_{\mathrm{w}} \times V_{\mathrm{swing}} \times V_{\mathrm{DD(driver)}} \times f \qquad (3.1)$$

where $\alpha$ is the switching probability, $C_{\mathrm{w}}$ is the wire capacitance, $V_{\mathrm{swing}}$ is the voltage swing on the wire, $V_{\mathrm{DD}}$ is the supply voltage of the driver and $f$ is the signaling frequency [9]. In this section, we will give a brief design guideline for low-power interconnection by minimizing each of $\alpha$, $C_{\mathrm{w}}$, $V_{\mathrm{swing}}$ and $f$ (Fig. 3.10) [58].

#### Channel coding to reduce the switching probability

There are many researches on the bus coding for reducing the switching probability such as *bus-invert* (BI) coding [51], gray-code [36], T0-code [4],

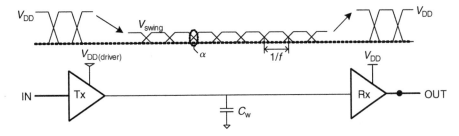

■ **FIGURE 3.10**

Interconnection model.

partial BI coding [46], probability-based mapping [42] and so on. However, there was also a report on the ineffectiveness of those on-chip bus coding techniques because the power dissipation on the (de)coder is comparable to the power saving obtained by the coding when the wire length is shorter than few tens-mm [28]. Furthermore, those bus coding schemes are effective to parallel busses but ineffective in the multiplexed channel used in packet switched networks.

As the link wires connecting *processing units* (PU) and switches are abundantly used, the wiring congestion will become one of the major challenges in the NoC design. To alleviate the wiring congestion, a narrow channel [45] or OCS channel [32, 48] are proposed. However, the switching probability on the serial link gets much higher than a parallel link as illustrated in Fig. 3.11. To solve the problem, a simple but effective transition-reducing encoding (SILENT) was reported for a serial link [33]. In this coding, the transition on parallel wires is encoded as symbol "1" before serialization as depicted in Fig. 3.12. By using the data correlation between successive data words, the probability of 1's occurrences on

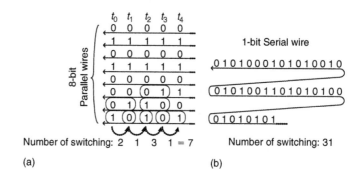

**■ FIGURE 3.11**

Switching probability of the same data pattern on (a) parallel and (b) serial wires.

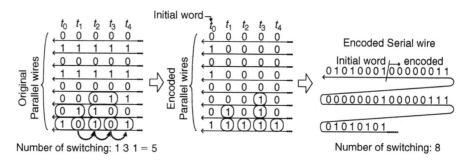

**■ FIGURE 3.12**

SILENT encoding.

the serial wire decreases so that the switching probability is also reduced. This encoding technique is more effective for the highly correlated data like those in the multimedia applications.

**Wire capacitance reducing techniques**

Wire capacitance can be reduced simply by wide spacing the wires and using different layers. Those approaches need more wiring resources horizontally and vertically. To minimize the wiring area, a serialized wire [32, 48] can be used. This serialization technique will be discussed in the Section 3.3 in more detail.

**Low-swing signaling**

Lowering the swing and supply voltage is the most effective way to reduce the power dissipation on interconnections. But the swing voltage and receiver circuits should be carefully designed against the noise sensitivity and delay degradation.

*Driver circuits*
There is a limited degree of freedom in the design of drivers. If a reduced supply voltage, $V_{DDL}$, is used, the power dissipation is reduced to $(V_{DDL}/V_{DD})^2$ (Fig. 3.13(a) and (b)). The $V_{DDL}$ can be provided from an off-chip or generated on a chip by DC–DC converting from the main supply, $V_{DD}$. Unless the additional power supply is applied, the low swing can be obtained by exploiting a transistor $V_{th}$ drop or pulse enabled driving (Fig. 3.13(c) and (d)). However, these designs are susceptible to process variation and noise. In the pulse-controlled driver, the $V_{swing}$ is determined by not only the pulse duration but also the $C_W$ which is hard to be estimated in design stage. The most widely used driver is the type of (b) [21, 48, 59]. By using a NMOS pull-up transistor instead of a PMOS transistor, faster rising-time on the output wire is obtained with smaller transistor size.

*Receiver circuits*
There are two design options in receivers with regard to the noise immunity; a single-ended level converter [17, 38, 40] or differential amplifier

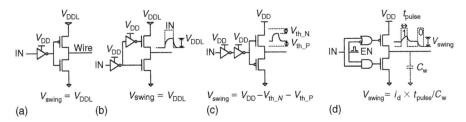

**■ FIGURE 3.13**

Driver circuits for low-swing signaling.

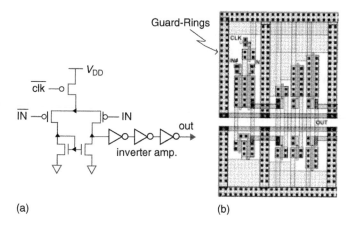

■ **FIGURE 3.14**

Clocked differential sense amplifier: (a) circuits and (b) layout.

[7, 20, 21, 32]. Differential signaling is more immune to noise due to its high common-mode rejection, allowing a further reduction in the swing voltage [21, 59]. Although the differential signaling requires double wires, the wiring congestion can be alleviated by using on-chip serial links [49]. Zhang evaluated many different receivers including a pseudo-differential amplifier in respect of energy, delay, swing and *signal-to-noise ratio* (SNR) [59].

In the on-chip interconnection network, the receiver circuits should be lightweight and occupy as small area as possible because it is used abundantly in most of the network interfaces. Figure 3.14 shows an example of a simple clocked sense amplifier with a three-stage CMOS inverter chain. PMOS transistors are used as the input gates in order to receive a low common-mode input signal. The sizes of the input gates and their bias currents are chosen to amplify the desired low-swing differential input ($W_P/L_P = 3/0.18\,\mu m$, $V_{swing} = 0.2\,V$, $V_{DD} = 1.6\,V$, area = $10 \times 15\,\mu m^2$ [32]).

*Static and dynamic wires*

There are two kinds of wires: static and dynamic wires. To speed up the response of wires, it is precharged to $V_{DDL}$ through PMOS transistors before each bit transition. After the precharging phase, the wire is conditionally discharged by the pull-down transistor of a driver. This dynamic signaling is used for multidrop busses having large fan-in and large fan-out. In the NoC, however, the link is point to point so that there is only single fan-in and fan-out. Therefore, the dynamic wire is not a good candidate for OCNs, especially when the wire has long latency. Furthermore, it is susceptible to noise.

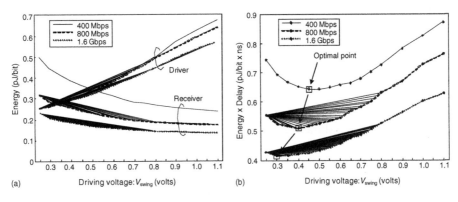

(a) Energy consumption and (b) energy and delay product versus $V_{swing}$ and $f$.

*Optimal voltage swing*

There was a report on the existence of an optimum voltage swing on long-wire signaling for the lowest-energy dissipation [52]. To find the optimum voltage swing, the interconnection system is simulated with extracted wire capacitance value from the real layout. Figure 3.15(a) shows an example of the energy consumption per bit transmission via the low-swing interconnection [32]. The required energy on the driver to create a certain voltage swing on the wires decreases linearly with the swing level, whereas the energy to amplify this signal back to its normal logical swing level increases superlinearly with the decrease of $V_{swing}$ level. The optimum $V_{swing}$ exists due to such opposite trends of the required energy in the transmitter and the receiver. Figure 3.15(b) shows energy and delay product versus $V_{swing}$. The optimal swing voltage is not unique but varies with the signal rates as demonstrated in the figure.

**Frequency/voltage scaling**

Lee et al. [32] presented a three-stepped frequency/voltage scaling based on the network workload and [57] introduced a self-calibrating circuit that controls the voltage swing on the link so that dynamic power consumption is minimal for required data rate.

## 3.2.3 Synchronization

### Clock distribution inside an NoC

As global synchronization of an SoC gets harder to achieve, the asynchronous design techniques become attractive solution. Building a globally asynchronous system, however, moves the global synchronization effort from a system design issue to a network design task. The NoC covers all over the chip area, and the issue of the global synchronization of the network is the same as that of the SoC.

Without global synchronization, a NoC can adopt one of three communication styles: *mesochronous*, *plesiochronous* and *asynchronous* communication. Applying asynchronous signaling to global interconnection or OCN system has been proposed by Villiger et al. [53], Chattopadhyay et al. [10] and Muttersbach et al. [39]. In the asynchronous communication, the data are transferred from a transmitter to a receiver based on a certain handshaking protocol like 2- or 4-phase signaling. In global wires, however, the handshaking protocol suffers from considerable performance degradation because it requires a signal to make a round trip for every signal transaction. Using asynchronous pipeline [53], the long link can be pipelined and throughput can be enhanced. However, it is incompatible with the high-speed wave-pipelining.

In plesiochronous signaling, each network device is fed by its own clock source, and each clock source has the same nominal frequency. However, deviations from nominal values may occur, and frequencies cannot be guaranteed to be exactly the same. Plesiochronous signaling is used in large-scale networks where sharing a common clock source between a transmitter and a receiver is not practical. Since the clock frequencies among network devices are not exactly the same, the receiver should have a mechanism to resolve the variable phase margin. In OCN, however, plesiochronous signaling seems to be a too extreme clocking condition, because sharing a common clock source is not a big problem and does not cost much, whereas the effort to resolve the undetermined phase and clock frequency difference is considerable.

The mesochronous communication seems to be an optimal solution for the OCN, where the network devices are fed by a common clock source but the clock phases of functional blocks may be different from each other due to asymmetric clock tree design and difference in load capacitance of the leaf cell. If the network has regular placement such as in a mesh topology, phase difference is deterministic. If the network has irregular shape thus the phase difference among the distributed network devices are indeterminable, synchronizers are required between the clock domains (Fig. 3.16).

### Synchronizers

Synchronizing mesochronous input signals requires a synchronizer. Several kinds of synchronizers including FIFO synchronizer, delay-line synchronizer and simple pipeline synchronizer have been proposed [12]. If an OCN is designed with intensive cares, the clock skew and signal propagation delay between the clock domains can be estimated precisely, and only single pipeline latch can synchronize the mesochronous signals. In reality, the post-fabrication factors, however, generate unexpected changes on the skew and the delay, which reduces the intended phase margin.

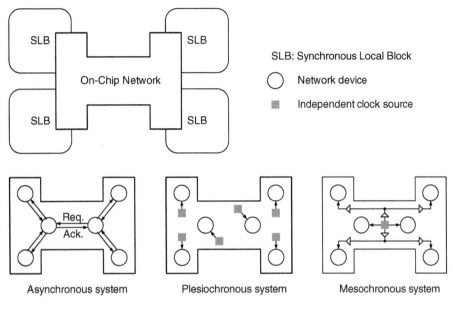

Clocking schemes of OCN.

In OCN, usually the phase difference has a range of finite values. In this case, simple pipeline synchronizer can be used to solve the unknown phase problem. There are three situations in which the values of the fixed phase difference can vary. The first is the variation in the physical distance due to irregular routing of the physical paths using multiplexers. The second is the variation in electrical distance due to supply voltage fluctuation, and the third is the reference timing variation due to clock frequency variation (Fig. 3.17).

A receiver can have more than one transmitter using a multiplexer. Then, the phase difference of the incoming signals at the receiver depends on which transmitter is sending the signal. Supply voltage control is used for power management in low-power application SoCs. In case of supply voltage variation, electrical distances between blocks vary, so that phase differences can be changed. The network clock frequency variation also changes the phase difference of clock signals. The variation of the clock signal is useful technique to manage overall power consumption according to the requirements of network bandwidth. For example, Lee et al. [32] adopt four-step network clock frequency management scheme for low-power application. When the clock frequency is changed, reference timing is changed accordingly, which shifts the phase, too.

The variation in values of phase difference is determined by the parameters such as clock frequency setting and packet path configuration, therefore, the variation is limited to several cases, and it can be quantized.

To deal with the unknown but quantized phase differences, programmable delay synchronizer is used by Lee et al. [50]. It is basically delay-line synchronizer. Additionally, it has a register file which stores appropriate delay values according to the network situations. At initialization period, the OCN performs delay calibration while varying the network configurations and conditions. According to the network situations, the *phase detector* (PD) finds the best delay time of the variable delays, and stores the value in the register file. After the initialization, the network restores a proper delay value from the register file according to the network situations (Fig. 3.18).

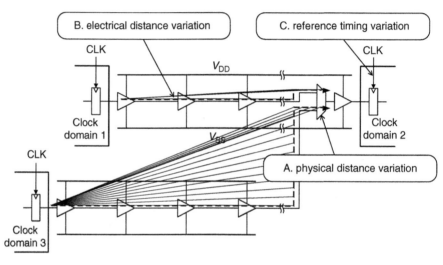

▪ **FIGURE 3.17**

The factors affecting the phase margin in mesochronous on-chip communication.

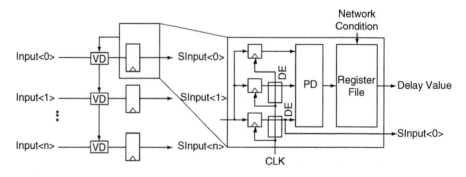

▪ **FIGURE 3.18**

Programmable delay synchronizer.

## 3.3 BUILDING BLOCKS

### 3.3.1 Switch Design

#### Switch fabric

The conventional switch consists of *input queue* (IQ), scheduler, switch fabric and *output queue* (OQ) as shown in Fig. 3.19(a). There are two kinds of switch fabric design: a cross-point and MUX-based switch fabric as presented in Fig. 3.19(b) and (c), respectively. The cross-point switch has pass transistors at each crossing junction of input and output wires. In this switch fabric, the capacitive loading driven by input driver is junction capacitance of pass transistors on input and output wires and the wire capacitance itself. The voltage swing on the output wire is reduced to $V_{DD} - V_{th\_N}$ because of the threshold voltage drop of the NMOS pass transistor thus the power dissipation is reduced. However, this design is hard to be synthesized. The fabric area is determined by the wiring area and not by the transistors so that its area cost can be the minimum. The MUX-based switch uses multiplexer for each output port. The capacitive loading driven by the input driver is the input gate capacitance of the multiplexers and input wire capacitance.

Table 3.1 presents the power, speed and area comparison of the two different designs. It is a simulation result using a capacitive wire model extracted from physical layout. The power consumption of a cross-point switch is much lower than that of a MUX-based design: 37%, 56% and 65% lower for $4 \times 4$, $8 \times 8$ and $16 \times 16$ switches, respectively. Furthermore, the

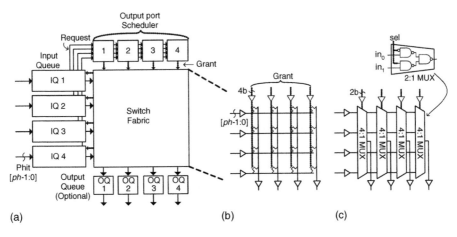

**■ FIGURE 3.19**

(a) Switch structure, (b) cross-point and (c) MUX-based switch fabric.

**TABLE 3.1** ■ Comparison of two designs of the switch fabric: power, delay and area.

| Switch size | Power (mW) | | Delay (ps) | | Area (mm$^2$) [49] | |
|---|---|---|---|---|---|---|
| | Cross-point | MUX based | Cross-point | MUX based | Cross-point | MUX based |
| $4 \times 4$ | 7.7 | 12.4 | 300 | 370 | 0.038 | 0.059 |
| $8 \times 8$ | 23.2 | 52.2 | 460 | 580 | 0.154 | 0.235 |
| $16 \times 16$ | 76.8 | 217.2 | 740 | 1000 | 0.614 | 0.941 |

delay of a MUX-based switch is longer than that of a cross-point switch because of the multiplexer gate-delay. The cross-point switch occupies 65% area compared to the MUX-based switch [49].

*Low-power technique for switch fabric*: Low-swing and current signaling technique can decrease the energy consumption of a crossbar [47, 56] because a crossbar has a large number of long and high-capacitance wires. Another effective way is to divide the fabric into multiple segments and partially activating a selected segment [32, 54].

### Switch scheduler

To arbitrate the output conflicts, a scheduler is used on each output port. The arbitration scheduling adds to the latency, power and area of the switch design. The latency of the arbiter becomes larger than that of switch fabric as the switch size gets bigger than $16 \times 16$ [49]. Furthermore, the area cost is not ignorable when we use a serialized link [32]. As shown in Fig. 3.20, the scheduler occupies similar area like the switch fabric when the phit width of a port is 10 bits. Therefore, the scheduler design is important as much as the switch fabric.

Round-robin scheduling algorithm is most widely used in the OCN [32] because of its fairness and no-starvation properties. The round-robin scheduler can be implemented by using two priority encoders [19] or MUX-tree-connected logic [34]. A pseudo-LRU algorithm and its implementation were also proposed for lower area and lower latency than those of the round-robin algorithm [30, 31].

### 3.3.2  Queuing Buffer and Memory Design

Queuing buffer is used in the input port of a switch and in the network interface too. The queuing buffer consumes the most area and power among composing building blocks in the OCN. The buffer circuits can be designed by two different memory units, either registers (flip-flops (F/Fs)) or static-RAM cells. Figure 3.21 shows four different register designs: (a) a conventional Shift-Register, (b) Push-In Shift-Out Register, (c) Push-In Bus-Out Register and (d) Push-In MUX-Out Register. A SRAM style design is also shown in Fig. 3.22.

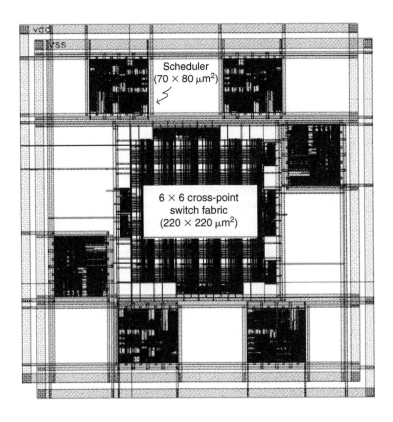

■ **FIGURE 3.20**

A 6 × 6 switch layout (phit width of a link: 10 bits).

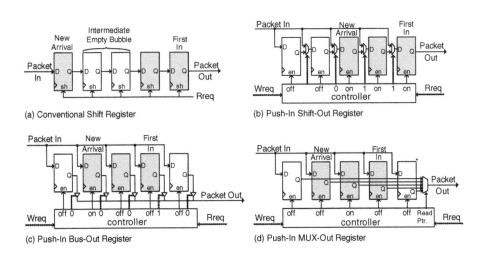

(a) Conventional Shift Register

(b) Push-In Shift-Out Register

(c) Push-In Bus-Out Register

(d) Push-In MUX-Out Register

■ **FIGURE 3.21**

Register designs for queuing buffer.

In a conventional Shift-Register, intermediate empty cells can exist when the packet in/out rates are different temporally in any case. Shifting all the registers at every packet-out consumes huge amount of power. Furthermore, the minimum latency in a queue is as long as the physical queue length rather than the backlog. Although this design is the simplest, it is not desirable to implement on a chip due to its longer latency and unnecessary power dissipation.

To remove the intermediate empty bubble, the arrival packet can be stored at the front empty place rather than the tail of a queue. This input style is called as "Push-In" as illustrated in Fig. 3.21(b). It can remove unnecessary latency and power consumption caused by the empty bubble. Only the occupied register cells are enabled. However, the Shifting Register style still consumes unnecessary power by shifting all of the occupied cells at every output packet. To avoid the shifting operation, the outputs of all registers are tied to a shared output bus line via tri-state buffers as shown in Fig. 3.21(c). The register holding the first-in packet is connected to the output bus by turning-on the tri-state buffer. In this design, only a cell, in which a new arrived packet is stored, is enabled. As the queuing capacity increases, the capacitance of the shared bus wire increases as well because of the parasitic capacitance of tri-state buffers, and the delay and power consumption are also enlarged. To eliminate this effect, output multiplexers can be used as shown in Fig. 3.21(d).

These register-based implementations have a definite limitation in their capacity because of the area and power constraints. As the queuing capacity rises up to a dozen of packets, the register-based implementation is not good in both respects of area and power [55]. Therefore, the queuing buffer should be designed based on a dual-port SRAM cell for large capacity queuing as shown in Fig. 3.22. The figure shows the cell circuit and its

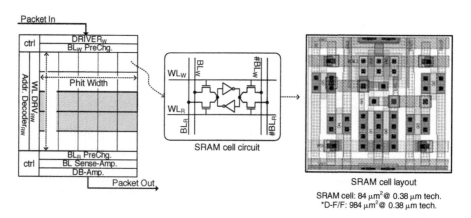

■ **FIGURE 3.22**

Dual-port SRAM design for queuing buffer.

layout also. An SRAM cell occupies only a tenth of a register (F/Fs) area. This area ratio will not be largely varied with technology.

### 3.3.3  Size Determination of Physical Transfer unit

**Size of physical transfer unit**

The *physical transfer unit* (phit) is a unit into which a packet is divided and transmitted through the OCN. Simply speaking, the phit size is the bitwidth of a link. Large phit size easily provides high-bandwidth link whereas the small phit size reduces network area. Assume that the phit size is 80-bits which can carry 32-bits address, 32-bits data and 16-bits header information in a single phit. Then, the area of $5 \times 5$ switch fabric which is a basic building block of mesh networks amounts to larger than $0.25\,mm^2$ when the metal pitch of the switch fabric is $1\,\mu m$. Considering that a one of the very popular micro-controller core, ARM7TDMI, occupies $0.53\,mm^2$ using $0.18\,\mu m$ technology [1], the switch fabric size will result in almost comparable to or slightly larger than that. The switch size includes queuing buffers, arbiter logic and driving buffers as well as the switch fabric. Therefore, the smaller phit size than 80-bits seems to be adequate for reasonable OCN area.

**On-chip serialization**

If the phit size is smaller than the interface bitwidth of a PU, serialization must be performed by the factor of:

$$\text{Serialization ratio}\,(R_{SER}) = \text{I/O bitwidth/phit size}$$

OCS reduces the phit size, the area and energy consumption of the switch as well as the energy consumption of the link. Figure 3.23 illustrates the concept of the OCS. An SERDES circuit is inserted at I/O of a PU, reducing the bitwidth of the I/O and the switch size. The reduction of switch size results in decrease of coupling capacitance of wires in the switch fabric, resulting in reduction of switch energy consumption. In the case of link, as the number of wires is reduced, wire space can be wide, which results

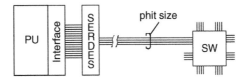

■ **FIGURE 3.23**

The concept of the OCS.

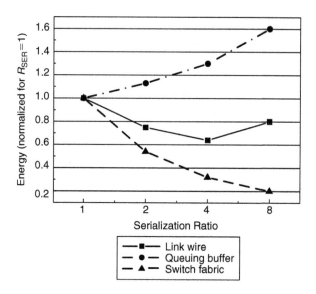

Energy variation of building blocks according to the $R_{SER}$.

in reduction of wire capacitance load or energy consumption. As the link bitwidth decreases, the driver size must be increased to come up with increased operating frequency. Also, the OCS increases switching activity factor because it breaks the correlation between consecutive transfer unit which is often observed in address field of memory access. The I/O bitwidth of a queuing buffer is the phit size. Because the I/O wires must be shared by the storage cells, the small phit size means larger number of storage cells or larger capacitance can be connected to each I/O wire. Therefore, as the $R_{SER}$ increases, the power consumption of a queuing buffer increases, as well. Figure 3.24 shows an example of energy analysis results for the building blocks based on 0.18 μm technology when the OCS is applied. The switch and link energy consumptions decrease effectively according to the $R_{SER}$. The link power increases again at $R_{SER} = 8$, however, due to the driving circuit overhead caused by the high-frequency operation.

**Serializer/deserializer circuits**

Figure 3.25 shows the two typical serializer circuits. The Shift-Register type serializer loads the parallel data through the 2:1 MUXs. After the load operation, the shift mechanism of the series F/Fs realizes high-speed serialization. The maximum clock frequency is given by:

$$f_{MAX} = \frac{1}{T_{MUX} + T_{SETUP} + T_{HOLD}}$$

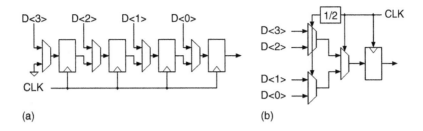

(a)    (b)

■ **FIGURE 3.25**

(a) Shift-Register type and (b) MUX-tree type serializers.

where the $T_{MUX}$ is the 2:1 MUX delay time, and the $T_{SETUP}$ and $T_{HOLD}$ are the setup and hold time of the F/F.

In the MUX-tree type serializer, one of the parallel data bit is selected through series of MUXs. The maximum clock frequency is given by:

$$f_{MAX} = \frac{1}{\log_2 R_{SER} \times T_{MUX} + T_{SETUP} + T_{HOLD}}$$

To achieve very high-speed operation, the MUX and F/Fs can be designed using *current-mode logic* (CML) [27], which increases power dissipation seriously while achieving state-of-the-art operation speed.

Both of the serializer types require as high-speed clock as the serialization speed. Lee et al. [50] introduce a new serializer structure, which uses physical delay constant of *delay elements* (DEs) as a timing reference instead of clock, and utilizes signal propagation phenomenon instead of the shifting mechanism. The serialization scheme is called *wave-front-train* or WAFT.

Figure 3.26 shows a 4:1 WAFT serializer circuitry. When EN is low, D<3:0> is waiting at QS<3:0>. The $V_{DD}$ input of MUXP, which is called a pilot signal, is also loaded to QP. The GND input of MUXO discharges the *serial output* (SOUT) while the serializer is disabled. If EN is asserted, QS<3:0> and the pilot signal start to propagate through the serial link wire. Each signal forms a wave-front of the SOUT signal, and the timing distance between the wave-fronts is the DE and MUX delay which is called as a unit delay. The series of wave-fronts propagates to the deserializer like a train.

When the SOUT signal arrives at the deserializer, it propagates through the deserializer until the pilot signal arrives at the end of the deserializer, or STOP node. As long as the unit delay times of the sender and the receiver are the same, D<3:0> arrives at its exact position when the pilot signal arrives at the STOP node. When the STOP signal is asserted, the MUXs feed back its output to its input, so that the output value is latched.

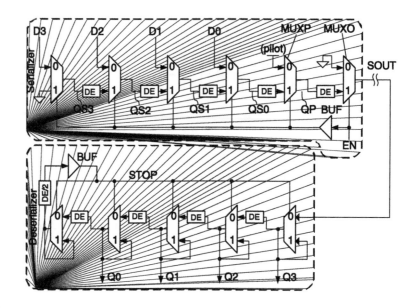

■ FIGURE 3.26

WAFT SERDES.

## 3.4  Summary

This chapter describes physical design issues for NoCs. A major design problem in NoCs is the interconnect realization, which requires modeling and performance metrics, such as the signal integrity and noise immunity. High-speed signaling, such as current-mode signaling, and wave-pipelining and low-power interconnection designs, such as low-swing signaling and transition-reducing coding, are widely used within the NoC physical layer. Furthermore, the clock and data synchronizing methods, as well as the physical design issues of NoC building blocks, such as switches, buffering memories and SERDES, are analyzed here in depth.

## REFERENCES

[1]  http://www.arm.com/products/CPUs/ARM7TDMI.html

[2]  H.B. Bakoglu, *Circuits, Interconnections, and Packaging for VLSI*, Addison-Wesley, Reading, MA, 1990.

[3]  R. Bashirullah, et al., "Current-Mode Signaling in Deep Submicrometer Global Interconnects," *IEEE Transactions on* VLSI Systems, Vol. 11, No. 3, June 2003, pp. 406–417.

[4] L. Benini, et al., "Asymptotic Zero-Transition Activity Encoding for Address Busses in Low-Power Microprocessor-Based Systems," *Proceedings of the Great Lakes Symposium on VLSI*, March 1997, pp. 77–82.

[5] T. Blalock, et al., "A High-Speed Sensing Scheme for 1T Dynamic RAM's Utilizing the Clamped Bit-Line Sense Amplifier," *IEEE Journal of Solid-State Circuits*, Vol. 27, April 1992, pp. 618–625.

[6] J.R. Brews, "Transmission Line Models for Lossy Waveguide Interconnections in VLSI," *IEEE Transactions on Electron Devices*, Vol. ED-33, No. 9, September 1986, pp. 1356–1365.

[7] G.C. Cardarilli, et al., "Low Voltage Swing Circuits for Low Dissipation Busses," *Proceedings of the International Symposium on Circuits and Systems*, June 1997, pp. 1868–1871.

[8] A.P. Chandrakasan and R.W. Brodersen, *Low Power Digital CMOS Design*, Kluwer Academic Publishers, Norwell, MA, 1995.

[9] A. Chandrakasan, et al., *Design of High-Performance Microprocessor Circuits*, IEEE Press, Piscataway, NJ, 1999, p. 360.

[10] A. Chattopadhyay, et al., "High Speed Asynchronous Structures for Interclocking Domain Communication," *International Conference on Electronics, Circuits and Systems*, 2002, pp. 517–520.

[11] W.J. Dally, "Computer Architecture is All About Interconnect," *Proceedings of the IEEE International Symposium on High-Performance Computer Architecture*, February 2002.

[12] W.J. Dally and J.H. Poulton, *Digital Systems Engineering*, Cambridge University Press, Cambridge, UK, 1998.

[13] I.B. Dhaou, V. Sundarajan, H. Tenhunen and K.K. Parhi, "Energy Efficient Signaling in Deep Submicron CMOS Technology," *Proceedings of the International Symposium on Circuit and Systems*, May 2001, pp. 319–324.

[14] I.B. Dhaou and H. Tenhunen, "Modeling Techniques for Energy-Efficient System-on-a-Chip Signaling," *IEEE Circuits and Devices*, Vol. 19, No. 1, January 2003, pp. 8–17.

[15] I.B. Dhaou, et al., "Current Mode, Low-Power, On-Chip Signaling in Deep-Submicron CMOS Technology," *IEEE Transactions on Circuit and Systems-I*, Vol. 50, No. 3, March 2003, pp. 397–406.

[16] E.S. Fetzer, et al., "A Fully Bypassed Six-Issue Integer Datapath and Register File on the Itanium-2 Microprocessor," *IEEE Journal of Solid-State Circuits*, Vol. 37, No. 11, November 2002, pp. 1433–1440.

[17] R. Golshan, et al., "A Novel Reduced Swing CMOS BUS Interface Circuit for High Speed Low Power VLSI Systems," *Proceedings of the IEEE International Symposium on Circuits and Systems*, May 1994, pp. 351–354.

[18] K.W. Goossen and R.B. Hammond, "Modeling of Picosecond Pulse Propagation in Microstrip Interconnections on Integrated Circuits," *IEEE Transactions on Microwave Theory Technology*, Vol. 37, No. 3, March 1989, pp. 469–478.

[19] P. Gupta, et al., "Designing and Implementing a Fast Crossbar Scheduler," *IEEE Micro*, Vol. 19, January–February 1999, pp. 20–28.

[20] M. Hiraki, et al., "Data-Dependent Logic Swing Internal Bus Architecture for Ultralow-Power LSI's," *IEEE Journal of Solid-State Circuits*, Vol. 30, April 1995, pp. 397–402.

[21] R. Ho, et al., "Efficient On-Chip Global Interconnects," *IEEE Symposium on VLSI Circuits*, Digest of Technical Papers, June 2003, pp. 271–274.

[22] M. Horowitz and W. Dally, "How Scaling Will Change Processor Architecture," *International Solid-State Circuits Conference*, Digest of Technical Papers, February 2004, pp. 132–133.

[23] Available: http://public.itrs.net/files/2003ITRS/Home2003.htm

[24] M. Izumikawa, et al., "A Current Direction Sense Technique for Multiport SRAM," *IEEE Journal of Solid-State Circuits*, Vol. 31, No. 4, April 1996, pp. 546–551.

[25] H.W. Johnson and M. Graham, *High-Speed Digital Design, A Handbook of Black Magic*, Prentice-Hall, Englewood Cliffs, NJ, 1993.

[26] A.P. Jose, et al., "Near Speed-of-Light On-Chip Interconnects Using Pulsed Current-Mode Signaling," *Symposium on VLSI Circuits*, Digest of Technical Papers, 2005, pp. 108–111.

[27] K. Kanda, et al., "40 Gb/s 4:1 MUX/1:4 DEMUX in 90 nm Standard CMOS," *International Solid-State Circuits Conference*, Digest of Technical Papers, 2005, pp. 152–153.

[28] C. Kretzschmar, et al., "Why Transition Coding for Power Minimization of On-Chip Buses Does Not Work," *Proceedings of the Design Automation and Test Europe Conference (DATE)*, February 2004, pp. 512–517.

[29] R. Kumar, "Interconnect and Noise Immunity Design for the Pentium 4 Processor," *Intel Technology Journal*, Vol. 5, Q1, 2001. http://developer.intel.com/technology/itj/archive/2001.htm

[30] K. Lee, et al., "A High-Speed and Lightweight On-Chip Crossbar Scheduler for On-Chip Interconnection Networks," *Proceedings of the IEEE European Solid-State Circuits Conference*, September 2003, pp. 453–456.

[31] K. Lee, et al., "A Distributed Crossbar Switch Scheduler for On-Chip Networks," *Proceedings of the IEEE Custom Integrated Circuits Conference*, September 2003, pp. 671–674.

[32] K. Lee, et al., "A 51 mW 1.6 GHz On-Chip Network for Low-Power Heterogeneous SoC Platform," *IEEE International Solid-State Circuits Conference*, Digest of Technical Papers, February 2004, pp. 152–153.

[33] K. Lee, et al., "SILENT: Serialized Low Energy Transmission Coding for On-Chip Interconnection Networks," *Proceedings of the International Conference on Computer Aided Design*, November 2004, pp. 448–451.

[34] K. Lee, et al., "Low-Power Network-on-Chip for High-Performance SoC Design," *IEEE Transactions on VLSI Systems*, February 2006, pp. 148–160.

[35] E.T. Lewis, "An Analysis of Interconnect Aline Capacitance and Coupling for VLSI Circuits," *Solid-State Electronics*, Vol. 27, No. 8/9, 1994, pp. 741–749.

[36] H. Mehta, et al., "Some Issues in Gray Code Addressing," *Proceedings of the Great Lakes Symposium on VLSI*, March 1996, pp. 178–181.

[37] M. Mizuno, et al., "Elastic Interconnects: Repeater-Inserted Long Wiring Capable of Compressing and Decompressing Data," *International Solid-State Circuits Conference*, Digest of Technical Papers, 2001, pp. 346–347.

[38] Y. Moisiadis, et al., "High Performance Level Restoration Circuits for Low-Power Reduced-Swing Interconnection Schemes," *Proceedings of the International Conference on Electronics Circuits and Systems*, December 2000, pp. 619–622.

[39] J. Muttersbach, et al., "Practical Design of Globally-Asynchronous Locally-Synchronous Systems," *Proceedings of the International Symposium on Advanced Research in Asynchronous Circuits and Systems*, 2000, pp. 52–59.

[40] Y. Nakagome, et al., "Sub-1-V Swing Internal Bus Architecture for Future Low-Power ULSI's," *IEEE Journal of Solid-State Circuits*, Vol. 28, April 1993, pp. 414–419.

[41] E. Schutt-Aine and S.M. Kang, Special Issue on Interconnection, *Proceedings of the IEEE*, Vol. 89, No. 4, April 2001.

[42] S. Ramprasad, et al., "A Coding Framework for Low-Power Address and Data Busses," *IEEE Transactions on VLSI Systems*, Vol. 7, June 1999, pp. 212–221.

[43] T. Sakurai and K. Tamura, "Simple Formulas for Two- and Three-Dimensional Capacitances," *IEEE Transactions on Electron Devices*, Vol. ED-30, No. 2, February 1983, pp. 183–185.

[44] E. Seevinck, et al., "Current-Mode Techniques for High-Speed VLSI Circuits with Application to Current Sense Amplifier for CMOS SRAM's," *IEEE Journal of Solid-State Circuits*, Vol. 26, April 1991, pp. 525–536.

[45] Y. Shin, et al., "Narrow Bus Encoding for Low-Power DSP Systems," *IEEE Transactions on VLSI Systems*, Vol. 9, October 2001, pp. 656–660.

[46] Y. Shin, et al., "Partial Bus-Invert Coding for Power Optimization of System Level Bus," *Proceedings of the International Symposium on Low Power Electronics and Design*, August 1998, pp. 127–129.

[47] M. Sinha, et al., "Current-Sensing for Crossbars," *Proceedings of the International ASIC/SOC Conference*, September 2001, pp. 25–29.

[48] S.-J. Lee, et al., "An 800 MHz Star-Connected On-Chip Network for Application to Systems on a Chip," *IEEE International Solid-State Circuits Conference*, Digest of Technical Papers, February 2003, pp. 468–469.

[49] S.-J. Lee, et al., "Packet-Switched On-Chip Interconnection Network for System-on-Chip Applications," *IEEE Transactions on Circuits and Systems II*, Vol. 52, June 2005, pp. 308–312.

[50] S.-J. Lee, et al., "Adaptive Network-on-Chip with Wave-Front Train Serialization Scheme," *IEEE Symposium on VLSI Circuits*, Digest of Technical Papers, June 2005, pp. 104–107.

[51] M.R. Stan, et al., "Bus-Invert Coding for Low-Power I/O," *IEEE Transactions on VLSI Systems*, Vol. 3, March 1995, pp. 49–58.

[52] C. Svensson, "Optimum Voltage Swing on On-Chip and Off-Chip Interconnect," *IEEE Journal of Solid-State Circuits*, Vol. 36, July 2001, pp. 1108–1112.

[53] T. Villiger, et al., "Self-Timed Ring for Globally-Asynchronous Locally-Synchronous Systems," *Proceedings of the International Symposium on Asynchronous Circuits and Systems*, 2003, pp. 141–150.

[54] H. Wang, et al., "Power-Driven Design of Router Microarchitectures in On-Chip Networks," *Proceedings of the IEEE/ACM International Symposium on Microarchitecture*, 2003, pp. 105–116.

[55] H. Wang, et al., "A Technology-Aware Energy-Oriented Topology Exploration for On-Chip Networks," *Proceedings of the Design, Automation and Test in Europe Conference*, March 2005, pp. 1238–1243.

[56] P. Wijetunga, et al., "High-Performance Crossbar Design for System-on-Chip," *Proceedings of the IEEE International Workshop on System-on-Chip for Real-Time Applications*, June 2003, pp. 138–143.

[57] F. Worm, et al., "A Robust Self-Calibrating Transmission Scheme for On-Chip Networks," *IEEE Transactions on VLSI Systems*, Vol. 13, January 2005, pp. 126–139.

[58] H. Yamauchi, et al., "An Asymptotically Zero Power Charge-Recycling Bus Architecture for Battery-Operated Ultrahigh Data Rate ULSI's," *IEEE Journal of Solid-State Circuits*, Vol. 30, April 1995, pp. 423–431.

[59] H. Zhang, et al., "Low-Swing On-Chip Signaling Techniques: Effectiveness and Robustness," *IEEE Transactions on VLSI Systems*, Vol. 8, June 2000, pp. 264–272.

# CHAPTER 4

# THE DATA-LINK LAYER IN NOC DESIGN*

The continued scaling in process technologies makes it imperative to consider reliability as a cross-cutting problem concerning not only test engineers but also system designers. In the *network-on-chips* (NoC) domain, the task to guarantee reliable data transfers across inherently unreliable physical links is performed by the data-link layer. In this context, communication reliability will be strongly impacted by the reliability of switch-to-switch connections. The main challenge is represented by the increased prominence of noise sources with shrinking feature sizes. Lower supply voltages, smaller nodal capacitances, a decrease of inter-wire spacing, the increasing role of coupling capacitance, the higher clock frequencies will make NoC communication increasingly sensitive to both internal (power supply noise, crosstalk noise, inter-symbol interference) and external (*electromagnetic interference* (EMI), thermal noise, noise induced by alpha particles) noise sources.

Traditional fault models are not proving capable of capturing the characteristics and the effects of failure mechanisms affecting on-chip communication links. Disturbances will in fact concern multiple adjacent wires, so that errors on these wires can no longer be considered as statistically independent, as is the case for many noise models traditionally used to assess performance and error resilience of on-chip communication schemes. Moreover, errors will be inherently bidirectional, thus exceeding the detection capability of traditional error control techniques such as the Berger code.

Common practices to enhance communication reliability include the insertion of repeaters, shielding of link wires, proper global wire configurations, spacing rules, error control coding-aware layout modifications, or unbalancing of link line drivers. The common feature of these techniques is that they require layout knowledge or electrical-level circuit modifications. In contrast, the data-link layer implements a technology-independent approach to the communication reliability problem, and redundant link encoding is proving capable of effectively achieving this goal.

A code protecting data on switch-to-switch links must allow for fast decoding because the decoding delay adds up to the critical path and limits

---

* This chapter was provided by Davide Bertozzi of University of Ferrara, Italy.

maximum operating frequency. Additionally, area constraints motivate a switch design with as few buffers as possible. The requirement of low-area fast decoding can be fulfilled by combinational logic circuits with low logic depth. Therefore, parity-based codes or *cyclic redundancy checks* (CRCs) are of primary interest for this application domain, such as Hamming codes, Hsiao codes or CRC codes. The implementation complexity of these error control schemes depends on the error recovery technique (packet dropping associated with time-out mechanisms, retransmission of corrupted data words with explicit feedback channels or error correction), on code-specific parameters (number of redundant link lines, generator polynomial of a CRC code, all related to the target error detection or correction capability of the code), and on implementation-related parameters (e.g., the wire voltage swing). However, their performance and energy dissipation is related not only to circuit complexity, but also to operating conditions such as wire load capacitance, traffic injection rate into the network and bit error rate.

As a consequence, communication reliability cannot be addressed in isolation, but power and performance implications need to be considered due to the limited power budgets of portable devices and to the stringent timing requirements of real-time systems. Several works in the open literature point out the conflicting requirements that many times *quality-of-service* (QoS) metrics, such as target probability of undetected errors or energy/performance constraints, pose on *error control code* (ECC) design. Several techniques have been investigated to span the power-reliability trade-off for on-chip communication links, for instance tuning the voltage swing of each coding scheme so to match its detection capability to the requirements on communication reliability. Alternatively, specific codespace features can be exploited. Recently, unified link coding architectures have been proposed to address crosstalk noise, power consumption and ensure general-purpose reliability in a unified framework.

More aggressive approaches have been proposed, which combine the principles of self-calibrating circuits with *dynamic voltage scaling* (DVS) applied to on-chip interconnects. In this way, conservative assumptions concerning application bandwidth requirements and operating conditions (e.g., noisy environment) can be avoided, resulting in more energy efficient implementations. Other techniques are even more revolutionary, since they introduce new fault-tolerant NoC communication protocols, leveraging probabilistic broadcast algorithms and multiple transmission schemes combined with error detection. Their goal is to achieve fault-tolerant communication at the cost of increased network traffic, since many researchers agree on the fact that bandwidth is largely available in today's NoCs.

Addressing communication reliability at the data-link layer depends on two fundamental conditions. First, the flow control mechanism implemented in a given NoC architecture must provide support for fault tolerance, otherwise solutions at a higher level of abstraction have to be devised. Second, the overall cost of link-level protection schemes in terms of buffering resources, overhead network traffic and performance degradation must be lower with respect to an end-to-end data protection scheme.

This chapter provides principles and guidelines for data-link layer design in the NoC domain. The reader will find an overview of state-of-the-art design techniques and of guiding principles which are emerging from research efforts reported in the open literature. In particular, basics of error control coding will be provided, with emphasis on those coding schemes (and their relevant properties) which are more frequently used by NoC designers. This chapter at first presents the theoretical framework which is at the core of data-link layer design and then introduces practical architectural solutions. In particular, the requirements the data-link layer poses on flow control implementation and the viability of a link-level error control with respect to end-to-end data protection will be carefully investigated by means of practical case studies and leveraging real NoC prototypes.

## 4.1 TASKS OF THE DATA-LINK LAYER

As mentioned in the introduction, the main purpose of data-link protocols is to abstract the physical layer as an unreliable digital link, where the probability of bit upsets is non-null. Since upset probabilities are increasing as technology scales down, this will be a critical and challenging task for NoC designers, and it is also a promising but almost unexplored research field. Therefore, communication reliability will be the main topic covered by this chapter.

However, it is worth observing that another important function of the data-link layer is to regulate access to shared communication resources. As a consequence, the data-link layer can be partitioned in two sublayers. The lower sublayer (i.e., closest to the physical channel) is called *media access control* (MAC), while the higher sublayer is the *data-link control* (DLC). The MAC regulates access to the medium, while the DLC increases channel reliability, for example, by providing error detection and/or correction capabilities.

Contention resolution is fundamentally a non-deterministic process [80], and is a key to avoid network congestion, that is, an interconnect working condition where an increase of offered bandwidth yields decreasing network performance. The development of fast arbiters has been extensively studied in computer networks [81–83], as one of the dominant factors for high-performance network switches. Also, fast and efficient switch arbiters are needed to switch packets in NoCs, and in many cases implementation schemes for baseline arbitration algorithms (such as fixed priority or round-robin) can be reused from the computer network community [18, 84].

Since NoCs will most likely replace state-of-the-art busses in highly integrated *multiprocess systems-on-chips* (MPSoCs), it is important to understand that relevant differences do exist between busses and NoC arbitration schemes. In a bus, master modules request access to the interconnect, and the arbiter grants access for the whole interconnect at once. Arbitration is global and centralized since there is only one arbiter component which has visibility of all access requests as well as the state of the interconnect. Moreover, when the path is given, the complete path from source to destination is exclusively reserved. We distinguish between split and non-split busses. In the latter bus architecture, arbitration takes place just once when a transaction is initiated. As a result, the bus is granted for both the request and the response phases of the transaction. This is the case of the *advanced microcontroller bus architecture* (AMBA) *advanced high-performance bus* (AHB) [85], where the communication medium is rearbitrated only when the slave response (as notified by signal *hready*) is correctly sampled at the master. However, the AHB protocol also supports split transactions, which require two arbitration rounds, one for the request and other for the response. More advanced communication architectures, such as the STBus [79], leverage two physical channels, each one with its own arbiter, handling requests and responses, respectively. This is an example of split bus architecture, featuring two separate arbitration rounds per transaction.

In NoCs, contention for communication resources is almost unavoidable. Although circuit switching enables efficient data transmission in a contention-free regime, it requires arbiters just at the same during the circuit set-up phase. In fact, programming packets are sent across the network to reserve a physical path from source to destination, and they might collide in order to get the reservation of the same resource. As a consequence, contention-free routing is the only way to avoid using arbitration logic. With this scheme, packet injection in the network is scheduled such that packets never contend for the same link at the same time. The pipelined time-division-multiplexing scheme used in Ref. [42] is one relevant example of global network scheduling technique. However, contention-free communication comes at a very high cost, and is usually implemented to provide QoS guarantees to time-constrained traffic.

Contrarily to busses, in an NoC the arbitration is fully distributed, since it is performed at every router, and is based on local information. In the most general router architecture scheme (assuming wormhole switching), an arbitration stage is required at each input port, in order to serialize crossbar switch access requests of ready-to-transmit virtual channels. Moreover, another arbitration stage is required to allocate slots on the central crossbar switch, in case multiple input ports are contending for the same output port. Packetization of information flow is instrumental in implementing fair sharing of communication resources. For instance, in the pipeline of wormhole routers the routing and output virtual channel allocation stages are performed for each packet, while crossbar switch allocation and traversal are performed on a flit-by-flit basis.

Current NoC designs typically use standard round-robin token passing schemes for bus arbitration [40, 42, 70]. Alternatively, priority queuing has been employed in QoS-oriented networks, where the arbiter favors packets from the non-empty highest-priority queue [41]. Arbitration strategies and implementation approaches to contention resolution have been extensively covered by Dally and Towles [18], and will not be addressed here, since they represent well-known and consolidated concepts and ideas. Instead, we devote the remaining part of this chapter to the exploration of recent and challenging issues in data-link layer design for NoCs, which are mostly related to communication reliability.

## 4.2   ON-CHIP COMMUNICATION RELIABILITY

The first objective of the data-link layer is to abstract physical communication links as higher-level communication channels. The main aspect of such abstraction process consists of dealing with the increased sensitivity of interconnects to on-chip noise sources, and of exposing a reliable communication service in spite of the underlying unreliable medium. Several design techniques [1–4] and analysis techniques [5–7] have been developed to help design for margin and minimize signal integrity problems. To the limit, interconnect-centric design methodologies have been devised, which try to address wiring issues at the beginning of chip design (e.g., wire plan for performance and for noise immunity [8]). Even so, process variations, manufacturing defects and unpredictable noise sources (e.g., an unexpected increase in cross-coupling capacitance between interconnects) may result in glitches and delay effects, causing logic errors and failure of the chip. The data-link layer should be able to control such residual errors, by implementing an error-tolerant communication scheme at a higher level of abstraction, increasing the communication reliability over an on-chip interconnect to the desired level.

One of the most widely used techniques to implement reliable systems consists of adding redundant information to the original data. As an example, error detecting and correcting codes append check bits to the data bits to enable detection and correction of erroneous bits. Error detecting codes are widely deployed for the implementation of *self-checking circuits* (SCCs), mainly because of design cost considerations and because they allow error recovery to be carried out either in hardware or in software. Basically, a functional unit provides an information flow protected by an error detecting code, so that a checker can continuously verify the correctness of the flow and provide an error indication as soon as it occurs. In the networking domain, the well-known disadvantage of this scheme lies in the feedback channel to relay the retransmission requests from the receiver back to the transmitter when errors are detected. For on-chip realizations, the feedback channel might be inherently provided by some flow control schemes (such as *ACKnowledge/Not ACKnowledge* (ACK/NACK) [9]), but the main problem is in the retransmission delay, which prolongs system energy dissipation and which impairs the performance of latency critical applications. Of course, the much lower frequency of retransmissions in the on-chip domain (orders of magnitude lower than the bit error rate of wireless channels) plays in favor of these schemes.

On the other hand, error correcting codes incur the minimum error recovery penalty, from a performance viewpoint, in that a correction logic at the receiver is able to reconstruct the error-free message. The correction delay is the only timing penalty here. However, for on-chip realizations, such delay might end up being in the critical path, therefore error correcting decoders with minimum logic depth must be used in this context. Moreover, the overhead for the correction circuitry, which is active at each clock cycle, might seriously degrade communication power metrics and increase area occupancy. Finally, in case an error pattern occurs, which exceeds the correction capability of the error correcting code, a decoding error is made. In the telecommunication domain, hybrid schemes are often used, where both error correction and error detection are deployed: most likely error patterns can be corrected, while the more infrequently occurring error patterns are detected and retransmitted. The viability of this approach for error-tolerant NoC-based communication architectures has not been proven yet.

Based on the observation that many faults in *very-large-scale integration* (VLSI) circuits cause unidirectional errors (i.e., 0-1 or 1-0 errors, provided the two kinds of errors do not occur simultaneously), the testing community has traditionally relied on the *Berger code*, which is the optimal separable all-unidirectional error detecting code. No other separable code can detect all-unidirectional errors with a fewer number of check bits [10]. The check bits are the binary representation of the number of 0s counted in the information bits. Unfortunately, the effects of many noise sources affecting *deep submicron* (DSM) on-chip interconnects can be better abstracted as multiple bidirectional errors rather than unidirectional errors. For instance, crosstalk causes bidirectional errors, when two coupled lines switch in the opposite direction and both transitions are delayed inducing sampling errors.

Many solutions have been proposed to overcome the detection capability limitation of traditional error control schemes with respect to multiple bidirectional errors [11, 12]. Very often, such techniques require layout knowledge or electrical-level design techniques, and a more general and abstract approach is desirable. The data-link layer in NoC design implements a technology-independent approach to the communication reliability problem through redundant link encoding. The main challenge is to select the most suitable coding framework for the NoC scenario, which involves identifying the most frequent error patterns, comparing the detection capability and the energy efficiency of pre-existing codes in dealing with these patterns or eventually coming up with new coding schemes.

## 4.3  FAULT MODELS FOR NoCs

Error-tolerant schemes for NoCs have to deal with three main issues: how the communication across a link is going to fail (the failure mode), how often this is going to happen and how to restore the corrupted information. The first information is captured in the *fault model*, which expresses the relevant behavior of the failure mode while hiding complex details on its physical generation mechanism.

Failures can be either static or dynamic. *Static failures* are present in the network when the system is powered on. *Dynamic failures* appear at random during the operation of the system. The fault models of both types of failures are generally referred to as *permanent*, that is, they remain in the system until it is repaired (e.g., stuck-at-faults). Alternatively, *transient faults* generally occur in the field. It has been showed that 80% of system failures are associated with transient faults [13–15]. Two well-known examples of transient faults are represented by *soft errors* and *timing faults* [16].

*Soft errors*, also called *single-event upsets* (SEUs), are radiation-induced transient errors caused by neutrons generated from cosmic rays and alpha particles generated by packaging material. An SEU is the consequence of a *single-event transient* (SET) created on a sensitive node by a particle striking an integrated circuit. When an SET occurs on a memory-cell node and flips the state of the cell, it is transformed to an SEU. Similarly, an SET occurring on a node of logic network is transformed to an SEU when a latch captures it. Traditionally, soft errors were regarded as a major concern only for DRAM and space applications, where radiation levels are higher. For designs manufactured at advanced technology nodes, such as 65 or 45 nm, devices are smaller, therefore the capture area into which a neutron would have to pass to cause trouble decreases, but the operating voltage is also going down, thus outweighing the reduction in area. A significantly lower charge deposed by a particle strike suffices to flip the logic value of a node, creating a transient pulse (SET). The transient pulse, after propagation through the logic network, can be captured by a latch to create an SEU. In the past, the probability of occurrence of an SEU in logic parts was drastically lower than in memories, since: (i) the propagation through logic gates can filter the induced transient pulse and (ii) a transient pulse propagated through a logic network can result in a logic error if it reaches the input of a latch in coincidence with the set-up or the hold period of the clock cycle. Unfortunately, at each technology generation, transient pulses become wider with respect to logic transition time of the logic gates (operating speeds are increasing), and end up being therefore propagated without attenuation through the logic network. In addition, as the clock frequency increases, the probability of latching a transient pulse increases as well. Due to these trends, the error rates in logic parts become as high as the error rates in memories.

Timing faults are the consequence of signal integrity problems such as crosstalk or ground bounce. The complexity of signal integrity verification in ultra-scaled silicon technologies increases the probability that a design, which under some circumstances exceeds its timing budget, is not detected as such by the signal integrity verification process. Timing faults can also be the consequence of the combination of a signal integrity issue with a fabrication defect. Process parameter variations and other defects (shorts, opens, etc.) often increase the path delays of a circuit. Eliminating such defects during fabrication testing requires to test the timing faults under the worst delay conditions, such as the excitation and propagation of the defect together with worst crosstalk and/or ground bounce conditions. Testing the huge number of paths in modern circuits for such complex test conditions makes the test generation process computationally infeasible and the test length unrealistic. Thus, circuits with timing faults might be declared fault-free by fabrication testing.

Failure modes and fault models for typical off-chip interconnection networks are reported in Table 4.1 [18]. Some failures, such as Gaussian noise and alpha-particle strikes, cause transient faults that result in one or more bits being in error but do not permanently impair machine operation. Others, such as electromigration of a line, cause a permanent failure of some module. The stuck-at fault involves that some logical node is stuck at logic one or zero. In presence of fail-stop faults, some components (link or router) stop functioning and inform adjacent modules that they are out-of-service. Some types of failures are Byzantine in that rather than stopping operation, the system continues to operate, but in a malicious manner,

**TABLE 4.1** ■ Failure modes and fault models traditionally considered for interconnection networks.

| Failure mode | Fault model | Typical value | Units |
|---|---|---|---|
| Gaussian noise on a channel | Transient bit error | $10^{-20}$ | BER (errors/bit) |
| Alpha-particle strikes on memory (per chip) | Soft error | $10^{-9}$ | SER ($s^{-1}$) |
| Alpha-particle strikes on logic (per chip) | Transient bit error | $10^{-10}$ | BER ($s^{-1}$) |
| Electromigration of a conductor | Stuck-at-fault | 1 | MTBF (FITs) |
| Threshold shift of a device | Stuck-at-fault | 1 | MTBF (FITs) |
| Connector corrosion open | Stuck-at-fault | 10 | MTBF (FITs) |
| Cold solder joint | Stuck-at-fault | 10 | MTBF (FITs) |
| Power supply failure | Fail-stop | $10^4$ | MTBF (FITs) |
| Software failure | Fail-stop or Byzantine | $10^4$ | MTBF (FITs) |

purposely violating protocols in an attempt to cause adjacent modules to fail. Fortunately, such faults are quite rare in practice, although tough to contain. Stuck-at-faults and fail-stop faults are examples of permanent faults and are usually described in terms of their *mean-time between failure* (MTBF) in units of time (often expressed in hours). Sometimes such failure rates are expressed in failures in $10^9$ h (FITs).

As interconnection networks move from off-chip to on-chip realizations, transient faults affecting switch-to-switch links are expected to increase, so that communication architectures cannot be considered to be 100% reliable anymore. Shrinking of feature sizes is determining an increased sensitivity to on-chip noise sources such as crosstalk, EMI, radiations, process variations and so on.

Present and future technologies show a decrease of the interwire spacing. Moreover, wires are relatively higher and thinner than in previous technologies, therefore *coupling capacitance* is playing a dominant role as a speed-limiting factor, also considering the Miller effect. Meanwhile, crosstalk noise is seriously threatening signal integrity, since the flip-flops at the receiver side of a link might sample incorrect data and propagate it throughout the system, thus compromising system's correct operation. Crosstalk noise might even induce multiple bidirectional errors, thus pushing the development of ECCs targeting these errors rather than unidirectional errors in an efficient way. Crosstalk on a link depends not only on physical and electrical parameters, but also on dynamic switching behavior (the sequence of bit patterns), therefore links should be designed for the worst-case crosstalk-induced delay or should leverage information about link data.

Long on-chip interconnects, typical for instance of customized, domain-specific irregular NoC topologies, are increasingly susceptible to EMI. Moreover, as the level of chip integration keeps growing, the chip itself generates electrical noise (e.g., due to high-frequency analog components in mixed-signal ICs or to high-frequency clock signals).

Although short switch-to-switch links in regular networks will be more immune to such noise sources, their load capacitance is expected to be much smaller than that of today's highly connected SoC busses, since they will be point-to-point connections and not shared resources among all communication initiators. In turn, this will make them more susceptible to transient faults because of cosmic radiation or alpha particles. Even if the injected charge will not suffice to flip the logic value of a wire, it might result in delay variations and it might generate transient pulses when affecting link drivers (receivers) and encoders (decoders). As a result, an increase of SEUs is expected.

At decreasing feature size, unavoidable *process variations* may lead to a greater variance of circuit parameters. Random fluctuation of device properties may for instance result in variations of transistor current drive capabilities and propagation delays [17]. Not only transistors, but also interconnects can be altered: a change in cross-coupling capacitance increases delay variation on a link. In multi-GHz circuits, even short delay variations may result in timing faults.

Another important noise source for on-chip communication is represented by *power supply noise* due to *simultaneously switching outputs (SSO)* [19, 20]. This noise has five primary undesirable effects:

(i)     It increases the propagation delay through the switching drivers and the neighboring circuitry sharing the same $V_{dd}$ and ground distribution networks.

(ii)    It induces resonance in the power distribution network, which further alters the values of $V_{dd}$ and ground.

(iii)   It causes false switching of a circuit, affecting the logical one or logical zero voltage threshold.

(iv)    It causes false switching of a circuit whose input nets are capacitively coupled to $V_{dd}$ or ground rails.

(v)     It alters the logical value stored in a high-impedance node of a dynamic circuit, making a transistor that was expected to be off temporarily conducting.

On-die *temperature fluctuations* are another important factor causing VLSI parameters uncertainty. Large heterogeneous SoCs are likely to consist of areas featuring different operational activity and power consumption. Thermal variations are likely to influence device properties as well as wire delays, noise and IR losses. In addition, the contribution of *leakage* to power consumption is increased in DSM technologies and is almost equal to the dynamic power [21]. This will cause an additional uncertainty in energy ranges, in that less active systems with multiple transistors might consume more static power than highly active systems consuming dynamic power. This consideration is important when designing complex drivers and receivers for on-chip communication, which is our focus here, since the encoder and decoder power overhead will have to increasingly account for static power dissipation.

Finally, as routing a single clock tree for large circuits will be increasingly difficult and is likely to lead to high power consumption, future network-centric ICs will have several clock domains and will be designed *globally asynchronous, locally synchronous* (GALS). For such designs, the communication can be affected by *synchronization errors*.

## 4.3.1   Bit Error Rate Model

Deriving an analytical model of the raw bit error rate and using it to forbid operating conditions and design parameters that do not meet reliability constraints does not seem a viable option today. In fact, there exists no accurate global model of raw bit error rate and this fact is unlikely to change as the complexity and interaction of physical phenomena affecting on-chip functionality and communication increases.

As a result, a number of design problems arise. For instance, it is very difficult to assess how efficiently an error correcting/detecting scheme can react to error conditions, power- and performancewise. In fact, these quality metrics are tightly related to the activation frequency of error recovery mechanisms (e.g., number of retransmissions and impact on application perceived performance).

Moreover, conservative design techniques for communication reliability can be hardly relaxed, since the number of self-induced errors cannot be precisely estimated. The most relevant example thereof is represented by the impact that lowering wire voltage swing might have on the bit error rate. A precise model linking the two effects would help designers to select the desired power-reliability trade-off point for their communication schemes.

In practice, the alternative is between implementing *self-calibrating* communication controllers, able to reduce the voltage swing until an increase in the bit error rate is detected, or assessing the reliability and power efficiency of new coding schemes by means of highly simplified (and hence less accurate) *bit error rate models*.

For the latter case, the most widely used modeling approach is the following [22]. Every time a transfer occurs across a wire, it can make an error with a certain probability $\varepsilon$, which depends on the knowledge of different noise sources and their dependence on the voltage swing. The following assumptions can be made to simplify the modeling problem: (i) for the purpose of statistical analysis, the sum of several uncorrelated noise sources affecting a link line can be modeled as a *Gaussian noise source*; and (ii) errors occurring on the different link lines are supposed to be *independent*.

As a result, the model assumes that a Gaussian distributed noise voltage $V_N$ with variance $\sigma_N^2$ is added to the signal waveform to represent the cumulative effect of all noise sources. Then, the probability of bit error is given by:

$$\varepsilon = Q\left(\frac{V_{sw}}{2\sigma_N}\right) \tag{4.1}$$

where $V_{sw}$ is the voltage swing and $Q(x)$ is the Gaussian pulse:

$$Q(x) = \int_x^\infty \frac{1}{\sqrt{2\pi}} e^{-\frac{y^2}{2}} \, dy \tag{4.2}$$

This model accounts for the decrease of noise margins (and hence for an increase of the line flipping probability or BER $\epsilon$, as shown in Fig. 4.1) caused by a decrease of the voltage swing across a wire, and allows

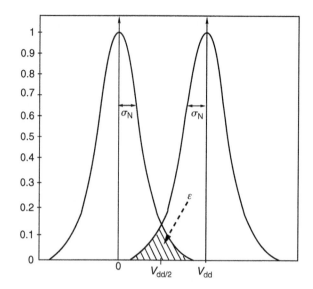

■ **FIGURE 4.1**

Model of the bit error probability $\varepsilon$ on a single wire. Lowering the voltage swing (reported on the x-axis) leads to an increased bit error rate (the overlapping of the Gaussian curves).

a simplified investigation of the energy-reliability trade-off for different error control schemes. It has been adopted in many research efforts to evaluate energy efficiency of reliable on-chip interconnects [23–29].

### 4.3.2   New Fault Model Notations

In the presence of crosstalk effects, statistical independence of errors on adjacent bus lines cannot be assumed any more. A new fault model notation for links which can represent multiple-wires, multiple-cycle faults has been proposed in Ref. [30].

All faults that are caused by the same physical effect belong to one fault of type $f_i$. Faults of different types are statistically independent. The *Fault Model* $\mathcal{FM}(f_i)$ describes faults of type $f_i$ in a given link architecture. It is based on the fault's probability of occurrence ($\alpha_i$), their characteristics (expressed in the matrix $P_i$) and a distance metric $DM_i$, influenced by the physical-link layout:

$$\mathcal{FM}: f_i \rightarrow (\alpha_i, P_i, DM_i)$$

$\alpha_i$ is the probability of occurrence of a fault of type $f_i$ per wire and cycle. Thus, $\alpha_i = 1/128$ leads to one fault of type $f_i$ occurring on a 128-bit wide link every cycle or every eight cycles on a 16-bit link.

When a fault of type $f_i$ occurs, it has a probability $p_i(\omega, d, e)$ to affect $\omega$ wires for a duration of $d$ time units (typically clock or transfer cycles) with the *effect e*. The effect of a fault is selected from a list of all possible effects, such as "logic value on wires inverted" or "wires forced to logic 0." All possible combinations of $p_i(\omega, d, e)$ are represented in the normalized matrix $P_i$ of three dimensions $\omega_{\max_i} \times (d_{\max_i} + 1) \times e_{\max}$. $\omega_{\max_i}$ and $d_{\max_i}$ are the maximum values for the number of wires affected and the fault's duration, respectively. These maximum values can be different for every fault type $f_i$. While $d_{\max_i}$ can take any positive value, $\omega_{\max_i}$ must not exceed the link width $\omega_{\max}$ of the interconnects between the switches: $\omega_{\max_i} \le \omega_{\max}$. $e_{\max}$ is the number of different effects which can be described.

For example, when setting $e_{\max} = 1$ and thus restricting the model to the effect "logic inversion," the normalized matrix $P_i$ is written as:

$$P_i = \begin{bmatrix} p_i(1, 0, Inv) & \dots & p_i(1, d_{\max_i}, Inv) \\ p_i(2, 0, Inv) & \dots & p_i(2, d_{\max_i}, Inv) \\ \dots & \dots & \dots \\ p_i(\omega_{\max_i}, 0, Inv) & \dots & p_i(\omega_{\max_i}, d_{\max_i}, Inv) \end{bmatrix}$$

where the increasing row number corresponds to increasing number of wires, and the increasing column number (from left to right) corresponds to increasing fault duration. A fault generally lasts at least one clock cycle ($d \ge 1$). The elements of $P_i$ with the index $d = 0$ are used to indicate the probability of permanent faults. Since the matrix elements describe the probabilities of different characteristics of one fault occurrence, their values must satisfy the following condition:

$$\sum_{a=1}^{\omega_{\max_i}} \sum_{b=0}^{d_{\max_i}} \sum_{c=1}^{e_{\max}} p_i(a, b, c) = 1 \tag{4.3}$$

The manifestation of a fault type $f_i$ affecting $\omega$ wires for $d$ cycles with the effect $e$ is determined by the probability of occurrence and the normalized value $p_i(\omega, d, e)$ to a probability of $\alpha_i p_i(\omega, d, e)$.

**Example 4.1.** $P_1$ characterizes a fault that affects multiple wires ($e = Inv$) for multiple cycles:

$$P_1 = \begin{bmatrix} 0 & 0.65 & 0.1 \\ 0 & 0.2 & 0.05 \end{bmatrix}$$

meaning that a fault of type $f_1$ will be confined to one wire and one clock cycle (or transfer) in 65% of all occurrences. Another 20% of these faults will disturb two wires for one clock cycle. Only in 15% of the cases can the effects be noted for two cycles: on one (10%) or two wires (5%), respectively. Since the first column is null, faults of type $f_1$ will never lead to permanent errors.

One parameter of the fault characteristics is the number of wires that are affected by a fault. This does not include the information which wires

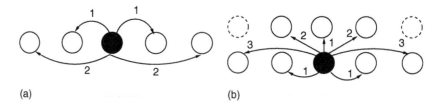

Example for distance relation between wires of a link running on one or two layers.

these are. Without loss of generality, we say that a fault affects $\omega$ adjacent wires and describes the relative distance between wires in a *distance metric* $DM_i$. This distance metric may be different for every fault type and bus layout. Zimmer and Jantsch [30] use a weighted, directed acyclic graph as shown in Fig. 4.2. The $\omega$ wires affected by a fault will be those with the lowest distance from the wire where the fault occurs. If multiple wires have the same distance, the choice among them is random.

For a complete representation of all faults that occur in a given environment, a set of fault models is necessary. This set is referred to as the *fault scenario* and is defined as:

$$S = \{\mathcal{FM}(f_1), \mathcal{FM}(f_2), ..., \mathcal{FM}(f_n)\}$$

When using a fault scenario in simulation, at every time step it will be checked for each wire if one or multiple faults have occurred (based on the probability of occurrence). If a fault of type $f_i$ has occurred, $P_i$ and $DM_i$ are evaluated to determine the fault's duration, effect and the affected wires.

Although *generalized fault model notations* like the one described above are being introduced, the problem of a lack of realistic fault scenarios for future DSM technologies remains, and a significant contribution of the research community is expected in this field.

### 4.3.3   Stochastic Failure Model

The failure model proposed in Refs [31, 32] assumes three common communication failure mechanisms for NoC-based systems. Because of crosstalk and EMI, the most common type of failures in DSM circuits will be data transmission errors (or upsets). If noise in the interconnect causes a message to be scrambled, a *data upset* error is said to occur, and it is characterized by a probability $p_{upset}$. Another assumption is that a message can be lost because of buffer overflow, and this is modeled by probabilities $p_{send\_miss}$ if it happens in a send buffer and $p_{recv\_miss}$ for a receive buffer. Since in Ref. [31] scrambled messages are treated as lost messages, the three probabilities can be combined into one, $p_{lost}$, which gives the odds that a message transmission fails, that is:

$$p_{lost} = p_{upset} + p_{send\_miss} + p_{recv\_miss} \qquad (4.4)$$

Moreover, it is assumed that tiles and links can be manufactured unreliably. More in general, the fault model depends on the following parameters:

- $p_{lost}$, the probability a message is lost (either because of data upsets or buffer overflow).

- $p_{tiles}$, the probability a tile is non-functional from manufacturing.

- $p_{links}$, the probability a link is defective from manufacturing.

If the links are experiencing arbitrary failures, the model also considers how the information transmitted is altered. If a message contains $n$ bits, the error vector is defined as: $e = (e_1, e_2, \ldots, e_n)$, where $e_i = 1$ if an error occurs in the $i$th transmitted bit and $e_i = 0$ otherwise. If all $2^n - 1$ non-null error vectors are equally likely to occur, we have the *random error vector model*. In this model, the probability of $e$ does not depend on the number of bit errors it contains, therefore:

$$p_{upset} = \sum_{e \neq 0} P[e] = (2^n - 1)p_v \approx 2^n p_v \rightarrow p_v \approx p_{upset}/2^n \qquad (4.5)$$

where $p_v$ is the probability of an error vector $e$. In contrast, in the random bit error model, $e_1, \ldots, e_n$ are independent of each other, so:

$$p_{upset} = 1 - \sum_{e=0} P[e] = 1 - (1 - p_b)^n \approx np_b \rightarrow p_b \approx p_{upset}/n \qquad (4.6)$$

where $p_b$ is the probability of a bit error.

The ultimate objective of this stochastic failure model is to emphasize the *non-deterministic nature* of DSM faults and it paves the way for a new class of NoC communication protocols called *on-chip stochastic communication*, introduced later in this chapter.

## 4.4  PRINCIPLES OF CODING THEORY

Redundant link encoding provides an effective and technology-independent approach to communication reliability. There are many different types of error detecting/correcting codes, but historically they have been classified into *block codes* and *convolutional codes* [33–35]. A block of $k$ data bits has to be appended by $n - k$ redundant parity bits in order to generate a codeword of an $(n, k)$ block code, where the parity bits are algebraically related to the $k$ data bits. Overall, the codeword consists of $n$ code bits. The ratio $R_C = k/n$ is called the *code rate*, where $0 < R_C \leq 1$.

On the other hand, convolutional codes are generated by the discrete time convolution of the input data sequence with the impulse response of the encoder. While block encoder operates on $k$-bit blocks of data bits, a convolutional encoder accepts a continuous sequence of input data bits.

Block codes and convolutional codes find potential applications in mobile radio systems. Some second generation digital cellular standards (e.g., GSM, IS-54) use convolutional codes, while others (e.g., PDC) use block codes. Although hard decision block decoders are easy to implement, there exist some very simple soft decision decoding algorithms (e.g., the Viterbi algorithm) for convolutional codes. As a result, convolutional codes are often preferred over block codes in the telecommunication domain.

In contrast, the NoC community has shown interest mainly for linear block codes because of the distinctive requirements of this domain. For link-level error control, the delay of encoders and decoders adds up to the delay of switch-to-switch links, thus limiting the clock frequency of the NoC datapath. Although in some state-of-the-art NoC prototypes the critical path is still in the control path [73], the propagation delay across on-chip interconnects degrades at each technology node. Therefore, it is likely that in future implementations the datapath will limit the maximum overall NoC clock frequency, and as few logic as possible has to be placed on this path. Moreover, it is already clear that most of the power in an on-chip network is drained by buffering resources. This requires error control codecs to mainly consist of combinational logic with low logic depth. Finally, since the power issue is already critical for early NoC prototypes, encoders and decoders should dissipate powers in the order of tens or hundreds of uW (microwatts). Common VLSI implementations of soft-output Viterbi decoders exhibit a few hundreds of mW power dissipation and more than $100\,k$ transistors in $0.18\,\mu m$ technology [72]. Such decoders can then be used as constituent decoders for Turbo codes in high-performance applications.

These considerations are pushing the adoption of very simple block codes for link-level error control in NoCs. Therefore, we will now focus on the main coding theory principles and go into the details of linear block codes only. Definitions and theorems have been selected from Refs [33–35], and re-ordered based on the objectives of this chapter. Proof of theorems will not be given while preserving readability. The interested reader can refer to those books for more details on coding theory.

Since the characteristics of the communication channel play an important role in the decoding decision rule, we will hereafter refer to a *binary symmetric channel* (BSC) for the sake of simplicity. A BSC is a memoryless communication channel, where the outcome of any one transmission is independent of the outcome of the previous transmissions. Moreover, the BSC has a channel alphabet consisting of two symbols $\{0, 1\}$, and channel probabilities:

$$P(1 \text{ received} \mid 0 \text{ sent}) = P(0 \text{ received} \mid 1 \text{ sent}) = p \qquad (4.7)$$

$$P(0 \text{ received} \mid 0 \text{ sent}) = P(1 \text{ received} \mid 1 \text{ sent}) = 1 - p \qquad (4.8)$$

where $p$ is the symbol (bit) error probability (also called the *crossover probability*).

We now wish to study decision rules for BSCs, that is, procedures that associate a codeword to a given received word or declare a decoding error. In particular, we want our decision rule not to depend on the *input distribution*, which consists of the probabilities that the various codewords are sent through the channel. This would involve some knowledge of the message being sent. Therefore, we are interested in a decision rule that, under the assumption of uniform input distribution, associates a codeword to a received word such that for no other codeword would it be more likely that the given word was received. This is the notion of *maximum likelihood decoding*.

**Example 4.2.** Suppose that codewords from the code $\{000, 111\}$ are sent through a BSC with crossover probability $p = 0.01$. If the string 100 is received, maximum likelihood decoding computes the following probabilities:

$$P(100 \text{ received} \mid 000 \text{ sent}) = P(1 \text{ received} \mid 0 \text{ sent})P(0 \text{ received} \mid 0 \text{ sent})^2$$

$$= p(1 - p)^2 = 0.009801$$

and

$$P(100 \text{ received} \mid 111 \text{ sent}) = P(1 \text{ received} \mid 1 \text{ sent})P(0 \text{ received} \mid 1 \text{ sent})^2$$

$$= (1 - p)p^2 = 0.000099$$

Since the first probability is larger than the second, the maximum likelihood decision rule decodes 100 as 000.

It can be showed that (see Ref. [34]):

**Theorem 1.** For a BSC, the maximum likelihood decision rule is to choose a codeword that differs in the fewest places with the received word $x$.

The *Hamming distance* is the traditional way of measuring the closeness of two strings.

**Definition 1.** Let $x$ and $y$ be strings of length $n$ over an alphabet $A$. The Hamming distance from $x$ to $y$, denoted by $d(x, y)$, is defined to be the number of places in which $x$ and $y$ differ.

Now suppose that $C$ is a code of length $n$. The codewords that are closest (in Hamming distance) to a given received word $x$ are referred to as *nearest neighbor codewords*. The *nearest neighbor decision rule* is the rule that decodes a received word $x$ as a nearest neighbor codeword.

A code $C$ is *u-error detecting* if, whenever a codeword incurs at least one but at most $u$ errors, the resulting string is not a codeword. A code $C$ is *v-error correcting* if nearest neighbor decoding is able to correct $v$ or fewer errors. Let us express these concepts more formally in terms of the minimum distance between codewords.

**Definition 2.** Let $C$ be a code with at least two codewords. The *minimum distance* $d(C)$ of $C$ is the smallest distance between distinct codewords.

In symbols:

$$d(C) = \min\{d(c,d) \mid c,d \in C, \ c \neq d\} \tag{4.9}$$

since $c \neq d$ implies that $d(c,d) \geq 1$, the minimum distance of a code must be at least 1.

In particular:

**Theorem 2.** A code $C$ is *u-error detecting* if and only if $d(C) \geq u + 1$.

**Theorem 3.** A code $C$ is *v-error correcting* if and only if $d(C) \geq 2v + 1$.

A key issue when considering error detection and correction is that both cannot take place at the same time and at maximum levels [34]. In other words, suppose a code $C$ has minimum distance $d$. Thus, it is $(d-1)$ – error detecting and $\lfloor (d-1)/2 \rfloor$ – error correcting. If we use $C$ for error detection only, it can detect up to $d-1$ errors. On the other hand, if we want $C$ to also correct errors whenever possible, then it can correct up to $\lfloor (d-1)/2 \rfloor$ errors, but may no longer be able to detect a situation where more than $\lfloor (d-1)/2 \rfloor$ but less than $d$ errors have occurred. For if more than $\lfloor (d-1)/2 \rfloor$ errors are made, nearest neighbor decoding might correct the received word to the wrong codeword, and thus the errors will go undetected. In a sense, employing *v-error correction* turns each codeword into a "magnet" that attracts any received word that is within a distance of $v$, even if the received word "came from" a more distant codeword. This issue is very important since, in practice, it is not uncommon to use a mixed strategy of both error correction and error detection.

**Definition 3.** Consider the following strategy for error correction/ detection. Let $v$ be a positive integer. If a word $x$ is received and if the closest codeword $c$ to $x$ is at a distance of at most $v$, and there is only one such codeword, then decode $x$ as $c$. If there is more than one codeword at minimum distance to $x$, or if the closest codeword has distance greater than $v$, then simply declare a word error. A code $C$ is simultaneously *v-error correcting* and *u-error detecting* if, whenever at least one but at most *v* errors are made, the strategy described above will correct these errors and if, whenever at least $v+1$ but at most $v+u$ errors are made, the strategy above simply reports a word error.

**Theorem 4.** A code $C$ is simultaneously *v-error correcting* and *u-error detecting* if and only if $d(C) \geq 2v + u + 1$.

If a codeword $c$ is transmitted, and if $d$ or more symbol errors are made, so that the received word $x$ has the property that $d(x, c) \geq d$, then $x$ will be closer to a different codeword, and so minimum distance decoding will definitely result in a decoding error.

Since the probability that a codeword of length $n$ is received correctly in a BSC is:

$$P(\text{no word error}) = (1 - p)^n \tag{4.10}$$

the probability of exactly $k$ symbol errors is:

$$P(k \text{ symbol errors}) = \binom{n}{k} p^k (1 - p)^{n-k} \tag{4.11}$$

therefore we get the following lower-bound on the probability of a decoding error:

$$P(\text{decoding error}) \geq \sum_{k=d}^{n} \binom{n}{k} p^k (1 - p)^{n-k} \tag{4.12}$$

On the other hand, the probability of correct decoding is at least as large as the probability of making $\lfloor (d - 1)/2 \rfloor$ or fewer errors (since the code is $\lfloor (d - 1)/2 \rfloor$ – error correcting), that is:

$$P(\text{decoding error}) = 1 - P(\text{correct decoding}) \leq 1 - \sum_{k=0}^{\lfloor \frac{d-1}{2} \rfloor} \binom{n}{k} p^k (1 - p)^{n-k} \tag{4.13}$$

## 4.4.1  Linear Block Codes

In block coding, the binary information sequence is segmented into message blocks of fixed length; each message block, denoted by $a$, consists of $k$ information digits. There are a total of $2^k$ distinct messages. The encoder, according to certain rules, transforms each input message $a$ into a binary $n$-tuple $c$ with $n > k$ ($n$ is said to be the codeword length and $n - k$ represents the number of check bits). This binary $n$-tuple $c$ is referred to as the codeword of the message $a$. Therefore, corresponding to the $2^k$ possible messages, there are $2^k$ codewords. This set of $2^k$ codewords is called a block code. For a block code to be useful, the $2^k$ codewords must be distinct. A binary block code is *linear* if and only if the modulo-2 sum of two codewords is also a codeword [33].

In linear block codes, codeword $c$ is generated from message $a$ through a linear mapping $c = aG$, where $G = [g_{ij}]_{k \times n}$ is a $k \times n$ matrix called the *generator matrix*. The matrix $G$ has full row rank $k$, and the code $C$ is generated by taking all linear combinations of the rows of the matrix $G$, where field operations are performed using modulo-2 arithmetic. In practice, the generator matrix allows to describe an $(n, k)$ linear code by simply giving a basis for $C$, which consists of $k$ linearly independent codewords in $C$, rather than having to list all of the $2^k$ individual codewords in the code. The rows of the generator matrix are the codewords in the basis of $C$. The whole task of designing a block code is to find the generator matrices that yield codes that are both powerful and easy to decode.

For any block code with generator matrix $G$, there exists an $(n-k) \times n$ *parity check matrix* $H = [h_{ij}]_{(n-k) \times n}$ such that $GH^T = 0_{k \times (n-k)}$. The matrix $H$ has full row rank $n-k$ and is orthogonal to all codewords, that is, $cH^T = 0_{n-k}$. The matrix $H$ is the generator matrix of a dual code $C^T$ consisting of $2^{n-k}$ codewords. The parity check matrix of $C^T$ is the matrix $G$.

## Syndromes

A *systematic block code* is one having a generator matrix such that the first $k$ coordinates of each codeword are equal to the $k$-bit input vector $a$, while the last $n-k$ coordinates are the parity check bits. By using elementary row operations, the generator matrix of any linear block code can be put into systematic form.

Suppose that the codeword $c$ is transmitted and the vector $y = c + e$ is received, where $e$ is defined as the error vector. The *syndrome* of the received vector $y$ is defined as:

$$s = yH^T \tag{4.14}$$

If $s = 0$, then $y$ is a codeword; conversely if $s \neq 0$, then an error must have occurred. Note that if $y$ is a codeword, then $s = 0$. Hence, $s = 0$ does not mean that no errors have occurred. They are just undetectable.

The syndrome only depends on the error vector because:

$$s = yH^T = cH^T + eH^T = 0 + eH^T = eH^T \tag{4.15}$$

In general, $s = eH^T$ is a system of $n-k$ equations in $n$ variables. Hence, for any given syndrome $s$, there are $2^k$ solutions for $e$. However, the most likely error pattern $e$ is the one that has minimum Hamming weight.

Since for a linear code the sum of any two codewords is another codeword, it follows that the number of undetectable error patterns is equal to $2^k - 1$, the number of non-zero codewords. In addition, since there are $2^n - 1$ possible non-zero error patterns, the number of detectable error patterns is:

$$2^n - 1 - (2^k - 1) = 2^n - 2^k \qquad (4.16)$$

where usually $2^k - 1$ is a small fraction of $2^n - 2^k$. For a $(7, 4)$ Hamming code (which will be introduced later), there are 15 undetectable error patterns, as compared with 112 detectable error patterns. Conversely, a linear block code is usually capable of correcting $2^{n-k}$ error patterns, which is equal to the number of syndromes.

It can therefore be observed that although a linear block code is able to detect all error patterns of $d - 1$ or fewer errors and to correct all error patterns of $\lfloor (d - 1)/2 \rfloor$ or fewer errors (where $d$ is the minimum distance), its detection capability is extended to many error patterns of more than $d - 1$ errors and its correction capability includes many error patterns of $\lfloor (d - 1)/2 \rfloor + 1$ or more errors.

Nearest neighbor decoding involves finding a codeword closest to the received word. If a code has no particular structure, it may be necessary to employ the brute force method of computing the distance from the received word to each codeword. This may be impractical, if not impossible, for large codes, therefore much better decoding techniques have been devised for linear block codes.

### Standard array decoding

This method is conceptually very simple. The *standard array* of an $(n, k)$ linear block code is constructed as follows:

1. Write out all $2^k$ codewords in a row starting with $c_0 = 0$.

2. From the remaining $2^n - 2^k$ $n$-tuples, select an error pattern $e_2$ of weight 1 and place it under $c_0$. Under each codeword put $c_i + e_2$, $i = 1, \ldots, 2^k - 1$.

3. Select a minimum weight error pattern $e_3$ from the remaining unused $n$-tuples and place it under $c_0 = 0$. Under each codeword put $c_i + e_3$, $i = 1, \ldots, 2^k - 1$.

4. Repeat the previous step until all $n$-tuples have been used.

For instance, consider the $(4, 2)$ code with generator matrix:

$$G = \begin{bmatrix} 1 & 1 & 0 & 0 \\ 0 & 1 & 0 & 1 \end{bmatrix}$$

The standard array is:

$$G = \begin{bmatrix} e_1 & 0000 & 1100 & 0101 & 1001 \\ e_2 & 0001 & 1101 & 0100 & 1000 \\ e_3 & 0010 & 1110 & 0111 & 1011 \\ e_4 & 0011 & 1111 & 0110 & 1010 \end{bmatrix}$$

The standard array consists of $2^{n-k}$ disjoint rows of $2^k$ elements. These rows are called *cosets* and the $i$th row has the elements:

$$F_i = \{e_i, e_i + c_1, \ldots, e_i + c_{2^k-1}\}$$

The first element $e_i$ is called the *coset leader*. The standard array also consists of $2^k$ disjoint columns. The $j$th column has the elements:

$$D_j = \{c_j, c_j + e_2, \ldots, c_j + e_{2^{n-k}}\}$$

To correct errors, the following procedure is used. When $y$ is received, find $y$ in the standard array. If $y$ is in row $i$ and column $j$, then the coset leader from row $i$, $e_i$, is the most likely error pattern to have occurred and $y$ is decoded into $y + e_i = c_j$. A code is capable of correcting all error patterns that are coset leaders. If the error pattern is not a coset leader then erroneous decoding will take place.

**Syndrome decoding**

*Syndrome decoding* relies on the fact that all $2^k$ $n$-tuples in the same coset of the standard array have the same syndrome. This is because the syndrome only depends on the coset leader as shown in equation (4.15). To perform syndrome decoding:

1. Compute the syndrome $s = yH^T$.

2. Locate the coset leader $e_l$ where $e_l H^T = s$.

3. Decode $y$ into $y + e_l = \hat{c}$.

This technique can be used for any linear block code. The calculation in Step 2 can be done by using a simple *look-up table* (LUT). However, for large $n - k$ it becomes impractical because $2^{n-k}$ syndromes and $2^{n-k}$ error patterns must be stored.

### 4.4.2  Hamming Codes

The family of Hamming codes was discovered independently by Marcel Golay in 1949 and Richard Hamming in 1950. Hamming codes are perfect codes [33] and have the advantage of being very easy to decode. However, they are just *single-error correcting* (SEC) codes, having minimum distance 3.

Let us study how these codes are constructed starting from their parity check matrices, and let us focus only on binary Hamming codes, although the generalization to $r$-ary Hamming codes is straightforward, where $r$ can be any prime power radix. Binary Hamming codes $\mathcal{H}_2(h)$ exist for any integer value $h$. A theorem introduces the principles of Hamming codes.

**Theorem 5.** Let $P$ be a parity check matrix for an $(n,k)$ linear code with minimum distance $d$. Then the minimum distance is the smallest integer $t$ for which there are $t$ linearly dependent columns in $P$.

Therefore, the parity check matrix of an $(n,k)$ linear code with minimum distance 3 has the property that no two of its columns are linearly dependent, that is, no column is a scalar multiple of another column, but some set of three columns is linearly dependent. We can easily construct such a parity check matrix with $h$ rows, and with maximum possible number of columns. We simply take the binary representation of those positive integers, in increasing order starting with 1, that have most significant value (i.e., the value in the leftmost non-zero position, columnwise) 1. For binary codes, this latter condition is obviously always met, therefore we do not need to worry about this. For instance, when $h = 3$ rows, we get:

$$H_2(3) = \begin{bmatrix} 0 & 0 & 0 & 1 & 1 & 1 & 1 \\ 0 & 1 & 1 & 0 & 0 & 1 & 1 \\ 1 & 0 & 1 & 0 & 1 & 0 & 1 \end{bmatrix}$$

whose columns are the binary representation of the numbers 1, 2, 3, 4, 5, 6 and 7. Matrices built based on this criteria are called *Hamming matrices*.

By observing the properties of the Hamming matrices, it can be proved that:

**Theorem 6.** The Hamming matrix $H_2(h)$ is a parity check matrix for a binary linear $(n,k)$ code $\mathcal{H}_2(h)$ with parameters $n = 2^h - 1, k = n - h, d = 3$.

The form of the Hamming matrices $H_2(h)$ allows for perhaps the most elegant decoding procedure of any code. In fact,

**Theorem 7.** If a codeword from the binary Hamming code $\mathcal{H}_2(h)$ suffers a single error, resulting in the received word $x$, then the syndrome $s = xH_2(h)^{\mathrm{T}}$ is the binary representation of the position in $x$ of the error.

The proof is straightforward. If the error string is $e_i$ (i.e., an error in the $i$th bit), then $s = e_iH_2(h)^{\mathrm{T}}$ is the $i$th column of $H_2(h)$, which is just the binary representation of the number $i$.

The code rate of the Hamming code is $\frac{k}{n} = 1 - \frac{h}{n}$, which tends to the maximum possible value of 1 as $n$ gets large. Unfortunately, the error correction rate is $\frac{\lfloor \frac{d-1}{2} \rfloor}{n} = \frac{1}{n}$, which tends to the worst possible value of 0 as $n$ gets large.

### 4.4.3  Cyclic Codes

*Cyclic codes* are a class of linear codes with the property that any codeword shifted cyclically (an end-around carry) will also result in a codeword. For example, if $c_{n-1}, c_{n-2}, \ldots, c_1, c_0$ is a codeword, then $c_{n-3}, \ldots, c_0, c_{n-1}, c_{n-2}$ is also a codeword. CRC codes are the most widely used cyclic codes.

Components of a codeword are usually treated as the coefficients of a polynomial in order to develop the algebraic properties of a cyclic code. The string $c_0, c_1, \ldots, c_{n-1}$ is associated with polynomial (with binary coefficients):

$$p(x) = c_0 + c_1 x + c_2 x^2 + \cdots + c^{n-1} x^{n-1}$$

The addition and scalar multiplication of strings corresponds to the analogous operations for polynomials. Thus, we may think of a codeword of length $n$ as a polynomial of degree less than $n$ and a binary linear code $C$ of length $n$ as a subspace of the set of all polynomials of degree less than $n$ with binary coefficients. In terms of polynomials, a right cyclic shift can be expressed by $xp(x)$, provided we replace $x^n$ by $x^0 (=1)$. This is exactly the same as dividing $xp(x)$ by $x^n - 1$ and keeping only the remainder, that is, taking the product $xp(x)$ modulo $x^n - 1$.

Let us denote by $\mathcal{R}_n(Z_p)$ the set of all polynomials of degree less than $n$, with coefficients from the set $Z_p$ of integers modulo $p$, and with the operations of addition of polynomials, scalar multiplication by an element of $Z_p$ and multiplication modulo $x^n - 1$. Note that taking the product modulo $x^n - 1$ is very easy, since it is just necessary to take the ordinary product and then replace $x^n$ by 1. For instance, in $\mathcal{R}_4(Z_2)$:

$$(x^3 + x^2 + 1)(x^2 + 1) = x^5 + x^4 + x^3 + 1 = x^4 \cdot x + x^4 + x^3 + 1$$

$$= 1 \cdot x + 1 + x^3 + 1 = x^3 + x$$

Moreover, since the polynomials in $\mathcal{R}_n(Z_p)$ have coefficients in a finite field, their properties are different from those of the polynomials with coefficients in the field of real numbers [34]. For instance, in $\mathcal{R}_4(Z_2)$, since any multiple of 2 is equal to 0 and since $-1 = +1$, we have:

$$(x - 1)^4 = x^4 - 4x^3 + 6x^2 - 4x + 1 = x^4 + 1 = 0$$

that is, the product of non-zero polynomials may equal the zero polynomial.

A binary linear code $C$ can be viewed as a subspace of the vector space $\mathcal{R}_n(Z_2)$. In addition, if $p(x) \in C$, then $x^k p(x)$ represents $k$ right cyclic shifts.

**Theorem 8.** A linear code $C \subseteq \mathcal{R}_n(Z_p)$ is cyclic if and only if $p(x) \in C$ implies that $f(x)p(x) \in C$ for any polynomial $f(x) \in \mathcal{R}_n(Z_p)$.

### Generator polynomial

The leading coefficient of a polynomial $p(x)$ is the coefficient of the largest power of $x$ that appears in the polynomial with non-zero coefficient. If the leading coefficient is equal to 1, then we say that $p(x)$ is *monic*.

Now suppose that $C$ is a cyclic code. Let $g(x)$ be a polynomial of smallest degree in $C$, among all non-zero polynomials in $C$. Without loss of generality, assume that $g(x)$ is monic. If $p(x) \in C$ then we may divide $p(x)$ by $g(x)$ to get:

$$p(x) = q(x)g(x) + r(x) \tag{4.17}$$

where the remainder $r(x)$ is either the zero polynomial or else has degree less than that of $g(x)$. Since $r(x) = p(x) - q(x)g(x) \in C$ and since $g(x)$ has the smallest degree of any non-zero polynomial in $C$, we deduce that $r(x)$ must be the zero polynomial. Therefore, $p(x)$ is a multiple of $g(x)$ and, more in general, any multiple of $g(x)$ is a codeword in $C$. An $(n,k)$ cyclic code is completely specified by its non-zero code polynomial of minimum degree $g(x)$ (called the *generator polynomial*), since the codeword for $k$ message bits can be obtained by multiplying the message polynomial by $g(x)$. It can also be proven that a cyclic code has one and only one generator polynomial.

The link between the generator polynomial and the generator matrix is given by the following theorem.

**Theorem 9.** Let $C$ be a non-zero cyclic code in $\mathcal{R}_n(Z_p)$ with generator polynomial $g(x) = g_0 + x g_1 + \cdots + x^k g_k$, of degree $k$. Then $C$ has generator matrix:

$$G = \begin{bmatrix} g_0 & g_1 & g_2 & \cdots & & g_k & 0 & 0 & \cdots & 0 \\ 0 & g_0 & g_1 & g_2 & \cdots & & g_k & 0 & \cdots & 0 \\ 0 & 0 & g_0 & g_1 & g_2 & \cdots & & g_k & \cdots & \\ \cdot & \cdot & \cdot & \cdot & \cdot & & \cdot & \cdot & \cdot & \cdot \\ 0 & 0 & \cdots & 0 & g_0 & g_1 & g_2 & \cdots & & g_k \end{bmatrix}$$

where each of the $n-k$ rows consists of a right cyclic shift of the row above.

Let us now investigate how to obtain the parity check matrix through the following two theorems.

**Theorem 10.** A monic polynomial $p(x) \in \mathcal{R}_n(Z_p)$ is the generator polynomial of a cyclic code in $\mathcal{R}_n(Z_p)$ if and only if it divides $x^n - 1$.

This theorem tells us that there is exactly one cyclic code for each factor of $x^n - 1$. Thus, we can find all cyclic codes in $\mathcal{R}_n(Z_p)$ by factoring $x^n - 1$. Moreover, we can write:

$$x^n - 1 = h(x)g(x), \quad h(x) \in \mathcal{R}_n(Z_p) \tag{4.18}$$

The polynomial $h(x)$, which has degree equal to the dimension of $C$, is called the *check polynomial* of $C$, and is unique. The following theorem proves the importance of this polynomial.

**Theorem 11.** Let $C$ be a cyclic code in $\mathcal{R}_n(Z_p)$, with check polynomial $h(x)$. Then a polynomial $p(x) \in \mathcal{R}_n(Z_p)$ is in $C$ if and only if $p(x)h(x) = 0$ in $\mathcal{R}_n(Z_p)$.

It can be easily proven that the $(n - k) \times n$ matrix built with the coefficients of the check polynomial:

$$H = \begin{bmatrix} h_k & h_{k-1} & h_{k-2} & \ldots & & h_0 & 0 & 0 & \ldots & 0 \\ 0 & h_k & h_{k-1} & h_{k-2} & \ldots & & h_0 & 0 & \ldots & 0 \\ 0 & 0 & h_k & h_{k-1} & h_{k-2} & \ldots & & h_0 & \ldots & \\ \cdot & \cdot & \cdot & \cdot & \cdot & \cdot & \cdot & \cdot & & \cdot \\ 0 & 0 & \ldots & 0 & h_k & h_{k-1} & h_{k-2} & \ldots & & h_0 \end{bmatrix}$$

is a parity check matrix for a cyclic code $C$.

### Burst errors

An $(n, k)$ cyclic code is capable of detecting any error burst of length $n - k$ or less, including the end-around bursts [33]. The length of a burst is the span from first to last error, inclusive. This notion of burst error is very important, since it is likely to better capture the nature of many on-chip errors, where the communication failure mechanism affects a certain number of contiguous lines. DSM technologies are particularly sensitive to these effects, in that the shrinking of geometries makes the relative distance between interconnects smaller, therefore even localized noise sources are likely to have an impact on multiple contiguous bus lines.

Early NoC implementations of burst error control capability are limited to error detection and not to error correction [24]. The reason lies almost entirely in the implementation complexity, as suggested by the following two theorems (proofs in Ref. [34]).

**Theorem 12.** If a linear $(n, k)$ code $C$ can detect all burst errors of length $b$ or less, then $n - k \geq b$.

**Theorem 13.** If a linear $(n, k)$ code $C$ can correct all burst errors of length $b$ or less, using a standard array, then $n - k \geq 2b$.

Since in practice $k$ is given by the number of parallel information lines in NoC links and $n$ could be the phit size, the target error burst length to detect translates into heavy requirements on the number of additional check bits a linear code must implement, especially when error correction is the error recovery technique of choice.

### 4.4.4　Cyclic Redundancy Check

CRC codes are perhaps the most famous and widely used class of cyclic codes for error detection in computer architectures and applications [74]. CRC codes only have error detection capabilities.

Choosing a good generator polynomial $g(x)$ is something of an art, and beyond the scope of this book [36]. Two simple observations: for an $r$-bit checksum (the number of check bits appended to the original message bits), $g(x)$ should be of degree $r$, because otherwise the first bit of the checksum would always be 0, which wastes a bit of the checksum. Similarly, the last coefficient should be 1 (i.e., $g(x)$ should not be divisible by $x$), because otherwise the last bit of the checksum would always be 0. The following facts about generator polynomials hold:

(i)　If $g(x)$ contains two or more terms, all single-bit errors can be detected.

(ii)　If $g(x)$ is not divisible by $x$ (i.e., if the last term is 1), and $e$ is the least positive integer such that $g(x)$ evenly divides $x^e + 1$ then all double errors that are within a frame of $e$ bits can be detected. A particularly good polynomial in this respect is $x^{15} + x^{14} + 1$, for which $e = 32,767$.

(iii)　If $x + 1$ is a factor of $g(x)$, all errors consisting of an odd number of bits can be detected.

(iv)　An $r$-bit CRC checksum can detect all burst errors of length $\leq r$.

Many common CRC codes detect also most weight 4 errors and most burst errors of length $r + 1$. In fact, the codewords of weight 4 form only a small fraction of the total number of words of weight 4. Moreover, the only bursts of length $r + 1$ that are codewords are the $n$ shifts of the generator polynomial $g(x)$. In practice, the most likely error patterns can be all detected by common CRC codes, although the distinctive constraints of on-chip communication may force to select only very simple generator polynomials (involving low check bits redundancy), so the error detection properties have to be discussed case by case, as in Ref. [25]. In addition, there is little published guidance and less quantitative data upon which to base trade-off decisions. Conventional wisdom is that the best way to select a CRC polynomial is to use one that is already commonly used. As a consequence, it turns out that many applications employ CRCs that provide far less error detection capability than they might achieve for a given number of check bits. Recently, guidelines for generator polynomial selection in embedded networks have been reported in Ref. [36].

### 4.4.5  Implementation of Hamming Codecs

Hamming encoders and decoders exhibit a very desirable property for on-chip implementations: they simply consist of low-depth networks of XOR gates. In order to better understand this, let us focus on the simple (7, 4) binary Hamming code example. The generator matrix for this code in *left standard form* (i.e., the first $k$ columns form a $k \times k$ identity matrix) is:

$$G = \begin{bmatrix} 1 & 0 & 0 & 0 & 0 & 1 & 1 \\ 0 & 1 & 0 & 0 & 1 & 0 & 1 \\ 0 & 0 & 1 & 0 & 1 & 1 & 0 \\ 0 & 0 & 0 & 1 & 1 & 1 & 1 \end{bmatrix}$$

To encode a source string $s = s_1 s_2 s_3 s_4$, we multiply $s$ by $G$:

$$c = sG = [s_1 \ s_2 \ s_3 \ s_4] \begin{bmatrix} 1 & 0 & 0 & 0 & 0 & 1 & 1 \\ 0 & 1 & 0 & 0 & 1 & 0 & 1 \\ 0 & 0 & 1 & 0 & 1 & 1 & 0 \\ 0 & 0 & 0 & 1 & 1 & 1 & 1 \end{bmatrix} \tag{4.19}$$

and we get:

$$c = [s_1 \quad s_2 \quad s_3 \quad s_4 \quad s_2 + s_3 + s_4 \quad s_1 + s_3 + s_4 \quad s_1 + s_2 + s_4] \tag{4.20}$$

The first four elements of $c$ are the elements of the source string $s$, while the others are the parity check bits $p_i$. We can therefore derive the parity equations that can be used to implement a Hamming encoder for this case:

$$p_1 = s_2 + s_3 + s_4 \tag{4.21}$$

$$p_2 = s_1 + s_3 + s_4 \tag{4.22}$$

$$p_3 = s_1 + s_2 + s_4 \tag{4.23}$$

where the + operations has to be meant as a logic XOR operation.

As to decoding, the general scheme of a Hamming decoder is showed in Fig. 4.3 [33]. The error correcting stage is optional, since the code could be used only for error detection. In this case, the same parity equations can be recomputed again and the result compared with the received parity check bits. An error indication can be generated should the computed and the received parity bits differ. Alternatively, the syndrome bits could be

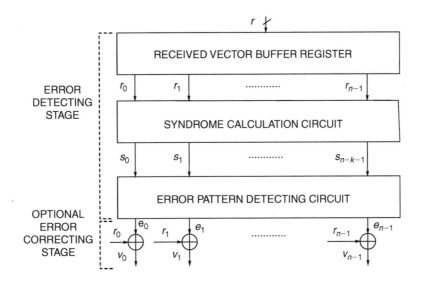

■ **FIGURE 4.3**

General scheme for a Hamming decoder.

computed. In both cases, a simple EXOR tree is required for the hardware implementation of the decoder. For instance, let us assume that the 7-bit codeword $c = 0101101$ is received. The syndrome is:

$$s = cH^\mathrm{T} = \begin{bmatrix} c_1 & c_2 & c_3 & c_4 & c_5 & c_6 & c_7 \end{bmatrix} \begin{bmatrix} 0 & 0 & 1 \\ 0 & 1 & 0 \\ 0 & 1 & 1 \\ 1 & 0 & 0 \\ 1 & 0 & 1 \\ 1 & 1 & 0 \\ 1 & 1 & 1 \end{bmatrix}$$

$$= \begin{bmatrix} b_4 + b_5 + b_6 + b_7 & b_2 + b_3 + b_6 + b_7 & b_1 + b_3 + b_5 + b_7 \end{bmatrix} \quad (4.24)$$

If the syndrome is zero, it means that no error has occurred. If a single error occurred, it must have been in the position denoted by the syndrome value. In our example, the syndrome is 100, indicating that an error occurred in the 4th bit of $c$. The error correcting logic can therefore reconstruct the right sequence 0100101.

## 4.4.6 Implementation of Cyclic Codecs

We now present a process for systematic encoding of cyclic codes. Unlike Hamming codes, cyclic codes are symmetric from an implementation viewpoint, meaning that encoder and decoder perform the same operations and have therefore the same complexity. The generator matrix $G$ is systematic if among its columns can be found the columns of a $k \times k$ identity matrix, in which case $G$ is said to be systematic on those columns. Notice that a generator matrix in left standard form (or *standard generator matrix*) is a special type of systematic generator matrix. If $G$ is a standard generator, then the first $k$ entries of the transmitted codeword $xG$ contain the message vector $x$. The linear encoding method we describe is based on the standard generator matrix, since the generator matrix found in Theorem 9 is almost never standard or even systematic.

The objective of the encoding process is to add parity check bits to the original message:

$$m = [m_0, \ldots, m_{k-1}] \Rightarrow c = [m_0, \ldots, m_{k-1}, -s_0, -s_1, \ldots, -s_{r-1}]$$

where $\sum_{j=0}^{r-1} s_j x^j$ is the remainder upon dividing $x^r m(x)$ by $g(x)$:

$$x^r m(x) = q(x)g(x) + s(x) \tag{4.25}$$

with $\deg(s(x)) < \deg(g(x)) = r$. To prove that this is the correct standard encoding, let us observe that:

$$x^r m(x) - s(x) = q(x)g(x) = b(x) \in C \tag{4.26}$$

with corresponding codeword:

$$b = [-s_0, -s_1, \ldots, -s_{r-1}, m_0, \ldots, m_{k-1}]$$

As this is a codeword of cyclic code $C$, every cyclic shift of it is also a codeword. In particular, codeword $c$ given above is found after $k$ right shifts. Thus $c$ is a codeword of $C$. Since $C$ is systematic on the first $k$ positions, this codeword is the only one with $m$ on those positions and so is the result of standard encoding. To construct the standard generator matrix itself, we encode the $k$ different $k$-tuple messages $[0, 0, \ldots, 0, 1, 0, \ldots, 0]$ of weight 1 corresponding to message polynomials $x^i$, for $0 \le i \le k - 1$. These are the rows of the standard generator matrix.

Let us apply this to the (7, 4) binary cyclic code with generator polynomial $x^3 + x + 1$ (so $r = 7 - 4 = 3$). We find for $i = 2$:

$$x^3 x^2 = (x^2 + 1)(x^3 + x + 1) + (x^2 + x + 1) \tag{4.27}$$

therefore the third row of the standard generator matrix is:

$$[m_0, m_1, m_2, m_3, -s_0, -s_1, -s_2] = [0, 0, 1, 0, 1, 1, 1]$$

Overall, the standard generator matrix is:

$$\begin{bmatrix} 1 & 0 & 0 & 0 & 1 & 1 & 0 \\ 0 & 1 & 0 & 0 & 0 & 1 & 1 \\ 0 & 0 & 1 & 0 & 1 & 1 & 1 \\ 0 & 0 & 0 & 1 & 1 & 0 & 1 \end{bmatrix}$$

and it can be easily checked that this is a Hamming code.

One practical way to perform cyclic decoding at the receiver side is to recompute the checksum $s(x)$ and to do a bitwise two complements addition with the computed checksum at the transmission side. In case of differences, an error is flagged. Alternatively, the code polynomial associated with the received string is divided by the generator polynomial. If the remainder is zero, no errors occurred, otherwise an error must have occurred. In all cases, there is no way of knowing how many errors occurred or which bits are in error.

*Linear shift register* (LSR) with serial data feed has been used since the 1960s to implement the CRC algorithm [37, 38], as depicted in Fig. 4.4. Like all hardware implementations, this method simply performs a division and then the remainder, which is the resulting checksum, is stored in the registers (delay-elements) after each clock cycle. The registers can then be read by use of enabling signals. Low-complexity and low power dissipation are the main advantages. This method gives much higher throughput than a software solution but cannot fulfill all the speed requirements of NoC nodes yet. Since fixed logic is used, there is no possibility of reconfiguring the architecture and changing the generator polynomial using this implementation. Several loop-connection schemes and reset alternatives exist.

In order to improve the computational speed in CRC generating hardware, *parallelism* has been introduced [75–78, 86, 87]. The speed-up factor

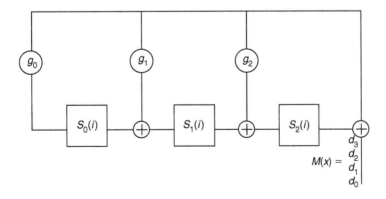

■ **FIGURE 4.4**

Linear feedback shift register.

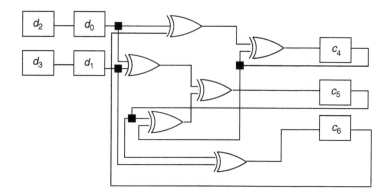

■ **FIGURE 4.5**

Serial–parallel checksum computation executing in two clock cycles.

is between 4 and 6 when using a parallelism of 8. By using fixed logic, implemented as parallelized hardware, this method can supply for CRC generation at wire speed and therefore it is the pre-dominant method used in computer networks. Parallel implementations allow to span the trade-off between speed and complexity. In fact, compact codec implementations can be obtained, that perform en/decoding in a finite number of clock cycles (faster but more complex with respect to the LSR solution). A *serial–parallel* hardware implementation of a (7, 4) cyclic code is illustrated by Fig. 4.5, executing in two clock cycles.

To the limit, a pure *combinational implementation* is feasible (e.g., using check equations [39]), which is very fast but prohibitively complex when a large number of check bits is required. This solution has the same flexibility problem of the previous one.

One way of implementing configurable hardware is by using LUTs as proposed by Albertango and Sisto [75], Glaise and Jacquart [87], and Ramabadran and Gaitonde [88] to store generator polynomial-dependent information. However, there is no possibility to adjust for a different number of check bits if the combinational logic is not reconfigurable. Furthermore, any change of polynomial and/or number of check bits would take a significant number of clock cycles to perform since all the contents of the LUT have to be recalculated and replaced. An architecture combining configurable and parallel hardware has been presented in Ref. [38].

## 4.5  THE POWER-RELIABILITY TRADE-OFF

When using error control coding for reliable on-chip communication, conflicting requirements on such coding schemes should be taken into close consideration to find the desired design point. The most critical trade-off concerns *communication reliability* versus *energy efficiency*.

Each coding scheme is able to meet the same constraint on communication reliability (in terms of MTBF) with different energy costs, according to its intrinsic characteristics, and this allows to search for the most efficient code from an energy viewpoint. The energy efficiency of a code can be expresses as average energy per useful bit, thus ascribing its coding and decoding energy overhead, and the energy consumed for switching the redundant check wires to the transfer of a single information bit.

On the other hand, switching of coupling capacitance between adjacent wires is becoming an important contributor to link power dissipation. Some approaches avoid such costly switching by introducing redundant check bits that replicate information wires. When information and check bits are transmitted on adjacent wires, the coupling capacitance does not contribute to power, since both wires are switching in the same direction, thus cutting down on coupling power. Unfortunately, the increased number of SSO in the system translates into an increased power supply noise that may seriously degrade system reliability and cause more EMI. Moreover, the layout footprint increases because of the doubling of the number of check lines.

These different aspects of the power-reliability trade-off for ECCs will be addressed in detail in the next subsections.

## 4.5.1  Conflicting Requirements on Voltage Swing

The wire voltage swing across switch-to-switch links in NoCs is a critical parameter for the energy-reliability trade-off. *Low-swing signaling* techniques have been traditionally used to decrease communication power in on-chip interconnects [48]. Unfortunately, decreasing the voltage swing reduces the noise margins of link receivers and makes on-chip wires inherently more sensitive to on-chip noise sources. From this point of view, reliable communication for unencoded links has to be ensured by means of full swing, power-inefficient data transfers.

The idea of using linear block codes to decrease wire voltage swing while still meeting communication reliability constraints was first proposed by Hedge and Shanbhag [22], and applied to state-of-the-art communication architectures in Ref. [23]. This approach consists of lowering the voltage swing beyond what would be required by reliability constraints, and to address the self-induced errors by exploiting the detection and recovery capability of linear block codes. Therefore, a certain amount of corrupted data transfers is allowed, provided the error detecting correcting code is able to deal with them (e.g., by means of retransmissions or error correction).

The effectiveness and energy efficiency of this approach is tightly related to the error detection capability of implemented codes (which determines the percentage of voltage swing reduction) and to the energy cost of error recovery procedures which, if activated frequently, might incur serious power and performance penalties. For linear block codes, error recovery technique and detection capability are mutually dependent. In fact, for these codes the probability of a decoding error is much larger than that of an undetected error. For instance, when a Hamming code is used for single-error correction, the decoder assumes that whenever an error occurs, it is single and not multiple, and can therefore proceed to correct it. Yet, it could have been a multiple error which is within the detection capability of the Hamming code, but the correction stage makes a decoding error in this case. If no restrictive assumptions are made on the nature of a detected error, correction cannot be carried out, but the residual word error probability improves a lot.

Moreover, implementing a Hamming decoder for error correction involves an amount of logic which is more than twice that required for error detecting Hamming decoders [25]. As a result, a larger number of errors can be detected with a simpler, detection-oriented decoder. However, whenever an error is detected, a retransmission has to be scheduled, which involves additional switching activity on the link lines and a longer execution time for the entire system (corresponding to the retransmission delay). Finally, in the NoC domain retransmissions require a feedback channel, which can be provided by the flow control architecture. Proper ACK/NACK feedback signals have in fact to be implemented to indicate the need for a retransmission, and this has also buffering implications, as will be described in Section 4.11.

All these considerations should be accounted for in a unique energy efficiency metric for a given coding scheme, able to capture the detection capability of a code, the voltage swing at which it can operate without violating reliability requirements, the number of redundant check lines, the encoder and decoder complexity, the overhead of the error recovery technique (e.g., correction logic, additional switching activity). In Fig. 4.6, the average energy per information bit is reported as a function of progressively tighter communication reliability requirements for different linear block coding schemes.

The plot is taken from Ref. [25], and is referred to a 0.25 μm technology with 2.5 V supply voltage. The considered wire load is equivalent to a few mm wires in that technology. Energy efficiency was derived by combining analytical modeling and functional simulation (for the backannotation of switching activity). The reliability requirement (in terms of residual word errors after the decoding stage) was changed into a requirement on the bit error rate, and then the needed voltage swing to sustain this rate was analytically derived by inverting equation (4.1). Finally, based on the knowledge of power supply and voltage swing, energy dissipation related to codec operation and to link switching activity was derived, leveraging the Synopsys synthesis and power estimation tool suite. Energy incurred by check bits and codec operation was ascribed to information lines, thus getting the average energy per useful bit.

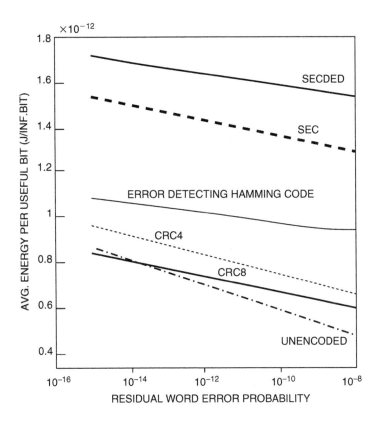

Energy efficiency of coding schemes for NoC-like point-to-point links. The x-axis indicates the requirement on communication reliability.

The importance of Fig. 4.6 lies in the trend it points out for lightly loaded on-chip interconnects, such as the switch-to-switch links of regular NoCs. When observing the plot, we have to recall that transitions on link lines play a minor role with respect to energy dissipation, since wires are short, while the contribution of codec complexity becomes relevant. In fact, a gap can be observed between SEC and single-error correcting–double-error detecting (SECDED) Hamming codes and the other schemes: correction circuitry at the decoder turns out to be power hungry. Error detecting Hamming code is energy efficient in spite of the retransmission overhead, since its decoder has three times less area than SECDED and half the area of SEC, with the same number of parity check lines (6, which grow to 7 for SECDED). However, two very simple CRC codes with redundancy 4 (CRC4) and 8 (CRC8), with generator polynomials described in Ref. [25], outperform Hamming codes, since in turn their decoders (with similar complexity) exhibit half the area of an error detecting Hamming decoder.

In this context, the unencoded link might seem a good solution, since it introduces no codec-related overhead but just operates at voltages progressively approaching the full swing as the reliability requirements become tighter. However, the crossing point in Fig. 4.6 lies in a region where the residual word error probability is close to $10^{-14}$. In the NoC domain, this value corresponds to a wrong data word every 2 days (with a worst-case injection rate of 1 flit/cycle), at a frequency of 500 MHz. The requirements will be much tighter than this, thus paving the way for *link coding schemes associated with tuning of voltage swing*.

The illustrated results point out that the detection capability of a code plays a major role in determining its energy efficiency because it is directly related to the wire voltage swing. As far as error recovery is concerned, error correction is beneficial in terms of recovery delay, but has two main drawbacks: it limits the detection capability of a code and it makes use of high-complexity energy-inefficient decoders. On the contrary, when the higher recovery delay (associated with the retransmission time) of retransmission mechanisms can be tolerated, they provide a higher energy efficiency, thanks to the lower swings and simpler codecs (pure error detecting circuits) they can use while preserving communication reliability.

The power-reliability trade-off has obviously an impact on system performance. An error detecting scheme with very high detection capability can be pushed to work at a very-low-voltage swing, at the cost of an increasing number of retransmissions. As a consequence, the longer execution time of applications has to be compared with their performance requirements. On the other hand, it must be observed that DSM ICs exhibit a lower degree of non-determinism with respect to other environments, such as wireless channels, therefore the impact of retransmissions might be tolerable. An accurate analysis of such an impact under realistic working conditions is still missing in the open literature.

The work in Ref. [28] points out however another effect of employing coding schemes with different error recovery delays. A single-error detecting code (PAR) and an SEC code are compared when the voltage swing is fixed but the traffic injection rate and the bit error rate vary. A $4 \times 4$ mesh network is considered, with 4.5 mm long inter-router links in a 0.18 µm technology. Error control schemes are evaluated by means of the model in Section 4.3.1 and assigning the noise voltage variance $\sigma$ to 0.5 and 0.36 V, respectively. The corresponding bit error rate varies from 0.035 (high, H in plots) to 0.0063 (low, L in plots), respectively.

Figure 4.7 reports network dynamic power consumption with low and high error rates using both PAR and SEC schemes. The SEC power consumption for high injection rates is more than PAR, but this time this is due to the higher acceptance rate for SEC. In fact, the latency involved in retransmissions under such levels of traffic injection limits the acceptance rate (i.e., the network throughput) of PAR. For low injection and low bit error rates, the power consumption for the SEC scheme is almost equal to PAR. However, the area consumed by the PAR implementation is lower than the SEC scheme, making it an attractive technique for error control in this case. This analysis stresses the role played by error recovery

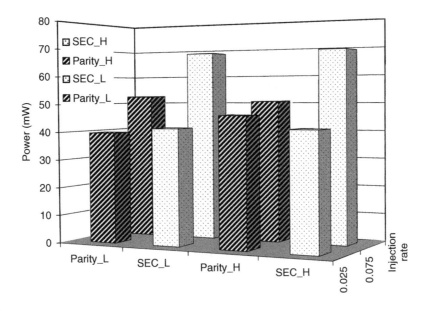

■ **FIGURE 4.7**

Power consumption of error detecting/correcting schemes as a function of the injection and bit error rates.

schemes in determining the ratio between offered and accepted rate in an NoC environment, and therefore the power-performance trade-off. Obviously, optimal code selection for a particular application domain depends on the combined effect of all factors addressed till now and on the desired trade-off point in the power-reliability-performance exploration space.

## 4.5.2 Conflicting Requirements on Switching Activity

While previous analysis has focused on power-sensitive parameters of a coding scheme from an implementation perspective, we will now focus on the ability of an error detecting/correcting code to be *intrinsically* low power because of the characteristics of its codespace, while maintaining the same error control capability. We will see that for some coding schemes, redundancy can be used to reduce coupling power and minimize the impact of certain noise sources, but at the cost of making the impact of other noise sources more significant.

For instance, in Ref. [43] a coding technique is introduced which, combined with a proper bus layout, allows a significant power reduction with respect to Hamming codes. A *dual rail* code adds $k + 1$ check bits to $k$ information bits, thus deriving a codeword of length $2k + 1$. The $k + 1$ check bits are given by the following relations: $c_i = d_i$, for $i = 0 \dots k - 1$; $c_k = d_0 \oplus d_1 \oplus \cdots \oplus d_{k-1}$. Since the first $k$ check bits are a copy of the correspondent data bits, they can be transmitted on adjacent wires, therefore the respective mutual capacitance does not need to be charged. As a consequence, the spacing between such wires can be kept minimal, while the spacing between wires carrying different values can be increased (intelligent spacing [43]). Even though the number of wires increases with respect to the Hamming code, the smaller number of mutual capacitances which must be charged, together with the intelligent spacing, leads to a power consumption reduction [43].

However, since the dual rail code reduces the power contribution due to link wires, the dissipation of the active elements of an encoded link (encoder, decoder, driver, receiver and repeaters) comes into play. When considering a 10-mm link line, with optional repeater stages every 2 mm, the Hamming code proves more power efficient in the buffered case and with links of more than 13 bits. This is due to the fact that the increase in power associated with the larger number of wires and buffers in the dual rail code is not balanced by the power saved by reducing the number of lateral coupling capacitances to be charged. In contrast, the dual rail code is more power efficient in the absence of repeater insertion, taking benefit of a lower link-related power and of a simpler codec [44].

The dual rail code is also more effective than Hamming codes in reducing crosstalk-induced link delay (CILD). In other words, the dual rail code has a larger redundancy which can be exploited to increase performance rather than increasing the code correction capability [46], limited to single errors. Hamming codes with wires at minimum spacing imply a high CILD. Increasing the spacing between adjacent wires increases the link footprint but produces better results. The increased efficiency of dual rail has been measured using the same link footprint. Interestingly, each inner link wire of a Hamming codeword might experience the worst-case CILD. In a dual rail code, each inner wire has only one neighbor switching oppositely, while the other neighbor switches in the same direction, speeding up the transition of the considered wire. As for the parity bit, it has one neighbor switching oppositely, while the other neighboring wire is the shield. Consequently, the wire carrying the parity bit should experience the worst-case transition. Furthermore, adding one extra check bit to the dual rail code (modified dual rail code) produces even better results, in terms of further reducing the effective coupling capacitance and, consequently, improving CILD [47].

Although the application of dual rail codes can guarantee resilience to CILD-related errors and reduce coupling power, it might aggravate the impact of other noise sources because of the additional switching signals. *In particular, code requirements for crosstalk and power minimization in terms of switching activity of adjacent wires are opposite to those for SSO noise reduction.* It has been proven in Ref. [45] that in the worst case, Hamming codes induce less SSO noise than dual rail and modified dual rail codes. As regards Hamming codes, the transition which induces the highest SSO noise is that involving the simultaneous switching of all data bits. The check bits switch based on the parity check matrix considered for different data lengths. For the test codes, the matrix was devised to minimize the number of 1s, thus reducing the encoder circuitry complexity. However, the switching of the check bits, being delayed with respect to that of the data bits due to the XOR blocks, turns out to be negligible with respect to the SSO noise. In fact, the worst case for such noise is when all bits switch simultaneously.

Instead, dual rail codes perform worse in terms of SSO noise than Hamming code. This is due to the larger number of SSO that these codes induce. In fact, the check bits are a copy of the data bits, and they switch at the same time, thus worsening the SSO problem due to an almost doubled number of switching lines. Only the transition of one check bit is delayed in this case. For a 32-bit bus, the SSO noise ranges from 20% for a Hamming code to 39% for a dual rail code. This test again stresses the conflicting requirements characterizing redundant error control coding.

## 4.6 UNIFIED CODING FRAMEWORK

The two techniques described above to span the energy-reliability trade-off (tuning voltage swing or employing duplication and parity schemes) can be seen as part of a more general coding framework [26], where different codes targeting different issues (e.g., crosstalk, power, general-purpose reliability) can be combined into a *unified coding system*, as depicted in Fig. 4.8.

*Crosstalk avoidance coding* (CAC) [49–52] aims at minimizing the worst-case communication delay across a wire by forbidding specific transitions. The CAC proposed in Ref. [51] reduces the worst-case delay by ensuring that a transition from one codeword to another one does not cause adjacent wires to transition in opposite directions (the *forbidden transition* (FT) condition). Shielding the wires of a bus by inserting grounded wires between adjacent wires is the simplest way to satisfy this condition. In Ref. [26] it is shown that there is no linear code that satisfies the FT condition while requiring fewer wires than shielding. The worst-case delay can also be reduced by avoiding bit patterns "010" and "101" from every codeword [50], which can be referred to as the *forbidden pattern* (FP) condition.

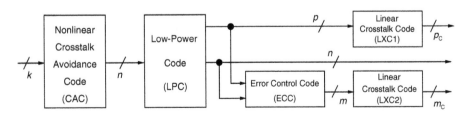

■ **FIGURE 4.8**

Unified coding framework.

*Duplication* is the most straightforward and expensive way to meet this condition. There is no linear FP code that satisfies the FP condition while requiring fewer wires than duplication [26]. The drawback of CAC is that it is not able to address other DSM noise sources other than capacitive crosstalk. CAC needs to be the outermost code as, in general, it involves nonlinear and disruptive mapping from data to codeword.

*Low-power coding* (LPC) has been traditionally used to reduce switching activity resulting in low-power busses. Early schemes ignored coupling capacitance and associated delay and power penalties [53, 54]. A simple but effective LPC is the *bus-invert code* [54], in which the data is inverted and an invert bit is sent to the decoder if the current data word differs from the previous data word in more than half the number of bits. The effectiveness of bus-invert coding decreases with increase in the bus width. Therefore, for wide busses, the bus is partitioned into several sub-busses each with its own invert bit. Codes that reduce both self and coupling transitions were then proposed by Sotiriadis [49], Kim et al. [55] and Zhang et al. [56]. The idea in Refs. [55, 56] was to conditionally invert the bus based on a metric that accounts for both self and coupling transitions, at the price of increased complexity. Alternatively, the transition pattern code proposed in Ref. [49] employs nonlinear mapping from data to codeword to achieve significant self and coupling activity reduction, but requires very complex codecs. Note that bus-invert coding is nonlinear. It has been shown in Ref. [49] that linear codes do not reduce transition activity. Unfortunately, LPCs only address the power consumption issue, neglecting delay and communication reliability problems. LPCs can follow CAC as long as LPC does not destroy the peak coupling transition constraint of CAC.

The additional information bits generated by LPC need to be encoded through a linear CAC to ensure that they do not suffer from crosstalk delay. ECCs need to be systematic to ensure that the reduction in transition activity and the peak coupling transition constraints are maintained. The additional parity bits generated by ECCs need to be encoded through a linear CAC to ensure that they do not suffer from crosstalk delay. LXC1 and LXC2 are linear CACs based on either shielding or duplication. Nonlinear CACs cannot be used here, since error correction has to be done prior to any other decoding at the receiver. In Fig. 4.8, a $k$-bit input is coded using CAC to get an $n$-bit codeword. The $n$-bit codeword is further encoded to reduce the average transitions through LPC resulting in $p$ additional low-power information bits. ECC generates $m$ parity bits for the $n + p$ code bits. The $m$ parity bits and $p$ low-power bits are further encoded for crosstalk avoidance to obtain $m_C$ and $p_C$ bits, respectively, that are sent over the bus along with $n$ code bits. The code rate is therefore $k/(n + p_C + m_C)$.

When designing the coding system depicted in Fig. 4.8, two main factors should be taken into close account:

- Not all possible combinations of codes are effective. For example, applying bus-invert coding to a FT codeword destroys its crosstalk avoidance property, while inverting an FP codeword maintains the FP condition. Moreover, in this latter case a duplication LXC1 code should be used to avoid crosstalk delay in the invert bits.

- By concatenating different codes, the codec complexity and delay is equal to the sum of codec overheads of the component codes, hence the joint code can be very complex. However, in many cases optimizations are feasible, making the complexity of the resulting unified coding system less than additive.

Let us now focus on some of these optimizations [26]. In Fig. 4.9(a) there is an example of *joint LPC and ECC*. Bus-invert coding is combined with a Hamming SEC code. However, the parity bits of the ECC (the XOR blocks) are determined using the original data bits, instead of waiting for invert bits of the LPC to be computed. In fact, a relevant property of XOR operation states that if an odd (even) number of the inputs of an XOR gate are inverted, then the output is inverted (unchanged). Therefore, once the invert bits are computed, the parity bits resulting from odd number of input bits are conditionally inverted using the invert bits. In this way, parity generation and invert bit computation can occur in parallel reducing the total delay to the maximum of the two plus the delay of an inverter (from 21% to 33% reduction in encoder delay from gate-level netlist estimations). This scheme is called *bus-invert Hamming* (BIH) coding.

An optimized version of a *joint CAC and ECC* code is reported in Fig. 4.9(b). The duplication scheme for avoiding crosstalk delay has a Hamming distance of two as any two distinct codewords differ in at least two bits. We can increase the Hamming distance to three by appending

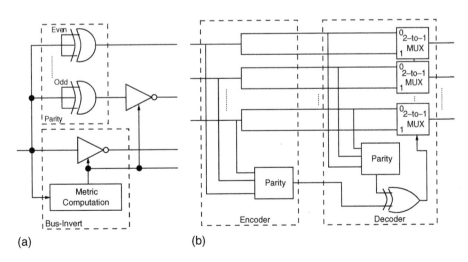

(a)                    (b)

■ **FIGURE 4.9**

(a) Joint LPC and ECC: BIH code with reduced encoder delay. (b) Joint CAC and ECC: DAP.

a single parity bit (*duplicate-add-parity code*, DAP). In order to decode, the parity bit is recreated at the decoder by using one set of the received data bits and compared with the received parity bit. If they match, the set of bits used to recreate the parity bit is chosen as the output, else the other set is chosen. Since a single error will at most affect one of the sets or the parity bit, it is correctable.

A *joint LPC (bus-invert coding), CAC (duplication)* and *ECC (1 parity check bit) code* can be easily designed by combining bus-invert coding with DAP (it will be denoted *DAPBI*). The encoder consists of the bus-invert logic operating in parallel with the parity check computation, and the parity bit is conditionally inverted, as in BIH coding. Information lines (inverted or not) are duplicated, then the decoder computes the parity and performs data set selection accordingly, as in DAP coding. In addition, the invert bit is duplicated (LXC1) to ensure error correction and crosstalk avoidance for the bit.

Finally, a technique can be devised to mask encoder delay in systematic coding. For instance, data bits in a Hamming code do not have crosstalk avoidance and, therefore, have a bus delay of $(1 + 4\lambda)\tau_0$, where $\tau_0$ is the delay of a crosstalk-free wire and $\lambda$ is the ratio of the coupling capacitance to the bulk capacitance. The parity bits obtained from the encoder have the same delay plus the encoder delay. However, this latter can be eliminated for long links by using half-shielding as LXC2 and thus reducing the link delay to $(1 + 3\lambda)\tau_0$. We call this the *HammingX code*. In DAP, we can use duplication as LXC2 for the single parity check bit, reducing its delay to $(1 + \lambda)\tau_0$. This is the *DAPX code*. The principle behind these codes is that the reduction in bus latency due to CAC can counterbalance the negative impact of codec latency, although the link must be long enough for this effect to bring performance benefits.

Let us now consider the design of a 32-bit bus in 0.13 μm technology under a reliability requirement of $10^{-20}$ (the target residual word error probability). The voltage swing can be tuned based on the detection capability of the ECC. Codes employing ECCs have swings that are lower than the nominal supply voltage: Hamming, HammingX, BIH, DAP, DAPX and DAPBI can operate at about 0.8 V rather than the nominal 1.2 V. Figure 4.10 shows the speed-up and energy savings over uncoded bus of the introduced coding schemes for increasing link lengths. For long links, the coding delay accounts for smaller portion of the total delay and, hence, speed-up achieved by the codes improve (see Fig. 4.10(a)). Several codes with CAC have speed-up of less than 1 at $L = 6$ mm. This indicates that, in the current technology, 6 mm is the critical length for these codes. As regards energy savings (Fig. 4.10(b)), BI-based codes achieve lower savings due to very high codec energy dissipation. Codes using ECC have some improvements due to the trade-off between supply voltage and reliability. DAPX turns out to be the most energy efficient code because of its low codec overhead. For this technology node, codec power and delay still play a dominant role, performance- and energywise, while bus average energy per transaction is

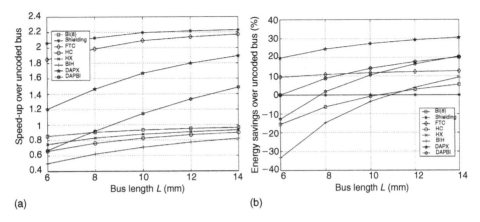

■ **FIGURE 4.10**

Code comparison for a 32-bit bus, performance- and energywise.

not able to make the difference even for wires that are a dozen mm long. Further details can be found in Ref. [26].

## 4.7    ADAPTIVE ERROR PROTECTION

### 4.7.1    Self-Calibrating NoCs

Another approach to reliable communication for NoCs consists of *self-calibrating link* design techniques, which aim at maintaining an acceptable design trade-off between energy, performance and reliability. This approach takes its steps from the fact that even though typical delays of new CMOS technologies tend to improve with respect to older ones, there is a larger spread of the probability distribution. Since traditional CMOS design relies on worst-case assumptions, many circuits will not exploit the full capabilities of new technologies. Instead of relying on over-conservative worst-case assumptions, self-calibrating circuits adjust their operating parameters to run-time actual conditions and will be needed to maintain acceptable design trade-off between energy, performance and reliability [57, 58].

Since self-calibrating techniques avoid worst-case assumptions as to silicon capabilities, they are complementary to DVS techniques, which avoid conservative assumptions about application requirements. For point-to-point links, DVS enables to tolerate uncertainty about bandwidth requirements (i.e., the frequency at which the link is operated), whereas self-calibrating techniques enable to tolerate uncertainty about silicon capabilities to operate at the required frequency (i.e., the voltage required

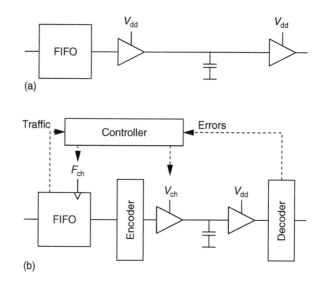

The basic idea of a self-calibrating, point-to-point, unidirectional on-chip interconnect: (a) the classic static scheme, with an FIFO buffer to decouple the two subsystems; (b) the self-calibrating scheme.

to operate at this frequency). The two techniques can be integrated into a layered architecture: the top layer determines the frequency required to meet application needs, which is then input to a self-calibrating voltage controller that determines, for the required frequency, which voltage satisfies the reliability constraint.

The application of this concept to a communication link [58, 59] is shown in Fig. 4.11. Voltage and frequency are set adaptively by a link controller. The encoding scheme has to detect errors resulting of over-aggressive link operation, or caused by additive noise. These requirements are challenging, since the link controller may cause bit error rates as large as 1, which renders the problem quite unique. The *automatic repeat request* (ARQ) and link controllers ensure that the system behaves correctly. The former guarantees an in-order data delivery, while the latter sets link voltage and frequency according to bandwidth and reliability requirements.

The link control problem can be formulated as a constrained minimization problem. The objective function to be minimized is average energy per useful bit, under two quality of service constraints related to performance and reliability. On one hand, the guarantee of an average transfer latency (which translates into guaranteeing a certain throughput) has to be provided. On the other hand, the optimizer has to ensure that the probability for a word declared correct by the decoder to be actually corrupted falls below a maximum tolerable limit. We call this probability the *residual word error rate*. The control knobs to minimize the objective function are the link voltage swing and operating frequency.

It is impossible for the controller to receive feedback on the words delivered but corrupted. Obviously, the controller can only be informed of detected word errors, and cannot distinguish words transferred correctly from those affected by undetected errors. As a result, the controller interprets retransmissions as a sign of unreliability, and has to react to run-time operating conditions by avoiding voltage–frequency pairs that cause retransmissions. Under these simplifying assumptions, the voltage and frequency selection problems become decoupled, and the optimization problem can be solved as follows:

(i) choose the slowest frequency that meets the bandwidth constraints;

(ii) for the derived frequency, determine the lowest voltage that does not cause detected word errors.

The reliability constraint does not appear in this policy because the maximum residual word error rate is determined by the encoding scheme only, and cannot be traded-off for energy. The first solution step is a typical *scheduling problem* encountered in DVS techniques. The second step defines instead a novel *self-calibrating control problem*.

In fact, the controller has to periodically decrease the voltage used for a particular frequency and observe whether this causes retransmissions. When doing so, it may happen that the frequency is too fast for the applied voltage. As a result, the received signal will be sampled even if many or all transitions failed, thus giving rise to timing errors. In general, naive threshold based algorithms, or extensions thereof, can be used to adjust a one-dimensional parameter in the presence or absence of detected word errors [58].

Interestingly, since the principle of self-calibrating operation may lead to temporarily generating a raw bit error rate equal to 1, the output of the timing error channel can be exactly equal to the previous input, because all transitions deterministically fail. Classic error detecting codes, such as CRCs, only add spatial redundancy. As a result, they do not detect timing errors in this situation, since the previous input was a codeword, and cannot be used in self-calibrating links. *Alternating-phase codes* are a simple extension of classic error detecting codes that incorporate temporal information about the data sequence within the redundant bits [60]. They detect all timing errors when the raw bit error rate is 1, which is a working region periodically visited by the link controller. In the same region, a CRC8 code has a residual word error rate tending to 1, while the CRC8 alternating-phase code detects an increasing number of errors.

Self-calibrating techniques can be seamlessly integrated in NoCs and enable to tolerate uncertainty on noise sources. This feature is of high interest for communication architectures automatically designed, as will be most likely the case for NoCs.

### 4.7.2  Coding Adaptation

An alternative adaptive scheme consists of *dynamically changing the coding strength* instead of scaling the supply voltage based on the observed error pattern [27]. The idea to change the coding scheme strength while retaining the same $V_{dd}$ can potentially lead to larger improvements in reliability per unit increase in energy consumption as compared to increasing the supply voltage using the same coding method.

The work in Ref. [27] reports an interesting case study. The error model in Section 4.3.1 is assumed, with $\sigma = 0.3$. Using a single parity check code (PAR) with a supply voltage of 2 V provides an undetectable error rate of 9.59E-5. When we employ a more powerful scheme (an Hamming code variant able to detect all single, double and triple errors, and a subset of higher bit errors) using the same 2 V supply voltage, the undetectable error rate is reduced to 2.75E-9. This latter approach consumes 1.18 mJ more energy (for 1 million transactions of the barnes application) than PAR for providing the additional reliability. In comparison, the voltage scaling approach needs the voltage of PAR to be increased from 2 to 2.75 V to reduce undetectable error rate to 2.75E-9, and consumes an additional 3.16 mJ of energy (as compared to PAR at 2 V) for the additional reliability. This example stresses the importance of adaptive coding schemes even in presence of voltage scaling.

However, it is more beneficial to consider code and supply voltage adaptation as complementary rather than competing approaches. Combining the two approaches could be particularly relevant given that the number of different coding schemes that can be implemented simultaneously and the number of supply voltages/supply scaling range are typically limited in real systems.

There are two important aspects in *code adaptation*: (i) detecting the variation in noise behavior and (ii) identifying the protection scheme to employ for the observed noise behavior. As in the previous section, an indicator of the variation in the noise behavior is the detected error rate observed at the decoder of the link. When detectable error rate is small, it requires a long period before the number of detectable errors changes and this makes it difficult to use the counted errors to affect changes in the protection scheme promptly. In addition, it should be considered that even small variations in detectable error rates indicate huge variations in undetectable error rates.

An alternate monitoring scheme consists of using a victim bus line that amplifies the number of detectable errors in order to detect changes in the noise behavior quickly. This victim line uses half the voltage swing as the normal link lines and is more susceptible to variations in noise. Consequently, the detectable error rate of the victim line is much higher than those of error detecting schemes. In the implementation in Ref. [27], the victim line transmits a repeated sequence of 0-1 with each data transmission and one per 10 cycles when the link is in idle state. Whenever, the decoder of this line receives the same bit value in two successive transmissions, it indicates a detected error.

The main goal of code adaptation is to keep the undetectable word error rate below a threshold when the noise variable $\sigma$ changes (modeling physical effects such as variations of flux distribution of alpha particles with altitude/latitude or changes in crosstalk patterns based on data activity in adjacent link lines). When more than one coding scheme can achieve the target undetectable word error rate, the scheme with the lowest energy consumption should be selected. The decoder should be notified whenever the coding scheme (e.g., switching between different variants of the Hamming code) changes. This task can be performed by transmitting a special code to all encoders and decoders using the information lines of the links.

Statically derived $\sigma$ switching thresholds could be used for code selection, moving to progressively more complex schemes whenever higher detection capabilities are required. In order to identify the switching points of $\sigma$, in Ref. [27] two counters are used: a saturating counter to track a sampling window and another counter to maintain the total number of detected errors on the victim line during this window. At the end of every sampling window, the error counter is reset.

## 4.8 DATA-LINK LAYER ARCHITECTURE: CASE STUDIES

Principles and guidelines for data-link layer design in NoCs have been presented so far. Now, the reader will be introduced to the viewpoint of the NoC architecture designer through two case studies. On one hand, the *Micro-Modem* solution to NoC communication reliability will be presented [61]. Then, the *layered* approach to NoC communication provided by the Nostrum framework [62] will be briefly described.

### 4.8.1 Micro-Modem

The Micro-Modem is a communication interface between routers which is responsible for transmitting and receiving digital data via the interconnect. It implements in hardware the physical layer and the data-link layer of the OSI model, while the higher layers are performed in the router and in the end nodes. The architecture is reported in Fig. 4.12 [61].

The parallel data from the source is stored in the input buffer which can be shared by the Modem and the source node. The packet is then transformed into a frame by addition of header and tail information, as depicted in Fig. 4.13. Header bits include check bits for error detecting/correcting codes or synchronization series and start/stop signals to

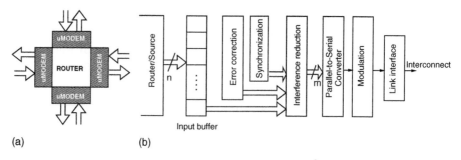

■ **FIGURE 4.12**

The Micro-Modem architecture. (a) Router interface and (b) Micro-Modem internal architecture.

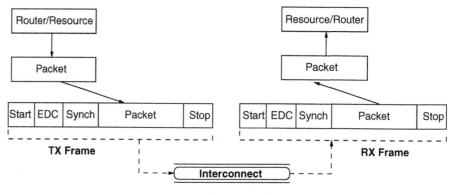

■ **FIGURE 4.13**

Frame building in the Micro-Modem.

solve the delay uncertainty. At this stage, ack/req signals can be formed for asynchronous and GALS structures and clock recovery mechanisms can be implemented.

The data undergoes additional processing to provide the *inter-signal interference* (ISI) and noise immunity. Signal shaping can be performed for minimization of sequent serial signals interference by using Raised Cosine Roll-off Filter. Interleaving of the data can be performed to enhance the immunity to burst noise and to improve the error correction ability of error detecting codes.

Then the frame is given the desired parallelism rate, or serialized [61, 63], and modulation is performed to allow data transport from multiple bit sources through a limited number of wires or a single wire. Link interface supplies the conditioning and buffering of the signal as it enters the wire, as well as swing restoration and timing optimization at the receiver side. At the wire level, various optimizations are performed to assure the minimal delay, such as repeater insertion.

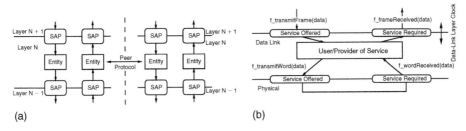

(a)                                                          (b)

■ **FIGURE 4.14**

(a) Layered communication protocol stack and (b) Behavior of the data-link layer.

It should be noted that the type of Modem and the way of its application are influencing the additional latency due to the signal processing in the transmitter and receiver. For low noise activity and short data links, the Modems should be applied only at the end-points of the communication infrastructure. In contrast, for high noise activity and long distance between communicating nodes, Modems should be applied at each router on the datapath, although resulting in potentially higher latency. Design techniques for low-power compact Micro-Modems should be researched and developed to make it an integrated part of reliable NoC design in DSM technologies.

### 4.8.2 Nostrum

The Nostrum architecture is a network-centric platform for SoC design. The *Nostrum backbone* is a layered communication platform used within the Nostrum architecture for on-chip communication [62, 64]. A communication protocol stack is defined within the Nostrum backbone, which consists of several layers with different responsibilities grouped based on the level of abstraction. Each layer is a user of services provided by the layer below. A concrete call for a service a layer provides is called a *function*. The functions a layer implements are grouped into *entities*, as depicted in Fig. 4.14(a). In order to provide these services, interfaces to the upper and lower layers have to be defined. These interfaces are known as *Service Access Points* (SAPs). The agreed data format within a layer is called the *Protocol Data Unit* (PDU).

Within the concept of the Nostrum backbone resides the three compulsory layers: the network layer (NL), the data-link layer (LL) and the physical layer (PL). Each layer has its own PDU format: words for the PL, frames for the LL and packets for the NL. Moreover, in order to synchronize the function calls and set up deadlines, clocks are associated with each layer.

Let us now focus only on the data-link layer. In Nostrum, it provides the reliable transfer of information across the physical link. Frames have to be sent to the remote peer taking care of synchronization, error control and flow control. The *Data-Link Layer Clock* (LLC) decides the speed at which frames are accepted from the SAP. On every rising edge of the LLC one frame is moved from the output of a switch to the corresponding input of the switch in the other end of the physical channel. The speed at which packets are sent out is decided by the Network Layer Clock. The ratio between the Network Layer Clock and the LLC decides the duration of packets in terms of LLC ticks.

In order to fit the SAP of the physical layer, frames of the data-link layer might get segmented or concatenated: the ultimate purpose is to allow changes in the physical implementation without affecting the upper layers. It also gives the possibility to elaborate different link widths and explore the trade-off between link width and operating frequency for a given requirement on communication bandwidth.

In turn, the SAP of the data-link layer accepts a function call for sending frames any time and executes on the next raising LLC. The time for transmission is decided by implementation and measured in LLC ticks. Also the possibility of error detection is introduced. Any single-bit error will be detected in a protection field, which is specified by means of start and end parameters. An entire packet can be protected, but at the cost of higher coding complexity and frame length. In contrast, only the most critical fields (e.g., the header) can be selected for protection. A pictorial overview of the Nostrum data-link layer is reported in Fig. 4.14(b). The protection range must be specified as an optional argument of the *f_transmitFrame* function.

## 4.9  ON-CHIP STOCHASTIC COMMUNICATION

Traditional acknowledgement/retransmission schemes can lead to severe loss of performance and even deadlock, particularly in presence of high bit error rates. Moving from this consideration, a novel paradigm for on-chip communication has been proposed, which is able to sustain low latencies even under severe levels of failures: *on-chip stochastic communication* [31, 32]. They key point behind this strategy is that at chip-level bandwidth is less expensive than in traditional networks, because of existing high-speed links and interconnection fabrics which can be used for the implementation of an NoC. Therefore, the designer can afford to have more packet transmissions than in the previous protocols in order to simplify the communication scheme and guarantee low latencies and fault tolerance.

At the heart of the new communication paradigm lays a *probabilistic broadcast algorithm* [66] which allows end-to-end communication between IP blocks using a *randomized gossip protocol* [65]. More precisely, if a tile (in a tile-based NoC architecture) has a message to send, it will forward the packet to a randomly chosen subset of the tiles in its neighborhoods, until the entire network becomes aware of it. The message is spread exponentially fast, and after $O(\log_2 n)$ stages it reaches all the $n$ tiles with high probability [65]. Every IP than selects from the set of received messages only the ones that have its own ID as the destination.

The performance of the stochastic communication scheme can be controlled using a few parameters. The message might reach its destination before the broadcast is completed, so the spreading could be terminated even earlier. In order to do this, a *time to live* (TTL) can be assigned to every message upon creation and decremented at every hop until it reaches 0; then the message can be destroyed. This optimization leads to important bandwidth and energy savings. Another important parameter is the probability that a message is transmitted over a link. It influences both the number of messages transmitted in the network and the number of rounds that a message requires to reach its destination. Therefore, this is a very powerful way to tune the trade-off between performance and energy consumption [32].

The way this stochastic communication scheme avoids link-level retransmissions is by combining error detection with the multiple transmissions mechanism. During transmission, packets are protected against data upsets by a CRC code; if an error is discovered, then the packet is discarded. Since a packet is retransmitted many times in the network, the receiver does not need to ask for retransmission, as it will receive the packet again anyway.

This communication paradigm based on probabilistic flooding achieves many desired features for NoCs, such as separation between computation and communication, fault tolerance, no retransmission-related latency and design flexibility. It is based on the assumption that simple deterministic algorithms do not cope efficiently with random failures (as captured by the noise model in Section 4.3.3), which might be better addressed by a stochastic approach to NoC communication. Stochastic communication has a very good average case behavior with respect to latency and energy dissipation. In the worst case, the protocol will not deliver the packet to destination, therefore it is more suitable for soft real-time constraints.

More in general, flooding is an effective fault-tolerant technique because if a path exists to destination, a message will almost certainly arrive. In practice, however, this level of fault tolerance may not be necessary. Resilience to a much smaller number of faults still offers increased chip yields, as well as resistance to transient faults, but also reduces unnecessary packet transmissions. In turn, this leads to higher network throughput and a more efficient use of the interconnect.

Two simple optimizations of the probabilistic flooding algorithm have been proposed in Ref. [67]. The first is *directed flooding*. While the original flooding algorithm [32] routes packets irrespective of where they are in the network, the directed flooding scheme makes use of an NoC's highly regular structure to direct a flood toward the destination. In this algorithm, the probability of passing a message to each outgoing link is not fixed, but instead varies based on the destination of the packet. At the algorithm's onset, packets have a high probability of being routed toward the destination node, allowing for packets to get in the general vicinity of the destination node quickly. As the packets approach the destination node, it is desirable for them to take varied paths, to mitigate the risk of all packets encountering the same defective nodes on their way to destination.

The second proposed algorithm is the *redundant random walk*, which sends only a predetermined number of copies of a message into the network not to significantly limit overall performance. These messages follow a non-deterministic path, where every node that receives one of them must forward it to one of its output links. The choice of output link is determined by a set of random probabilities for each port, where the sum of all probabilities is 1. In general, each message in the network will tend toward the same destination, but will follow slightly different paths to reach it. In all algorithms, an initial flooding phase of two cycles is performed to seed the area and ensure that all functional nodes within two hops receive a copy of the message. For the random walk algorithm, this allows to easily generate a finite number of packets, while limiting the impact of any local faults.

The comparison in Ref. [67] comes to the following conclusions:

- The flooding algorithms cause significant communication overhead, which limits their applicability to a very low message injection rate.

- The random walk algorithm typically requires an order of magnitude less communication overhead than would be possible for the flooding algorithms.

- The flooding algorithms tend to be more fault resistant. As error rates go beyond 10% random walk throughput begins to drop. However, this is a very high number of errors in practice.

- Since the flooding algorithms have a much higher quantity of message hops, they are likely to have higher packet drop rates than would be likely for random walk, due to buffer overflow.

Overall, flooding protocols, while simple and lightweight, rely on too many message replications to achieve fault tolerance. While this can be compensated in part by using a directed flooding with lower complexity, a much more drastic saving in overhead can be obtained by using fault-tolerant algorithms that focus on a limited replication of messages following individual non-uniform paths such as the random walk.

## 4.10    LINK-LEVEL VERSUS END-TO-END ERROR PROTECTION

Till now this chapter has focused on link-level error control techniques. However, end-to-end data protection is a viable alternative solution as well. Selecting the location at which error protection schemes should be implemented is a critical issue and not much has been done to investigate it. The open literature reports just a few works on this topic, which are however based on high-level simulation [29, 68], therefore results have to be consolidated by means of more accurate analysis frameworks. An insight on these results will be provided in this section.

Three error recovery schemes have been analyzed in Ref. [68]: *end-to-end*, *switch-to-switch* (or *link-level*) or *hybrid error protection*. The implementation overhead, power dissipation and performance implications of these schemes have been investigated in a reference NoC architecture, based on: (i) input-queued routers with credit-based flow control; (ii) static source routing and (iii) wormhole switching. For maximum throughput, the number of queuing buffers required at each switch input should be at least $2N_L + 1$ flits, where $N_L$ is the number of cycles required to cross the link between adjacent switches. The reason is that in credit-based flow control, it takes one cycle to generate a credit, $N_L$ cycles for the credit to reach the preceding switch and $N_L$ cycles for a flit to reach a switch from the preceding switch. Here are the details about the different error protection schemes:

- *End-to-end error protection*: In the *end-to-end* (*ee*) error-detection scheme, parity (*ee-par*) or cyclic redundancy check (*ee-crc*) bits have to be added to packets. A CRC or parity encoder is added to the sender *network interface* (NI) and decoders to the receiver NI. The sender NI has one or more packet buffers in which it stores packets that have been transmitted. The receiver NI sends a *nack/ack* signal back to the sender, depending on whether the data contained an error or not. In a transaction with response (e.g., reads, non-posted writes), the feedback signal can be piggybacked on the response packet. To account for errors in the *ack/nack* packets, a time-out mechanism for retransmission at the sender should be implemented. To detect reception of duplicated packets, packet sequence identifiers could be used. Since the header flit carries critical information (such as routing information), it should be protected by parity or CRC codes, which the switch checks at each hop traversal. If a switch detects an error on a packet's header flit, it could drop the packet. Finally, redundancy could be used also to protect the flit-type bits (which identify header, body or tail flits). Concept switch architecture is reported in Fig. 4.15(a).

- *Link-level error protection*: *Switch-to-switch* schemes add the error detection hardware at each switch input and retransmit data between adjacent switches. There are two types of switch-to-switch schemes: parity or CRC at flit level and at packet level. Switch architecture for this case is reported in Fig. 4.15(b). The additional buffers added at each switch input store packets until an *ack/nack* signal comes from the next switch or NI. The number of buffers required to support switch-to-switch retransmissions depends on whether error detection occurs at packet or flit level.

In the *flit-level protection scheme* (ssf), the sender NI or source switch adds the parity or CRC bits to each flit. At each switch input, there are two sets of buffers: queuing buffers for credit-based flow control, as in the basic switch architecture, and retransmission buffers to support the switch-to-switch retransmission mechanism. Like queuing buffers, retransmission buffers at each switch input require a capacity of $2N_L + 1$ flits for full-throughput operation. In Fig. 4.15(b), redundancy (*triple modular redundancy*, TMR) is used to handle errors on the control lines (such as the *ack* line).

In the *switch-to-switch packet-level* (ssp) error detection scheme, parity or CRC bits are added to the packet's tail flit. Because error checking occurs only when the tail flit reaches the next switch, the number of retransmission buffers required at each switch input is $2N_L + f$, where $f$ is the number of flits in the packet. The *ssp* scheme also requires header flit protection, like the *ee* scheme.

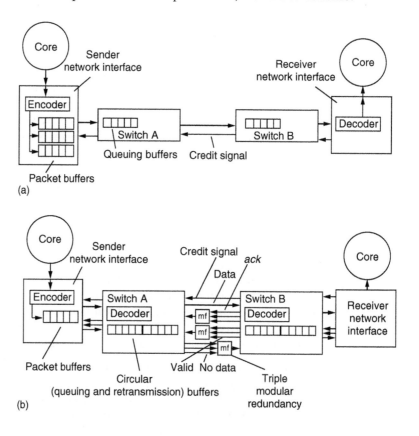

(a) End-to-end and (b) link-level retransmission architectures.

■ **FIGURE 4.15**

- *Hybrid error protection*: In the hybrid single-error correcting, multiple-error detecting scheme ($ec + ed$), the receiver corrects any single-bit error on a flit, but for multiple-bit errors, it requests end-to-end retransmission of data from the sender NI.

The error protection efficiency of the different coding schemes is investigated in a $4 \times 4$ mesh network with 16 cores and 16 switches. Packets are 4 flits long and flit size is 64 bit. System operating voltage is assumed to be fixed and equal to 0.85 V. High-level simulation of the system was performed under varying injection rates for a uniform traffic pattern. As Fig. 4.16 shows, with a low flit-error rate and a low injection rate, the various schemes' average packet latencies are almost the same. However, as the error rate and/or flit injection rate increase, the end-to-end retransmission scheme incurs a larger latency penalty than the other schemes. The link-level packet-based retransmission scheme has a slightly higher packet latency than the flit-based switch-to-switch retransmission scheme because the latter detects errors on packets earlier. As expected, the hybrid scheme outperforms the other ones.

More surprisingly, end-to-end protection schemes turn out to be more energy efficient, as depicted in Fig. 4.17. Energy models were derived in Ref. [68] from circuit schematics of individual switch components for 70 nm technology using the Berkeley Predictive Technology Model [69]. Soft-error-tolerant design methodologies were used to make the NoC components error-free, and the relative power overhead was taken into account (an increase by about 8–10%). Link lengths were assumed to be two cycles. Simulation was performed with uniform traffic pattern, with

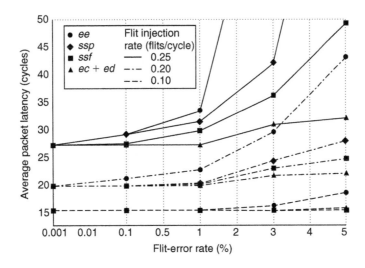

**■ FIGURE 4.16**

Packet latencies for end-to-end versus link-level error control schemes.

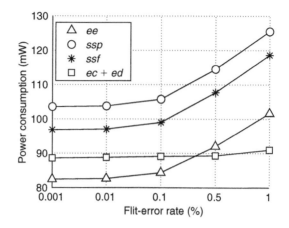

▪ **FIGURE 4.17**

Power consumption of error recovery schemes (0.1 flit/cycle).

each core injecting 0.1 flit/cycle. For *ee* and *ec + ed*, two packet buffers per NI were assumed.

Results show that power consumption of switch-based (*ssf* and *ssp*) error detecting schemes is higher than that of end-to-end (*ee* and *ec + ed*) retransmission schemes. The reason is twofold. First, the buffering required for retransmission in the *ssf* and *ssp* schemes for this setup is larger than in the *ee* and *ec + ed* schemes. Second, because of the uniform traffic pattern, traffic through each switch is greater (since the average number of hops is higher), thus increasing *ssf* and *ssp* retransmission overhead.

For the *ee* and *ec + ed* schemes, the major power overhead components are the packet-buffering needs at the NIs and the increase in network traffic caused by *ack/nack* packets. For the *ssf* and *ssp* schemes, the major power contributor are the retransmission buffers that are required at each switch. More in general, it was showed in Ref. [68] that for small link lengths and large packet-buffering requirements of the *ee* scheme, the *ssf* scheme is more power efficient than the *ee* scheme. On the other hand, when the link lengths are large, the *ee* scheme is more power efficient. Moreover, as the average hop count for data transfer increases, the *ssf* power overhead increases rapidly because more traffic passes through each switch, thereby consuming more power in the switch retransmission buffers. Finally, for reasonable flit sizes, it was shown that flit-based schemes are more power efficient than packet-based schemes, even under an increasing number of flits per packet. As regards area, the overhead of the different error protection schemes (switch area plus additional hardware for error recovery) was showed to be comparable.

The work in Ref. [29] comes to almost the same conclusions, but starting from a very different reference NoC-based platform. In fact, the Nostrum architecture is used, featuring deflective routing, minimal switch internal buffering and very high link-level bandwidth with 128-bit links. A regular two dimentional mesh topology is analyzed. In this design point, authors show that power consumption is dominated by switch-to-switch links while the switches and the NIs contribute with a mere few percent to the power consumption. In this scenario, link-level low-power encoding techniques (e.g., bus invert) spend several times more power than no encoding at all, if normalized for the same performance, which is done by adjusting supply voltage and frequency. Data protection schemes are also considered in order to reduce power if voltage levels can be reduced and certain faults can be tolerated. A link-level SECDED error control code is compared with a more powerful SECDED code used for end-to-end data protection. This latter code is more powerful to compensate for the accumulation of errors over multiple hops. As in the previous case study, the header is protected in both cases to avoid a corrupted header after each hop, and a random traffic pattern is assumed. When considering only link power (which dominates network power in this case), the end-to-end approach is showed to be consistently more power efficient than the other approach over the entire reasonable voltage swing range.

Although some interesting points are emerging from these early explorations, results have to be consolidated for more realistic traffic patterns and power models, in addition to a refined analysis of implementation costs on more accurate NoC models.

## 4.11 FLOW CONTROL

*Flow control* determines how network resources are allocated to packets traversing the network, and can be seen either as a problem of resource allocation or one of the contention resolutions [18]. Carrying out a certain error recovery strategy as an effect of a transmission error depends on the support the flow control mechanism provides for it. For instance, retransmitting a corrupted data word involves stopping the flow of packets from a source switch, signaling the request for retransmission and reallocating buffering and bandwidth resources for the original data word. Unfortunately, not all flow control strategies for NoCs are capable of reallocating resources as an effect of transmission errors, but can only manage link congestion (e.g., on–off or credit-based flow control). In this case, either communication reliability is guaranteed by means of error correction or it has to be managed at higher layers than the data link in the communication protocol stack.

Of course, the level of fault-tolerance support in flow control mechanisms is reflected into the trade-offs they span between area, power and error control capability. In general, in the NoC domain error events cannot be considered to be so frequent to violate performance specifications of applications. Therefore, the main interest is in pointing out whether the overhead for implementing combined flow and error control in hardware is such to degrade application perceived performance in the normal error-free operating mode.

This section sheds light on this trade-off for a number of relevant flow control schemes (STALL/GO, T-Error, ACK/NACK) in the context of pipelined switch-to-switch links and distributed buffering [9]. Xpipes NoC [70] is used as the reference architecture for performance exploration.

### 4.11.1    Flow Control Protocols

Each flow control protocol offers different fault-tolerance features at different performance/power/area points, as sketched in Table 4.2. STALL/GO is a low-overhead scheme which assumes reliable flit delivery. T-Error is much more complex, and provides logic to detect timing errors in data transmission; this support is however only partial, and usually exploited to improve performance rather than to add reliability. Finally, ACK/NACK is designed to support thorough fault detection and handling by means of retransmissions.

STALL/GO is a very simple realization of an ON/OFF flow control protocol (Fig. 4.18). It requires just two control wires: one going forward and flagging data availability, and other going backward and signaling either a condition of buffers filled ("STALL") or of buffers free ("GO"). STALL/GO can be implemented with distributed buffering along the link; namely, every repeater can be designed as a very simple two-stage FIFO. The sender only needs two buffers to cope with stalls in the very first link repeater, thus

**TABLE 4.2** ■ Flow control protocols at a glance.

|  | STALL/GO | T-Error | ACK/NACK |
|---|---|---|---|
| Buffer area | $2N + 2$ | $>3M + 2$ | $3N + k$ |
| Logic area | Low | High | Medium |
| Performance | Good | Good | Depends |
| Power (est.) | Low | Medium/High | High |
| Fault tolerance | Unavailable | Partial | Supported |

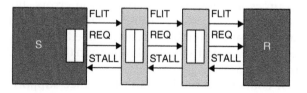

**■ FIGURE 4.18**

STALL/GO protocol implementation.

resulting in an overall buffer requirement of $2N + 2$ registers, with minimal control logic. Power is minimized since any congestion issue simply results in no unneeded transitions over the data wires. Performance is also good, since the maximum sustained throughput in absence of congestion is of 1 flit/cycle by design, and recovery from congestion is instantaneous (stalled flits get queued along the link toward the receiver, ready for flow resumption).

In the NoC domain with pipelined links, STALL/GO indirectly reflects the performance of credit-based policies, since they exhibit equivalent behavior.

The main drawback of STALL/GO is that no provision whatsoever is available for fault handling. Should any flit get corrupted, some complex higher-level protocol must be triggered.

The T-Error protocol (Fig. 4.19 [71]) aggressively deals with communication over physical links, either stretching the distance among repeaters or increasing the operating frequency with respect to a conventional design. As a result, timing errors become likely on the link. Faults are handled by a repeater architecture leveraging upon a second delayed clock to resample input data, to detect any inconsistency and to emit a VALID control signal (Fig. 4.20). If the surrounding logic is to be kept unchanged, as we assume in this chapter, a resynchronization stage must be added between the end of the link and the receiving switch. This logic handles the offset among the original and the delayed clocks, thus realigning the timing of DATA and VALID wires; this incurs a one cycle latency penalty.

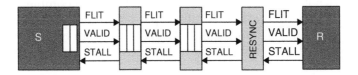

■ **FIGURE 4.19**

T-Error protocol implementation.

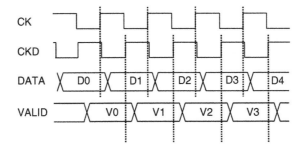

■ **FIGURE 4.20**

T-Error concept waveforms.

The timing budget provided by the T-Error architecture can also be exploited to achieve greater system reliability, by configuring the links with spacing and frequency as conservative as in traditional protocols. However, T-Error lacks a really thorough fault handling: for example, errors with large time constants would not be detected. Mission-critical systems, or systems in noisy environments, may need to rely on higher-level fault correction protocols.

The area requirements of T-Error include three buffers in each repeater and two at the sender, plus the receiver device and quite a bit of overhead in control logic. A conservative estimate of the resulting area is $3M + 2$, with $M$ being up to 50% lower than $N$ if T-Error features are used to stretch the link spacing. Unnecessary flit retransmissions upon congestion are avoided, but a power overhead is still present due to the control logic. Performance is of course dependent on the amount of self-induced errors.

The main idea behind the ACK/NACK flow control protocol (Fig. 4.21 [70]) is that transmission errors may happen during a transaction. For this reason, while flits are sent on a link, a copy is kept locally in a buffer at the sender. When flits are received, either an ACK or NACK is sent back. Upon receipt of an ACK, the sender deletes the local copy of the flit; upon receipt of an NACK, the sender rewinds its output queue and starts resending flits starting from the corrupted one, with a GO-BACK-$N$ policy. This means that any other flit possibly in flight in the time window among the sending of the corrupted flit and its resending will be discarded and resent. Other retransmission policies are feasible, but they exhibit higher logic complexity. ACK/NACK can either be implemented as end-to-end over a whole fabric or as switch-to-switch; due to complex issues with possible flit misrouting upon faults in packet headers, the latter solution is considered in this study. Fault tolerance is built in by design, provided encoders and decoders for ECCs are implemented at the source and destination, respectively.

In an ACK/NACK flow control, a sustained throughput of 1 flit/cycle can be achieved in presence of proper buffering resources. Repeaters on the link can be simple registers, while, with $N$ repeaters, $2N + k$ buffers are required at the source to guarantee maximum throughput, since

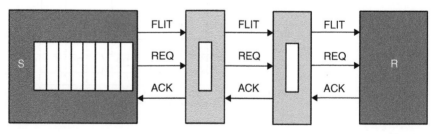

■ **FIGURE 4.21**

ACK/NACK protocol implementation.

ACK/NACK feedback at the sender is only sampled after a round-trip delay since the original flit injection. The value of $k$ depends on the latency of the logic at the sending and receiving ends. Overall, the minimum buffer requirement to avoid incurring bandwidth penalties in an NACK-free environment is therefore $3N + k$.

ACK/NACK provides ideal throughput and latency until no NACKs are issued. If NACKs were only due to sporadic errors, the impact on performance would be negligible. However, if NACKs have to be issued also upon congestion events, the round-trip delay in the notification causes a performance hit which is very pronounced especially with long pipelined links. Moreover, flit bouncing between sender and receiver devices causes a waste of power.

## 4.12 PERFORMANCE EXPLORATION

In the following, we are interested in the performance of flow control protocols during normal operation. The assumption is that with a conservatively clocked design, errors can be made rare enough to have a negligible performance impact. This holds also for T-Error. When used to increase system reliability, by deploying it with conservative link parameters, the fault-free communication will be assumed. When used to aggressively space link repeaters, thus artificially causing and handling a non-trivial amount of data corruption in exchange for better performance, self-induced corruption will be accounted for.

Protocol analysis is performed on a star-like topology, where up to eight clusters of three processors and their private memories can be instantiated. At the heart of each cluster is a $7 \times 7$ port xpipes switch. Therefore, up to 24 processors can be deployed in the platform. Shared slaves exist and are attached to a central $11 \times 11$ switch (see Fig. 4.22). The central switch is connected to the computation clusters by means of pipelined links, whose length can be customized. Depending on the amount of instantiated processors, the simultaneous traffic on the links toward this central switch

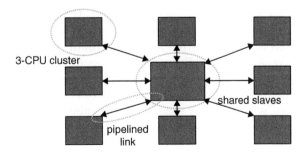

■ **FIGURE 4.22**

The star-like topology under test.

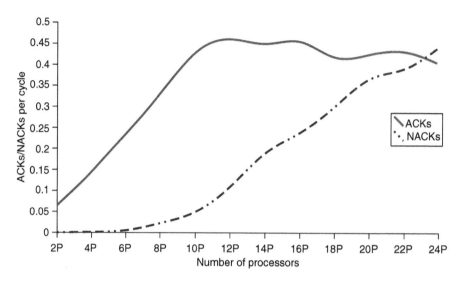

■ **FIGURE 4.23**

Link congestion under increasing traffic.

increases, resulting in congestion. All of the processors are executing the same benchmark, which encompasses local computation (which happens within the peripheral clusters) and performs communication and synchronization functions by accessing the shared slaves on the central switch.

An analysis of the link congestion trend under increasing pressure by IP cores can be found in Fig. 4.23, which plots the density of ACKs and NACKs (ACKs and NACKs over clock cycles) when using ACK/NACK flow control in the system. The chart is only plotting figures for the links connecting the central switch to the clusters; the links are here assumed to be very long, with six repeaters. As can be seen, with more processors, initially the ACK density increases thanks to the increase in offered bandwidth. Just before hitting the value 0.5 (one ACK every two cycles), growing congestion and the intrinsic inefficiency of this flow control protocol impose a bandwidth ceiling, while a growing amount of NACKs can be observed.

The average latency for communication transactions across the congested links are plotted as a function of the length of the pipelined links in Fig. 4.24. For T-Error, results are referred to two scenarios. The first accounts for 50% less repeater stages (four instead of six and two instead of three) and assumes a 5% error rate. The second scenario is more conservative, with as many repeaters as in other protocols and no errors. In both cases, a resynchronization stage is needed before the receiving switch. In the direct connection scenario, where it is assumed that operation over the links can be safely achieved in a single clock cycle, T-Error logic is unneeded and this case reduces to the direct STALL/GO connection. Under all circumstances, both variants of T-Error and STALL/GO exhibit

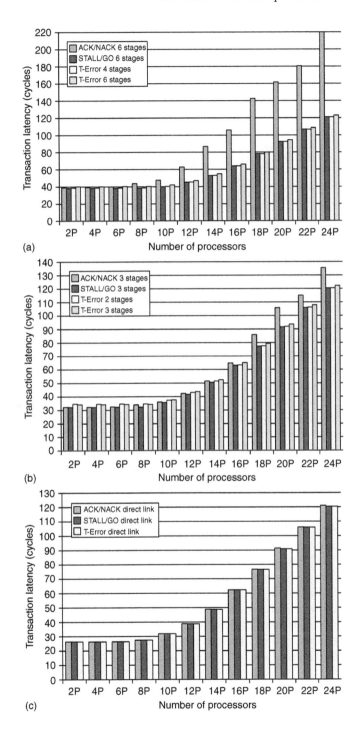

■ **FIGURE 4.24**

Communication latency over congested links of different lengths.

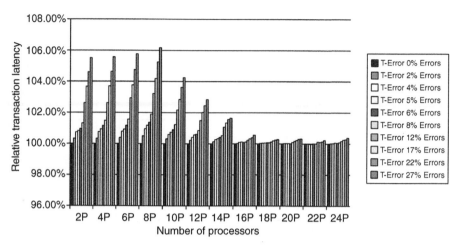

T-Error performance under varying error probabilities.

similar performance. The latency advantage gained over STALL/GO by the aggressive deployment of T-Error is mostly offset by the need for a resynchronization stage and by some penalty upon transmission errors; for short links and in low-congestion environments, T-Error can even perform worse. The conservative T-Error links perform almost on par with the aggressive ones because they trade repeater stages for error-free operation. The conservative T-Error links always perform slightly worse than the corresponding STALL/GO schemes due to resynchronizer latency, but the transaction overhead is negligible. As expected, if links are very long (six repeaters: Fig. 4.24(a)), the round-trip delay imposed by the ACK/NACK protocol proves to be a major drawback, and latencies increase steeply with congestion. With shorter links (three repeaters: Fig. 4.24(b)), the ACK/NACK overhead decreases, and substantial performance parity is achieved if the switches are directly attached (Fig. 4.24(c)).

The aggressive T-Error variant achieves lower communication latencies by accepting a certain amount of transmission errors as a trade-off. In Fig. 4.25, T-Error performance is explored in presence of different self-induced error probabilities. The plot is reporting figures for a long link, which is assumed to have a 0–27% error rate percentage. This number expresses the percentage of errors per clock cycle (not per transmitted flit), and is for the whole link. Latencies are normalized against the ideal error-free case. As the plot shows, under heavy congestion, T-Error is by design able to almost completely mask errors, because error penalties can be hidden behind congestion-triggered stalls. Under light traffic, transmission faults have a more noticeable impact, with 6% worse transmission latency when comparing a 27% link error probability against the ideal case. Still, T-Error is very good at minimizing the impact of faults on performance.

### 4.12.1   Considerations

STALL/GO proves to be a low-overhead efficient design choice showing remarkable performance, but unfortunately is fault sensitive.

T-Error can be either deployed to improve link performance or to improve system reliability by catching timing errors. In the former design, average latencies are on par with STALL/GO, but no error detection capability is present; in the latter case, speed degrades slightly, in exchange for a partial but significant reliability boost. In both alternative schemes, some area and power overhead are incurred. Overall, using T-Error to decrease the number of pipeline stages does not bring significant performance benefits, while the partial detection capability can be effectively exploited in a conservative design. Another possible option is the conversion of the timing margin budget into a frequency overclock; while this choice holds good potential, it requires the surrounding NoC components (switches, NIs) to be designed to work at the same extremely high frequencies.

ACK/NACK pays its most extensive fault handling support with significant power and area overheads. Performance penalties are also noticed in presence of heavy congestion and long pipelined links. However, it is reasonable to expect that in current and imminent design technologies, links will need no more than three repeaters, the very-high-latency-link scenario being representative of a more distant future. In low-congestion or short-link scenarios, the application-perceived latency overhead of ACK/NACK turns out to be negligible.

## 4.13   SUMMARY

This chapter provides a comprehensive overview of data-link layer design issues in the NoC domain. This layer is in charge of implementing a technology-independent approach to communication reliability, exposing a reliable communication channel in spite of the underlying unreliable medium. The increased sensitivity of interconnects to on-chip noise sources will make an efficient data-link layer architecture critical for the success of NoCs.

Redundant error control coding is proving a viable option for communication reliability across switch-to-switch links. Given the area, delay and power constraints of NoCs, the interest is in pure combinational codes with low logic depth. Therefore, the single parity check code, several variants of the Hamming code and CRC codes are the reference solutions for state-of-the-art NoC prototypes. Code design for a particular application domain has been shown to be a difficult task, since code selection and tuning of code parameters has (often conflicting) implications on power, performance and reliability. New link architectures are emerging for better power-reliability trade-offs, leveraging self-calibration and adaptive operating frequencies and voltage swings. More aggressive approaches are represented by new communication paradigms, based on probabilistic broadcast algorithms, which however largely reduce available bandwidth. One of the most critical issues in assessing effectiveness and energy efficiency of fault-tolerant communication schemes is represented by the lack of accurate bit error rate models, which presently rely on highly simplifying assumptions. However, new fault model notations are being introduced.

This chapter has also showed that addressing communication reliability at the data-link layer depends on two fundamental conditions. First, the flow control mechanism implemented in a given NoC architecture must provide support for fault tolerance, otherwise solutions at higher abstraction layers have to be devised. The cost for this support is explored in detail. Second, the overall cost in terms of buffering resources, network traffic overhead and performance degradation must be lower than that incurred by end-to-end protection schemes. Overall, the chapter sheds light on the basic principles and the main trends in data-link layer design for NoCs, and moves from the awareness that research in this domain is still in the early stage.

# REFERENCES

[1] H. Zhou and D.F. Wang, "Global Routing with Crosstalk Constraints," *Proceeding of the 1998 Design and Automation Conference 35th DAC*, 1998, pp. 374–377.

[2] Z. Chen and I. Koren, "Crosstalk Minimization in Three-Layer HVH Channel Routing," *Proceeding of the IEEE International Symposium on Defect and Fault Tolerance in VLSI System*, 1997, pp. 38–42.

[3] K.L. Shepard, "Design Methodologies for Noise in Digital Integrated Circuits," *Proceedings of the Design Automation Conference*, 1998, pp. 94–99.

[4] T. Stohr, H. Alt, A. Hetzel and J. Koehl, "Analysis, Reduction and Avoidance of Crosstalk on VLSI Chips," *Proceedings of the International Symposium Physical Design*, 1998, pp. 211–218.

[5] K. Rahmat, J. Neves and J. Lee, "Methods for Calculating Coupling Noise in Early Design: a Comparative Analysis," *Proceedings of the International Conference on Computer Design VLSI in Computers and Processors*, 1998, pp. 76–81.

[6] H. Kawaguchi and T. Sakurai, "Delay and Noise Formulas for Capacitively Coupled Distributed RC Lines," *Proceedings of the Asian and South Pacific Design Automation Conference*, 1998, pp. 35–43.

[7] A. Vittal and M. Marek-Sadowska, "Crosstalk Reduction for VLSI," *IEEE Transaction on Computer-Aided Design of Integrated Circuits and Systems*, Vol. 16, No. 3, 1997, pp. 290–298.

[8] T.Y. Ho, Y.W. Chang, S.J. Chen and D.T. Lee, "Crosstalk- and Performance-Driven Multilevel Full-Chip Routing," *IEEE Transactions on Computer-Aided*

*Design of Integrated Circuits and Systems*, Vol. 24, No. 6, June 2005, pp. 869–878.

[9] A. Pullini, F. Angiolini, D. Bertozzi and L. Benini, "Fault Tolerance Overhead in Network-on-Chip Flow Control Schemes," *Proceedings of the 18th Annual Symposium on Integrated Circuits and System Design (SBCCI 2005)*, pp. 224–229.

[10] D. Pradhan, *Fault Tolerant Computing: Theory and Techniques*, Prentice Hall, Englewood Cliffs, NJ, 1986.

[11] D. Das and N. Touba, "Weight-Based Codes and their Applications to Concurrent Error Detection of Multilevel Circuits," *Proceedings of the VLSI Test Symposium*, 1999, pp. 370–376.

[12] M. Favalli and C. Metra, "Optimization of Error Detecting Codes for the Detection of Crosstalk Originated Errors," *Proceedings of the DATE*, March 2001, pp. 290–296.

[13] H. Ball and F. Hardy, "Effects and Detection of Intermittent Failures in Digital Systems," *Proceedings of the Fall Joint Computer Conference, American Federation of Information Processing Societies Conference*, 1969, pp. 329–335.

[14] R. Iyer and D. Rosetti, "A Statistical Load Dependency of CPU Errors at SLAC," *Proceedings of the Fault Tolerance Computing Systems FTCS*, 1982, pp. 363–372.

[15] A. Maheshwari, W. Burleson and R. Tessier, "Trading Off Transient Fault Tolerance and Power Consumption in Deep Submicron (DSM) VLSI Circuits," *IEEE Transaction on Very Large Scale Integration (VLSI) Systems*, Vol. 12, No. 3, March 2004, pp. 299–311.

[16] M. Nicolaidis, "IP for embedded robustness," *Proceedings of the DATE*, 2002, pp. 240–241.

[17] H.P. Wong, D.J. Frank, P.M. Solomon, C.H.J. Wann and J.J.Welser, "Nanoscale CMOS," *Proceedings of the IEEE*, Vol. 87, No. 4, April 1999, pp. 537–570.

[18] W.J. Dally and B. Towles, *Principles and Practices of Interconnection Networks*, Morgan Kaufmann, San Francisco, CA, 2004.

[19] G. Bai and I.N. Hajj, "Simultaneous Switching Noise and Resonance Analysis of on-Chip Power Distribution Network," *Proceedings of the IEEE International Symposium on Quality Electronic Design*, 2002, pp. 163–168.

[20] C.L. Chen and B.W. Curran, "Switching Codes for Delta-I Noise Reduction," *IEEE Transaction on Computers*, September 1996, pp. 1017–1021.

[21] G.E. Moore, "No Exponential is Forever: But 'Forever' Can Be Delayed!," *Proceedings of the ISSCC*, Vol. 1, 2003, pp. 20–23.

[22] R. Hedge and N.R. Shanbhag, "Toward Achieving Energy Efficiency in Presence of Deep Submicron Noise," *IEEE Transactions on – VLSI Systems*, Vol. 8, No. 4, August 2000, pp. 379–391.

[23] D. Bertozzi, L. Benini and G. De Micheli, "Low Power Error Resilient Encoding for on-Chip Data Buses," *Proceedings of the DATE*, 2002, pp. 102–109.

[24] D. Bertozzi, L. Benini and B. Riccó, "Energy-Efficient and Reliable Low-swing Signaling for On-Chip Buses Based on Redundant Coding," *ISLPED*, I-93–I-96, 2002.

[25] D. Bertozzi, L. Benini and G. De Micheli, "Error Control Schemes for On-Chip Communication Links: The Energy-Reliability Tradeoff," *IEEE Transaction on CAD of Integrated Circuits and Systems*, Vol. 24, No. 6, June 2005, pp. 818–831.

[26] S.R. Sridhara and N.R. Shanbhag, "Coding for System-on-Chip Networks: A Unified Framework," *IEEE Transactions on VLSI Systems*, Vol. 13, No. 6, June 2005, pp. 655–667.

[27] L. Li, N. Vijaykrishnan, M. Kandemir and M.J. Irwin, "Adaptive Error Protection for Energy Efficiency," *International Conference on Computer Aided Design (ICCAD 2003)*, 2003, pp. 2–7.

[28] P. Vellanki, N. Banerjee and K.S. Chatha, "Quality-of-Service and Error Control Techniques for Mesh-Based Network-on-Chip Architectures," *INTEGRATION, the VLSI Journal*, Vol. 38, 2005, pp. 353–382.

[29] A. Jantsch, R. Lauter and A. Vitkowski, "Power Analysis of Link Level and End-to-End Data Protection in Networks on Chip," *International Symposium on Circuits and Systems*, Vol. 2, 2005, pp. 1770–1773.

[30] H. Zimmer and A. Jantsch, "A Fault Model Notation and Error-Control Scheme for Switch-to-Switch Buses in a Network-on-Chip," *CODES ISSS 2003*, pp. 188–193.

[31] T. Dumitras, S. Kerner and R. Marculescu, "Towards On-Chip Fault-Tolerant Communication," *Proceedings of the ASP-DAC 2003*, 2003, pp. 225–232.

[32] R. Marculescu, "Networks-on-Chip: The Quest for On-Chip Fault-Tolerant Communication," *Proceedings of the IEEE Computer Society Annual Symposium on VLSI*, ISVLSI'03, 2003, pp. 8–12.

[33] S. Lin and D.J. Costello, *Error Control Coding: Fundamentals and Applications*, Prentice-Hall, Englewood Cliffs, NJ, 1983.

[34] S. Roman, *Introduction to Coding and Information Theory*, Springer Verlag, Berlin, Germany, 1997.

[35] G.L. Stueber, *Principles of Mobile Communication*, 2nd edition, Kluwer Academic Publishers, Norwell, MA, 2001.

[36] P. Koopman and T. Chakravarty, "Cyclic redundancy code (CRC) polynomial selection for embedded systems," *The International Conference on Dependable Systems and Networks*, 2004, pp. 1–10.

[37] W.W. Peterson and D.T. Brown, "Cyclic Codes for Error Detection," *Proceedings of the IRE*, January 1961, pp. 228–235.

[38] U. Nordqvist, T. Henrikson and D. Liu, "Configurable CRC Generator," *Proceedings. DDECS*, Brno, Czeck Republic, April 2002, pp. 192–199.

[39] A. Sobski and A. Albicki, "Partitioned and Parallel Cyclic Redundancy Checking," *Proceedings of the 36th Symposium Circuits and Systems*, Vol. 1, August 1993, pp. 538–541.

[40] A. Mello, L. Moeller, N. Calazans and F. Moraes, "MultiNoC: A Multiprocessing System Enabled by a Network on Chip," *Design, Automation and Test in Europe (DATE'05)*, Vol. 3, 2005, pp. 234–239.

[41] E. Bolotin, I. Cidon, R. Ginosar and A. Kolodny, "QNoC: QoS Architecture and Design Process for Network on Chip," Special issue on Networks on Chip, *The Journal of Systems Architecture*, Vol. 50, No. 2–3, February 2004, pp. 105–128.

[42] E. Rijpkema, K. Goossens, A. Radulescu, J. Dielissen, J. van Meerbergen, P. Wielage and E. Waterlander, "Trade Offs in the Design of a Router with Both Guaranteed and Best-Effort Services for Networks on Chip," *IEE Proceedings: Computers and Digital Techniques*, Vol. 150, No. 5, 2003, pp. 294–302.

[43] D. Rossi, V.E.S. van Dijk, R.P. Kleihorst, A.H. Nieuwland and C. Metra, "Coding Scheme for Low Power Consumption Fault Tolerant Bus,"

*IEEE Proceedings of the International On Line Testing Workshop*, 2002, pp. 8–12.

[44] D. Rossi, V.E.S. van Dijk, R.P. Kleihorst, A.K. Nieuwland and C.Metra, "Power Consumption of Fault-Tolerant Bus: The Active Elements," *Proceedings of the IEEE International On Line Testing Symposium*, 2003, pp. 61–67.

[45] D. Rossi, A. Muccio, A.K. Nieuwland, A. Katoch and C. Metra, "Impact of ECCs on Simultaneously Switching Output Noise for On-Chip Busses of High Reliability Systems," *Proceedings of the 10th IEEE International On-Line Testing Symposium*, 2004, pp. 135–140.

[46] D. Rossi, A.K. Nieuwland, A. Katoch and C. Metra, "Exploiting ECC Redundancy to Minimize Crosstalk Impact," *IEEE Design and Test of Computers*, 2005, pp. 59–70.

[47] D. Rossi, A.K. Nieuwland, A. Katoch and C. Metra, "New ECC for Crosstalk Impact Minimization," *IEEE Design and Test of Computers*, 2005, pp. 340–348.

[48] H. Zhang, V. George and J.M. Rabaey, "Low-Swing On-Chip Signaling Techniques: Effectiveness and Robustness," *IEEE Transactions on Very Large Scale Integration (VLSI) Systems*, Vol. 8, No. 3, 2000, pp. 264–272.

[49] P.P. Sotiriadis, *Interconnect Modelling and Optimization in Deep Submicron Technologies*, Ph.D. dissertation, Massachusetts Institute of Technology, Cambridge, May 2002.

[50] C. Duan, A. Tirumala and S.P. Khatri, "Analysis and Avoidance of Crosstalk in On-Chip Buses," *Proceedings of the Hot Interconnects*, 2001, pp. 133–138.

[51] B. Victor and K. Keutzer, "emphBus Encoding to Prevent Crosstalk Delay," *Proceedings of the ICCAD'01*, 2001, pp. 57–63.

[52] S.R. Sridhara, A. Ahmed and N.R. Shanbhag, "Area and Energy-Efficient Crosstalk Avoidance Codes for On-Chip Buses," *Proceedings of the ICCAD'04*, 2004, pp. 12–17.

[53] S. Ramprasad, N.R. Shanbhag and I.N. Hajj, "A Coding Framework for Low-Power Address and Data Busses," *IEEE Transactions on VLSI Systems*, Vol. 7, No. 2, June 1999, pp. 212–221.

[54] M.R. Stan and W.P. Burleson, "Bus-Invert Coding for Low-Power I/O," *IEEE Transactions on VLSI Systems*, Vol. 3, No. 1, March 1995, pp. 49–58.

[55] K. Kim, K. Baek, N. Shanbhag, C. Liu and S. Kang, "Coupling-Driven Signal Encoding Scheme for Low-Power Interface Design," *Proceedings of the ICCAD'00*, 2000, pp. 318–321.

[56] Y. Zhang, J. Lach, K. Skadron and M.R. Stan, "Odd/Even Bus Invert With Two-Phase Transfer for Buses with Coupling," *Proceedings of the ISLPED'02*, 2002, pp. 80–83.

[57] T. Austin, D. Blaauw, T. Mudge and K. Flautner, "Making Typical Silicon Matter with Razor," *IEEE Computer*, Vol. 37, No. 3, March 2004, pp. 57–65.

[58] F. Worm, P. Ienne, P. Thiran and G. De Micheli, "A Robust Self-Calibrating Transmission Scheme for On-Chip Networks," *IEEE Transactions on Very Large Scale Integration (VLSI) Systems*, Vol. 13, No. 1, January 2005, pp. 126–139.

[59] F. Worm, P. Thiran, G. De Micheli and P. Ienne, "Self-Calibrating Networks-on-Chip," *Proceedings of the IEEE International Symposium on Circuits and Systems (2005)*, 2005, pp. 2361–2364.

[60] F. Worm, P. Thiran and P. Ienne, "Soft Self-Synchronising Codes for Self-Calibrating Communication," *Proceedings of the International Conference on CAD*, November 2004, pp. 440–447.

[61] A. Morgenshtein, E. Bolotin, I. Cidon, A. Kolodny and R. Ginosar, "Micro-Modem – Reliability Solution for NoC Communications," *ICECS*, 2004, pp. 483–486.

[62] M. Millberg, "The Nostrum Protocol Stack and Suggested Services Provided by the Nostrum Backbone," Internal Report in Electronic System Design, TRITA-IMIT-LECSR02:01, LECS, IMIT, KTH, Stockholm, Sweden, 2003.

[63] S.J. Lee, K. Kim, H. Kim, N. Cho and H.J. Yoo, "Adaptive Network-on-Chip with Wave-Front Train Serialization Scheme," *International Symposium on VLSI Circuits (VLSI)*, Kyoto, June 2005, pp. 104–107.

[64] M. Millberg, E. Nilsson, R. Thid and A. Jantsch, "The Nostrum Backbone – A Communication Protocol Stack for Networks-on-Chip, *Proceedings of the International Conference on VLSI Design*, India, January 2004, pp. 693–696.

[65] R. Karp et al., "Randomized Rumor Spreading," *Proceedings of the IEEE Symposium on Foundations of Computer Science*, 2000, pp. 565–574.

[66] K. Birman et al., "Bimodal Multicast," *ACM Transactions on Computer Systems*, Vol. 17, No. 2, 1999, pp. 41–88.

[67] M. Pirretti, G.M. Link, R. Brooks, N. Vijaykrishnan, M. Kandemir and M.J. Irwin, "Fault Tolerant Algorithm for Network-on-Chip Interconnect," *Proceedings of the IEEE Computer Society Annual Symposium on VLSI Emerging Trends in VLSI Systems Design*, ISVLSI'04, 2004, pp. 46–51.

[68] S. Murali, T. Theocharides, N. Vijaykrishnan, M.J. Irwin, L. Benini and G. De Micheli, "Analysis of Error Recovery Schemes for Networks on Chips," *IEEE Design and Test of Computers*, 2005, pp. 434–442.

[69] *Berkeley Predictive Technology Model*, http://www-device.eecs.berkeley.edu/ptm/

[70] S. Stergiou, F. Angiolini, S. Carta, L. Raffo, D. Bertozzi, L. Benini and G. De Micheli, "Xpipes Lite: A Synthesis Oriented Design Library For Networks on Chips," *Proceedings of the Conference on Design, Automation and Test in Europe*, Vol. 2, March 2005, pp. 1188–1193.

[71] R.R. Tamhankar, S. Murali and G. De Micheli, "Performance Driven Reliable Link Design for Networks on Chips," *Proceedings of the ASP-DAC'05*, 2005, pp. 749–754.

[72] E. Yeo, S.A. Augsburger, W.R. Davis and B. Nikolic, "A 500-Mb/s Soft-Output Viterbi Decoder," *IEEE Journal of Solid-State Circuits*, Vol. 38, No. 7, July 2003, pp. 1234–1241.

[73] F. Angiolini, D. Bertozzi, L. Raffo, P. Meloni, S. Carta and L. Benini, "Xpipes Network on Chip: A Synthesis Perspective," *Parallel Computing Minisymposium*, 12–16 September 2005, Malaga (Spain).

[74] G.D. Nguyen, "Error-Detection Codes: Algorithms and Fast Implementation," *IEEE Transactions on Computers*, Vol. 54, No. 1, January 2005, pp. 1–11.

[75] G. Albertango and R. Sisto, "Parallel CRC Generation," *IEEE Micro*, Vol. 10, No. 5, October 1990, pp. 63–71.

[76] T.B. Pei and C. Zukowski, "High-Speed Parallel CRC Cicuits in VLSI," *IEEE Transactions on Communications*, Vol. 40, No. 4, 1992, pp. 653–657.

[77] R.F. Hobson and K.L, Cheung, "A High-Performance CMOS 32-Bit Parallel CRC Engine," *IEEE Journal Solid State Circuits*, Vol. 34, No. 2, February 1999, pp. 233–235.

[78] R. Nair, G. Ryan and F. Farzaneh, "A Symbol Based Algorithm for Implementation of Cyclic Redundancy Check (CRC)," *Proceedings of the VHDL International Users' Forum*, 1997, pp. 82–87.

[79] http://www.st.com/stonline/prodpres/dedicate/soc/cores/stbus.htm

[80] L. Benini and G. De Micheli, "Networks on Chip: A New Paradigm for Component-Based MPSoC Design," in A. Jerrraya and W. Wolf (editors), *Multiprocessors Systems on Chips*, Morgan Kaufmann, San Francisco, CA, 2004, pp. 49–80.

[81] H.J. Chao, C.H. Lam and X. Guo, "A Fast Arbitration Scheme for Terabit Packet Switches," *Proceedings of the IEEE Global Telecommunications Conference*, 1999, pp. 1236–1243.

[82] N. Mckeown, P. Varaiya and J. Warland, "The iSLIP Scheduling Algorithm for Input-Queued Switch," *IEEE Transactions on Networks*, 1999, pp. 188–201.

[83] H.J. Chao and J.S. Park, "Centralized Contention Resolution Schemes for a Larger-Capacity Optical ATM Switch," *Proceedings of the IEEE ATM Workshop*, 1998, pp. 11–16.

[84] E.S. Shin, V.J. Mooney III and G.F. Riley, "Round-Robin Arbiter Design and Generation," *Proceedings of the 15th International Symposium on System Synthesis (ISSS '02)*, 2002, pp. 243–248.

[85] AMBA AHB Specification, Rev.2.0, ARM, 1999.

[86] G. Castagnoli, S. Brauer and M. Herrmann, "Optimization of Cyclic Redundancy-Check Codes with 24 and 32 Parity Bits," *IEEE Transactions on Communications*, Vol. 41, No. 6, June 1993, pp. 883–892.

[87] R.J. Glaise and X. Jacquart, "Fast CRC Calculation," *1993 IEEE International Conference on Computer Design: VLSI in Computers and Processors*, 1993, pp. 602–605.

[88] T.V. Ramabadran and S.S. Gaitonde, "A Tutorial on CRC Computations," *IEEE Micro*, Vol. 8, No. 4, August 1988, pp. 62–75.

# CHAPTER 5

# NETWORK AND TRANSPORT LAYERS IN NETWORKS ON CHIP*

The main goal of the network and transport layers is to support the end-to-end communication between the modules at the specified *quality of service* (QoS) using a power- and resource-efficient sharing of the interconnect resources. This seemingly simple goal forces the designer to address all classical (and some new) network layer issues in the context of an individual chip design.

In order to define the end-to-end QoS, we need to define the on-chip communication requirements in terms of the module-to-module traffic types, rates, statistical behavior, and predictability. For each such end-to-end *flow* we also need to define the service it receives such as loss and delay. We also need to define the interrelation between these flows, such as priorities, and actions that should be taken when communication resources are scarce. Finally, we need to decide on the appropriate network architecture that supports the above QoS requirements. For this the following *network on chip* (NoC) characteristics must be defined: switching technique, NoC topology, addressing and routing scheme, and end-to-end congestion and flow control schemes.

First, the appropriate data switching mechanism (e.g., circuit switching, packet switching, etc.) needs to be selected for the multiplexing of multiple flows from different sources and different requirements in a single network. Then, the appropriate network topology needs to be selected and optimized to physically connect the different modules (including basic graph topology, links and router speeds, and specific layout issues). Addressing and routing schemes need to be devised in order to allow circuit or packets traversing the network to be routed to diverse destinations including possible multicast or broadcast of information. Names meaningful to applications (such as memory and I/O addresses) need to be translated into routing-efficient labels. Since the proper delivery of certain types of traffic (or signals) on the chip is crucial for performance and is categorized in different ways, the concurrent support of multiple QoS

---

* This chapter was provided by Israel Cidon, of Technion, Israel and by Kees Goossens, of Philips Research, The Netherlands.

requirements is essential. Since modules perform a variety of information exchanges, there is a need for end-to-end mechanisms to guarantee in order and assure delivery, end-to-end connection and flow control for rate matching and receiver buffer management and resource access control for multiple resource access resolution. In order to accommodate excessive traffic conditions and to quantify the behavior of the network under extreme conditions, NoC should also support network-level congestion control. One also needs to address reliability in the face of communication soft errors that may corrupt transmitted data.

## 5.1 NETWORK AND TRANSPORT LAYERS IN NoCs

The network layer deals with the QoS, switching technique, topology, and addressing and routing schemes. The transport layers address the congestion and flow control issues. However, we will not make the distinction in the remainder of this chapter.

Since similar network and transport layer problems have been extensively studied in the networking and system interconnect realm, one may be tempted to employ well-developed networking solutions in the NoC context. However, a direct adaptation of such network solutions to NoCs is impossible, due to the different communication and performance requirements, cost considerations, and architectural constraints [21].

First, the requirements are different. Unlike many off-chip networks, the NoC is at the heart of chips that support real-time operations and are embedded in critical systems from life-support gear to vehicular and aerospace equipment. Embedded systems also often deal with intrinsically real-time data, such as high-quality audio and video. Therefore, NoCs QoS requirements call for a high degree of predictability and robustness and cannot tolerate incidental glitches and anomalies [32]. As a specific example, network design must meet strict QoS requirements for certain types of traffic, such as interrupt signals or fetching real-time instructions and data from caches to *digital signal processors (DSPs)*. Such requirements may be specified in very strict terms, such as the number of clock cycles to accomplish a certain transaction.

Second, the cost considerations are different. The primary considerations in *very large-scale integration* (VLSI) are minimizing area and power dissipation. The area cost of an NoC is composed of routers and network interfaces (i.e., logic cost) and the cost of wires/links that interconnect them (area used by metal lines, spacing, shielding, repeaters, etc.). Similarly, power consumption can be separated to dynamic and static power consumed by the NoC logic (in routers and network interfaces) and links (in wires and repeaters). In both cases, both temporal power to reduce heat dissipation, and total energy consumption for saving battery power for nomadic systems [12] are to be minimized. Moreover, since an NoC connects modules or relatively small subsystems, these area and power costs should be kept much lower compared to networks that connect large systems.

Third, there are also many architectural constraints and freedoms unique to the NoC environment. For example, on-chip network topologies are quite restricted – they are planar, often organized as (full or partial) grids, and do not need to support the dynamic addition or removal of components. A new important dimension of freedom in NoC is the ability to alter the physical layout of the network routers and links along with the chip module placements, enabling the designer to optimize the NoC geography according to traffic and layout constraints [13, 77, 79].

Furthermore, each NoC is synthesized anew for each design [43, 61], eliminating the need for standardization of network protocols. That is, protocols and architectures employed in a new chip design do not have to be compatible with those used in previous designs. Consequently, NoCs do not possess the rigid standardization constraints of traditional networks. Except for retaining module reuse compatibility, NoCs can be customized to their specific chip environment and need not assume backward, upward, and different vendor compatibility as well as regulatory constraints. In that sense the NoC environment is much more open to multiple choice and architectural innovations. Nevertheless, the use of standard network interfaces, such as *Open Core Protocol* (OCP) [85] and *Advanced extensible Interface* (AXI) [2] is important, in order to allow the reuse of modules, across chip designs.

## 5.1.1 Classifying NoC Models

As mentioned in Chapter 1, Networks on chip can be classified into different families. We elaborate more on this concept here, and relate to a classification of chip designs based on Ref. [21] and on its impact on NoC design. *Systems on chip* (SoCs) span a vast spectrum of objectives and implementations.

On one extreme of the spectrum we identify the *application-specific* SoCs or ICs (ASICs), encompassing custom-made designs with a particular pre-known application, for example, a multichannel 3G base-station, a video camera, or a set-top box. In such chips, the network usage is known *a priori* and can be classified prior to the NoC design time.

On the opposite side of the spectrum, general-purpose *chip multiprocessors (CMPs)*, capture general-purpose chip designs supporting parallel processing, where the chip and interconnect usage is unpredictable and only determined at run time. Here we can find general multi-core processors, parallel DSP arrays, etc.

In between these two extremes we can find designs whose traffic patterns can be partially predicted, or may have several distinct usage patterns. These systems are the outgrowth of *application-specific standard*

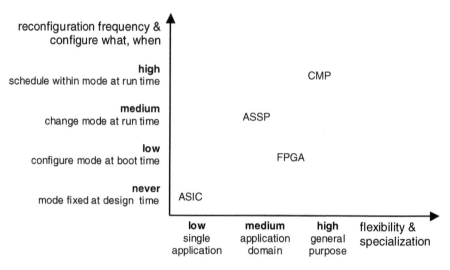

■ **FIGURE 5.1**

An NoC classification.

*products* (ASSP); also called *application-specific instruction set processors* (ASIPs), which are processors whose architecture has been tuned to an application. SoCs may contain multiple processors and mix general-purpose and application-specific processors. In general, it is commonplace to refer to *platforms* as to multiprocessors dedicated to a family of applications.

The *field-programmable gate arrays (FPGAs)* define multipurpose chips that include generic hardware resources like logic arrays, flip-flops, RAM, processors, and special purpose *Internet Protocol* (IP) modules (Ethernet, USB, DSP) that can be *configured* using a programmable interconnect grid, into specific systems. The design of such systems leaves a large degree of freedom for the FPGA designer (programmer).

We illustrate the NoC design spectrum in Fig. 5.1.

### ASIC

The distinguishing property of our ASIC NoC definition is the ability to predict the network usage patterns before the network is designed. Since a typical SoC has a specific functionality, this functionality can be simulated, and the inter-module communication patterns and requirements can be inferred at design time. Consequently, a specific NoC can be synthesized to satisfy these exact needs. Virtually all network parameters, including network layout, link capacity allocation, buffer sizes, packet headers, partial routing tables, and the number and requirements of service classes, can be custom synthesized to meet the particular SoCs requirements with no "open-ended" resources spent on "general-purpose" requirements. Specifically, most SoCs do not need to support all possible module-to-module communication patterns as well as all possible QoS classes for all modules or SoC vicinities.

ASICs are optimized for the application, and hence will have the smallest area and use the least power. This comes at the cost of reduced flexibility. Related to this is the cost of designing an ASIC, which is incurred most often of all categories, namely for each application.

### ASSP and platforms

A slightly more complex scenario is dealing with SoCs that are designed to support multiple applications, or that can take different chip "incarnations" at either production or power-up time. In such SoCs different modules and different software may be operating in each distinct SoC incarnation. A major design objective is to build a single SoC for a broad range of related applications [98]. In such a case the NoC is designed to support all possible applications. However, for each specific incarnation the NoC can be configured to operate at minimum power consumption. Therefore, non-required elements can be turned off and dynamic voltage and frequency scaling (DVS/DFS) techniques can be used to tune the NoC for an optimal performance versus power trade off [78].

ASSPs/platforms are tailored to an application domain (e.g., high-end set-top boxes), rather than to a particular application. They are less area and power efficient than an ASIC for each application. However, the increased flexibility is rewarded by reuse of the chip for multiple applications. As a result, the cost of designing the ASSP/platforms is paid only once per application domain.

### FPGAs

An FPGAs NoC is designed in two phases: (i) the *layout phase*, which occurs when the FPGA chip is designed and (ii) the *configuration phase*, occurring when a system design is programmed into the FPGA. The design made in the former should be flexible enough so as to allow for a large variety of configurations during the latter. Unlike in the ASIC model, little is known about traffic patterns when the network is laid out, and hence, custom optimizations are impossible. In the configuration phase the traffic patterns become available, but the physical wires and network resources are already fixed in place.

The NoC infrastructure can be designed into the fixed FPGA fabric during the layout phase. But it can also be programmed, just like any other functionality, during the configuration phase. In the former case, the communication patterns of the application to be programmed in the FPGA are not yet known, and the NoC must therefore be a general one (like for CMPs, described below). But the implementation of this NoC can be highly optimized, for example using custom layout. The reverse holds for the latter, where the application and its communication patterns are known, and an NoC optimized for the application can be programmed in the FPGA. The

NoC implementation will therefore use the generic FPGA infrastructure (look-up tables, switch boxes, etc.), just like any other module functionality. As a result, the NoC implementation will be much less optimized. Of course, a mixture of these two models is also possible. For example, an NoC backbone (routers and network interface kernels) is designed in the FPGA, which is augmented with network interface shells [90] during configuration.

FPGAs are optimized for hardware-programmable flexibility. Hence, they are very area and power inefficient compared to an ASIC for each application. However, their computational efficiency is good compared to ASICs and ASSPs. Because the FPGA is designed once and then programmed, it is able to run all applications. The design cost of an FPGA is amortized over a larger number of applications than ASICs and ASSPs.

### CMPs

In a general-purpose CMP, the traffic pattern is completely unpredictable until run-time and even during different phases of the same program because of very different program behaviors for different external inputs (e.g., dynamic games, simulations, etc.). In this regard, many of the challenges of CMP NoC design resemble those we normally see in traditional networks, for example, static versus dynamic routing, congestion control, connection rate fairness, etc. Unfortunately, traditional mechanisms for dealing with these issues may be prohibitively expensive to implement in silicon. On the other hand, the relatively small dimension of the complete network, combined with the fact that the entire chip is often controlled by a single operating system, may make the problems amenable to centralized solutions.

CMPs are optimized for software-programmable flexibility. Hence, they are very area and power inefficient compared to an ASIC or ASSP for each application. The CMP is designed once (in theory) and is able to run all applications. Of the four categories, the design cost of a CMP is amortized over the largest number of applications.

In the sequel we present the different aspects of the network layer architectures and their design choices: the QoS and traffic requirement characteristics, the switching paradigm, NoC topologies, the addressing and routing mechanisms, handling congestion control, and end-to-end transport layer issues.

## 5.2    NoC QoS

There are many important metrics for NoCs. In this chapter, we devote particular attention to QoS, because this requirement has a direct impact on switching and routing. QoS is a common networking term that applies to the specification of network services that are required and/or provided for certain *traffic classes* (i.e., traffic offered to the NoC by modules). QoS specification can be expressed by performance metrics like loss, rate, delay, delay variation (jitter), etc. and can be categorized by absolute (worst case) bounds, average values, percentiles etc. [4]. For example, an Internet packet-based voice traffic may require a throughput of 64 Kbps, less than 1% average packet loss and a packet one way delay bound below 150 ms. QoS definitions can also apply to different service granularities. In addition to the QoS specification of voice packets, the whole voice call can require a limit on its call blocking probability (failure rate) or a specification of the call establishment times.

If certain traffic classes require an assured (guaranteed) QoS specification, there is an inherent limit to the amount of traffic that can be allocated to such a class. In other words, when there are more potential users than the network can support, some of them will be refused access to the network. Therefore, in our voice examples, the network is designed for a limited voice offered load. In case there is a risk that the user load may overpass this limit, special provisions need to be taken in order to block the excess offered load (e.g., a busy signal for telephone users).

The NoC applications (such as DSP, consumer electronics, etc.), usually stress the need for stringent bandwidth and/or latency requirements. NoC traffic can consist of urgent tasks such as code fetch following a cache miss, CMP synchronization signals or periodic refreshed data that need a guaranteed latency. Other urgent tasks may require similar latency with a high probability. Yet streaming traffic, such as audio and video, has strict bandwidth and jitter requirements, but tolerates long latencies [46]. On the other hand, certain traffic classes may not be planned for a guaranteed service (as they may produce an unspecified amount of traffic) but may require a graceful degradation of service as well as a fair allocation of the available resources as offered load increases.

The mixture of different service classes in the same network poses a special challenge to the NoC designer as each class may affect the performance of other classes. In other words, QoS also specifies how to allocate network resources at times of conflict over the network resources. Therefore, QoS mechanisms usually combine mechanisms for traffic discrimination (such as applying delay or loss priorities [18]), mechanisms for traffic separation (such as fair queueing mechanisms [48]), and mechanisms for traffic policing (enforcing limitations of traffic generation at the source modules). We return to some of these issues in Sections 5.4, Switching techniques and 5.7, Congestion control and flow control.

While the available QoS depends on the overall NoC architectural components such as topology, link and router capacities, routing algorithms, and congestion control mechanisms, it is common to separate the issue of QoS from other network issues. The rationale is the following: given a complete network architecture we would like to be able to serve multiple traffic classes over the same network each with a different QoS specification. To achieve this, we need special mechanisms that isolate and differentiate these traffic classes within the network. These specific mechanisms are described in this section.

## 5.2.1    Traffic Classes and Service Classes

The definition of traffic classes is an open-ended task and NoC designers can specify and support any number of classes. However, practice has shown that the specification, management and implementation of a large number of classes are complex and ambiguous task. First, it is hard to predict ahead of time all the future possible NoC usages. Here we should differentiate between different NoCs in our design spectrum where ASIC and ASSP NoCs are much more predictable than CMP NoCs. Second, the exact interrelationships between the classes are hard to define. Finally, the mechanisms that allocate the exact performance to each class in a multiple performance metrics domain (loss, rate, delay, jitter, etc.) are complex and costly.

**Traffic classes and service classes in computer networks**

Consequently, most off-chip network standards have aggregated the large number of possible traffic classes (traffic offered to the NoC by the modules) into a few predefined *service classes* (offered by the NoC to the modules). Let us illustrate the way the *asynchronous transfer mode* (ATM) and IP networking standards have classified the different traffic classes.

The ATM network defines five possible standard classes [73]:

1. *Constant bit rate (CBR)*: associated with traditional time division multiplexing (TDM) services where delay and loss are fixed and small.

2. *Variable bit rate – real time (VBR-RT)*: associated with variable rate (usually compressed) real-time services that produce a predictable average rate and require low delay and jitter and can tolerate small loss.

3. *Variable bit rate – non-real time (VBR-NRT)*: associated with less interactive real-time services such as streaming or high-priority data services.

4. *Available bit rate (ABR)*: includes most data services. The term available bit rate means that under bandwidth shortage conditions, the bandwidth allocated to this class should be allocated fairly among competing applications within this class.

5. *Best effort (BE)*: low-revenue and low-priority data that can serve as "bandwidth gap filler" such as file sharing or background synchronization traffic.

In a typical ATM implementation, the CBR and VBR service classes are accomplished via bandwidth reservation before use: a router priority mechanism among these classes and external user traffic policing guarantee no oversubscription. The ABR is accomplished using a network to user feedback control loop to assure a fair allocation of resources among all sources that share a congestion link. The BE is given the lowest priority and can take what is left. We discuss these issues for NoCs in the Section 5.4 on switching, and Section 5.7 on end-to-end control.

IP networks have two main QoS standards. IntServ [16] is a per (end-to-end) flow QoS standard and therefore require a rather complex implementation in the Internet. DiffServ [81] defines the differentiated service field with six possible service classes which are similar to the ATM ideas.

### Traffic classes and service classes in NoCs

While NoC traffic classes can be mapped to ATM, IntServ, or DiffServ service classes, the direction translation of general computer networks to NoCs is not appropriate, as described in the introduction. Moreover, the specific NoC service classes should be defined based on the vast experience in supporting ASIC and CMP communications over busses and dedicated interconnection infrastructures. Another key factor that separates NoC-based designs from a general network environment is the intrinsic characterization of on-chip communication. In addition, to general message-passing type communication, modules exchange low-level signals and distributed-shared-memory transactions that are typical to SoC flows. Examples are timing and synchronization signals, control words, cache invalidations, and interrupts that require an immediate and time-controlled transfer. Memory transactions such as read/write (RD/WR), code fetch and pre-fetch, semaphore-based operation, and DMA transfer may vary in their timing priorities and urgency. Finally, I/O traffic and off-chip memory access needs to cross chip boundaries via drivers and external pin bottlenecks.

Each designer can identify service classes for a specific NoC implementation. We describe three NoC traffic and service classes in the remainder of this section:

■ Goossens et al. [46] characterize different traffic classes in ASIC and ASIP SoCs. The heterogeneity of processing modules results in a variety of traffic classes, based on data rate, latency, and jitter characteristics (see Table 5.1).

Control traffic originates from control tasks that are usually mapped on one or more processors, and which must obtain status

**TABLE 5.1** ■ A classification of traffic classes [46].

| Example | Data rate | Latency | Jitter |
|---|---|---|---|
| Control traffic | Low | Low | Low |
| Cache misses | Medium | Low | Tolerant |
| Cache pre-fetch | High | Tolerant | Tolerant |
| Hard real-time video | High | Tolerant | Low |
| Soft real-time video | High | Tolerant | Tolerant |
| Audio and MPEG2 bitstreams | Medium | Tolerant | Low |
| Graphics | Tolerant | Tolerant | Tolerant |

information from modules and program them. It has a low data rate, but requires low latency to minimize the system response time, for example, when the application mode changes.

Multi-tasking processors, such as high-performance VLIW processors, do not have sufficient local memory to contain all instructions (code) and data of the multiple tasks. Instruction and data caches are therefore used to automatically swap in and swap out the appropriate instructions and data. This leads to medium (but instantaneously high) data rates, and requires low latency. On the other hand instructions are also speculatively pre-fetched ahead of time resulting in higher traffic that is both latency and jitter tolerant.

Dedicated video-processing modules usually operate on and generate streaming (sequential) traffic with high data rates. They are composed in deep chains without critical feedback loops, and their low-latency requirement can therefore be made less critical by using buffers to avoid underflow. The resulting traffic has a high data rate but is latency tolerant. Medium-data-rate latency-tolerant traffic is generated, for example, by audio and MPEG2-processing modules.

Jitter (latency variation) can be handled similarly, and we use the distinction between low-jitter (hard real time) and jitter-tolerant (soft real time) traffic. Modules with the latter traffic, such as the memory-based video scaler, have an average data-rate requirement but can be stopped when there is no data, and make up by processing at a higher rate later, or by averaging out data bursts. By contrast, low-jitter modules do not tolerate variations in data rates, because they cannot make up for any lost processing cycles. Examples are video-processing modules operating at actual video frequencies, where line and field blanking cannot be used as slack.

■ Another rich example of a service classes set was given in Ref. [13] to illustrate a common NoC environment:

   – *Signaling* covers urgent messages and very short packets that are given the highest priority in the network to assure shortest latency. This service level is suitable for interrupts and control signals and alleviates the need for dedicating special, single-use wires for them. Some of the signals may also take the form of a complete transaction (such as semaphore operations).

- *Real-time* service level guarantees bandwidth and latency to real-time applications, such as streamed audio and video processing or the timely refresh of an LCD screen. While these operations need a guaranteed time for completion, the time limit itself may be quite large compared to other operations.

- *RD/WR* service level provides bus semantics and is hence designed to support short memory and register accesses. This class may be subdivided according to the urgency of the operation (fetching code, capturing resources). Extremely urgent R/W transactions may utilize the signaling service class.

- *Block transfer* service level is used for transfers of long messages and large blocks of data, such as cache refill and DMA transfers.

■ Several NoC studies have observed the need to classify traffic and differentiate the service according to pre-specified service classes. The Æthereal [44] and Mango [8] architectures mainly address ASIC and ASSP SoCs for consumer electronics and have separated the services into two distinct classes: guaranteed throughput (GT) and BE. The GT class (similar to the above ATM CBR) accomplishes a TDM like service by limiting the GT traffic to a limited number of periodic flows and prioritizing the GT over the BE. With the right router architecture, this bounds the delay transfer through the network [60].

It should be emphasized that like other NoC design issues, each specific NoC implementation may define its unique set of service classes and also define how to map various traffic classes to service classes. Classifying traffic class may prove to be a challenging issue that involves knowledge of the specific modules' internals. For example, it may turn out that a seemingly single traffic class (memory read operation) needs to be split into several QoS subclasses. For example, fetching code (a memory read operation) to a processor module is sometimes much more urgent (an instruction fetch was missed in a local cache) and sometimes much less urgent (a pre-fetch operation with no current miss) than fetching data (another memory read operation) at the same processor. In such a case the module that originates the memory read instructions may map these similar operations to different classes of service, yet they look identical to an external observer who is not aware of the original cause of these transactions. Connection (identifiers) can be used to indicate to the NoC to which traffic class a transaction belongs [56, 91].

The SoC type impacts the traffic classes that may be expected, and hence the NoC service classes that the NoC must offer. ASICs, ASSPs, and FPGAs implement specific, hard real time, often relatively static applications known in advance [98]. Hence, we can expect service classes that are tailored for assured (guaranteed) real-time data streaming. CMPs, on the other hand, execute a variety of dynamic soft real-time applications that are unknown in advance, and BE service may be the dominant service class.

## 5.3  NoC TOPOLOGY

NoC architectures and topologies have been described in Chapter 2. In this section, we summarize for convenience the main characteristics of NoC topologies, and we relate them to cost metrics such as area, performance, and power consumption.

Network topology has been intensively studied in the context of high-performance networks [33] and parallel computers architectures [23]. Here, we introduce the concepts germane to NoCs, and refer to [23, 33] for more extensive classifications.

NoCs differ from general networks because they are realized on a plane, even though new technologies, like the emerging die-stacking and three-dimensional integration techniques may change this in the future. Moreover, links between routers can travel only in X or Y direction, in a limited number of planes (the number of metal layers of the IC process). As a result, many NoCs have topologies that can be easily mapped to a plane, such as low-dimensional (1–3) meshes and tori. We list here some important topologies for NoCs, and discuss area cost and performance for each group:

- *Crossbar*: When all routers are connected to all other routers, the NoC is fully connected. The result is single crossbar [66, 67]. It does not scale up to large number of network interfaces.

- *n-dimensional k-ary mesh* (or grid): The two-dimensional 2-ary mesh is a popular NoC topology because routers can be preplaced in layout, and because all links have the same (limited) length. The number of network interfaces per router is usually one, but can also be higher. Examples include QNoC [13] and Nostrum [74]. The area of meshes (the number of routers and network interfaces) grows linearly with the number of cores. Meshes have a relatively large average distance between network interfaces, affecting power dissipation negatively. Moreover, their limited bisection bandwidth reduces performance under high load. Care has to be taken to avoid accumulation of traffic in the center of the mesh (creating a *hot spot*) [82].

- *k-ary n-cube (torus)*: The *k*-ary 1-cube (one-dimensional torus or ring) is the simplest NoC, and is used in Proteo [92]. The area and power dissipation costs (related to the number of routers and network interfaces, and the average distance) of the ring grows linearly with the number of cores. Performance decreases as the size of the NoC grows because the bisection bandwidth is very limited.

  The 3-ary 2-cube (two dimensional) torus adds wrap-around links to the two-dimensional mesh. To reduce the length of these links, the torus can be folded. Dally and Towles [30] use this topology. The area of tori is roughly the same as for meshes (some links are added), but the power dissipation and performance are better because the average distance is less than in meshes.

- *Express cube*: Meshes and tori can be extended with bypass links to increase the performance (bisection bandwidth and reduced average distance), for a higher area cost. The resulting express cubes [25] are essentially used in FPGAs.

- *d-dimensional k-ary (fat) trees*: These have *d* levels, in which each router has *k* children, network interfaces are attached only to the leaves. Examples are SPIN [50, 87]. The bisection bandwidth of tree is very low, due to concentration of all traffic at the root of the tree. To solve this problem the root can be duplicated. The resulting *fat trees* [68] (or folded butterfly) have a large bisection bandwidth (and hence performance), but associated high area cost. For larger number of nodes, the layout of the fat tree is more difficult, in comparison with meshes or tori.

- *Irregular*: NoCs are appropriate when the NoC can be optimized to a particular application domain or set of applications. Synthesizing application-specific NoCs in general at the desired cost–performance point is a challenging problem. Examples of NoCs that allow irregular topologies are Xpipes [5] and Æthereal [44]. Optimized irregular NoCs can also be obtained by removing unnecessary routers and links from a regular NoC. For example, a mesh can optimized to a partial mesh [13], which has sufficient performance for the application at hand, but at lower cost.

**NoC types and topologies**

There is no strict correspondence between the different types of SoC (application specific, reconfigurable, or general purpose) and their topologies. However, we can discern some general trends.

ASIC and ASSP NoCs tend to be irregular (reduced meshes or completely optimized), because much is known of the application (domain) and NoCs can be highly tailored [98]. The area and power dissipation costs can be reduced, while performance is still guaranteed. For example, the simple irregular tree topology is already used in commercial products.

FPGAs are composed of small-grained tightly coupled computation and storage units (look-up tables, RAMs, etc.). These units communicate mostly locally, and require high bandwidth and low latency for communication. As a result, express cubes (meshes with bypasses) are mostly used in FPGAs. The NoC is configured infrequently, and has high performance in the steady state.

NoCs for CMPs tend to consist of large-grained loosely coupled (usually homogeneous) computation subsystems (called tiles) [99, 102]. Tiles usually contain a local interconnect such as a bus or switch for frequent local communication, and use the NoC for less frequent global communication. As a result, two-dimensional meshes or tori (with limited bisection bandwidth) are the preferred NoC topology. Because applications are unknown at design time, NoCs are usually not statically configured but schedule traffic at run time.

## 5.4  SWITCHING TECHNIQUES

Once the topology of an NoC has been decided on, the switching technique, or how data flows through the routers, must be determined. This involved defining the granularity of data transfer, and the switching technique.

Data is transferred on a link, which has a fixed width, measured in bits. The unit of data transferred in a single cycle on a link is called the *phit* (physical unit). Two routers synchronize each data transfer, to ensure that buffers do not overflow, for example. Link-level flow control is used for this, and can be based on hand shaking or the use of credits [63]. The unit of synchronization is called a *flit* (flow control unit), and it is at least as large as a phit. Finally, multiple flits constitute a *packet*, several of which may make up *messages* that modules connected to the NoC send to each other. (Note that messages can be used for different NoC programming paradigms, including message passing and distributed-shared memory, cf. Chapter 7.) Fig. 5.2 shows this structure. To increase the packetization efficiency [42] message and packet boundaries do not need to be aligned, as shown.

Different NoCs use different phit, flit, packet, and message sizes. The phit and flit sizes reflect different design choices, such as link speed versus router arbitration speed [88]. For example, Æthereal uses phits of 32 bits, flits of 3 phits (or words), and packets and messages of unbounded length. Nostrum [74] uses phits of 128 bits, and flits equal to 1 phit. SPIN [50] uses phits and flits of 36 bits, and packets can be unbounded in length.

Phits are relevant for the link layer (Chapter 4), and will not be further discussed. The switching technique determines how flits and packets

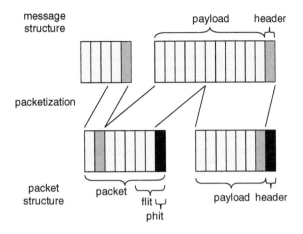

Phits, flits, packets, and messages.

are transported and stored by the routers, as described below. Chapter 7 further discusses messages. Note that the switching technique determines how data flows through the NoC, but not along which route. This is the subject of the next Section 5.5.

There are two basic modes of transporting flits: *circuit switching* and *packet switching*. Essentially, in circuit switching a circuit with a fixed physical path is set up between sender and receiver, and all the flits of the message are sent on this circuit. In contrast, in packet switching the packets constituting a message make their way independently from sender to receiver, perhaps along different routes, and with different delays. We discuss these techniques, and several variations, in more detail below.

To determine a switching technique for an NoC, a number of issues must be balanced, such as the granularity of the data to be sent, and the frequency with which it is sent; the cost and complexity of the router; the dynamism and number of concurrent flows to be supported; and the resulting performance (bandwidth, latency) of the NoC. Different types of NoC (ASIC, ASSP, FPGA, or CMP) often use different switching techniques, as we shall see. The switching technique strongly influences QoS, as mentioned in Section 5.2 and further elaborated in Section 5.7.3. In fact, to offer different QoS levels, NoCs can use multiple switching techniques at the same time [6, 44, 75].

## 5.4.1 Circuit Switching

Messages from one module to another are sent in their entirety when using circuit switching. First, a physical path, that is, a series of links and routers, from sending to receiving module is determined and reserved for the circuit. Logically, the head flit of the message makes its way from

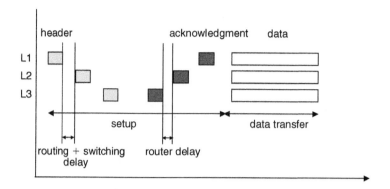

■ **FIGURE 5.3**

Circuit switching.

the sender to receiver, reserving links along the way. If it arrives at the receiver without conflicts (all links were available) an acknowledgment is sent back to the sender, who commences data transfer on its reception. If a link is reserved by another circuit, a negative acknowledgment is sent to the sender. Figure 5.3 shows the set-up and use of a circuit over time; R1, R2, and R3 are routers along the path of the circuit. The shaded boxes show when the inter-router wires are occupied.

After transmission of the message, the circuit is torn down, as part of the tail flit. Pure circuit switching has not been used much in NoCs, and the principal examples are SOCBus [105, 106].

Circuit switching has a high initial latency due to the set-up phase that has to complete before data transmission starts. (Scouting routing [33] can reduce this time.) Data transmission is very efficient, however, because the full link bandwidth is available to the circuit, and results in minimal latency. Data does not have to be buffered in the routers (only pipelined perhaps), reducing the area of routers. However, circuit switching does not scale well as the diameter of the NoC grows [31, 105] because links are occupied also when data is not being transmitted (during the set-up and tear-down phases).

Circuit switching is appropriate when data is sent very often, or when the communication pattern between senders and receivers is relatively static. The circuit can be left in place in these cases. When the amount of data to be transmitted is large (making the set-up phase less relevant) circuit switching also works well. ASICs and ASSPs have relatively static communication patterns, and FPGAs also send data (bits or words) every cycle on the circuit. FPGAs (currently) exclusively use circuit switching, and ASIC and ASSP NoCs often do [44, 98, 106].

**Virtual channels and virtual circuits**

Circuit switching reserves physical links between routers. Multiple virtual links (more commonly called *virtual channels* [29]) can be multiplexed on a

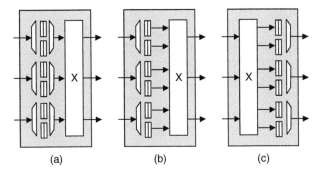

■ **FIGURE 5.4**

Virtual-circuit switching with multiple (2) virtual channels.

single physical link (channel). Virtual channels can be used to make circuit switching more flexible (see below), to increase performance by reducing blocking of links (described in Section 5.4.2), and to avoid deadlock (see Section 5.5).

Circuit switching reserves physical links between routers. *Virtual circuits* can be created by multiplexing circuits on links. The number of virtual channels that can be supported by a link depends on the number of buffers on the link. Two basic schemes have been used: a buffer per virtual circuit, or one buffer per link. Intermediate strategies are possible, but are more complex. We describe each scheme in turn, after which we discuss alternatives for circuit set-up and tear-down.

*Virtual circuits with multiple buffers per link*

A virtual circuit requires a buffer in each router it passes through. The spatial distribution of virtual circuits in the NoC therefore determines how many buffers are required in each router. This requirement can be used to determine the number of buffers for each router (virtual-circuit buffering). The number of buffers can then vary per router. Alternatively, the number of buffers per router can be given as constant and taken equal to the number of virtual channels (virtual-channel buffering). In this case, the number of virtual circuits using a given link is limited by the number of virtual channels on that link (and the routing algorithm, to avoid links of which all buffers are occupied). In both cases, virtual circuits are created by using virtual links, and it is only the determination of the number of the buffers in the routers that makes the difference. Figure 5.4(a) shows the outline of a router with multiple buffers at the inputs of the switch. Figure 5.4(b) shows how increasing the size of the switch from $N \times N$ (where $N$ is the router degree) to $(V \times N) \times N$ (where $V$ is the number of virtual channels) reduce the contention on the switch [44, 62]. Figure 5.4(c) shows virtual-channel buffering with output queueing [8].

In any case, many systems contain hot spots, where many virtual-circuits converge, such as the external memory interface [46]. There will

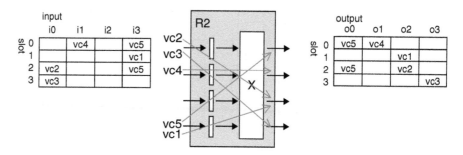

- input i0 transports two virtual circuits (vc2, vc3)
- input i1 transports one virtual circuit (vc4)
- input i3 transports two virtual circuits (v1, vc5)

■ **FIGURE 5.5**

Virtual circuit switching with a single buffer per link and TDM.

not be enough virtual channels for accommodate all flows. Virtual-circuit buffered router implementations become problematic either due to multiplexing the large number of small buffers, or due to the expensive (large) shared buffer implementations, such as SRAMs. Routers that implement virtual circuits with a single buffer per link obviously do not suffer from this problem, as described below.

The multiplexing of the individual virtual channels on a single link requires scheduling at each link/router, and results in an end-to-end schedule of the virtual circuits. Virtual circuits with input queuing are used by Refs [62, 106]. The scheduling of access to links and access to the (blocking) crossbar in the router interfere, making bandwidth and latency guarantees difficult to achieve. For this reason, the Mango NoC uses virtual-channel buffering implemented by output queuing with a non-blocking crossbar [8] (see Fig. 5.4(c)). This, together with an appropriate link scheduling scheme [9] enables it to guarantee bandwidth and latency on its virtual circuits.

*Virtual circuits with a single buffer per link*
Circuits can also be time multiplexed with one buffer per link, that is, without requiring buffering per virtual circuit. Essentially this is achieved by statically scheduling (using TDM) the usage of all links in the NoC by all virtual circuits. Figure 5.5 shows a $4 \times 4$ router with four TDM slots. Each input has a one-flit buffer even though it can be used by up to four virtual circuits. The TDM tables show how virtual circuits are mapped from inputs to output in time. The example is described in more detail in Section 5.7.1 and is consistent with Fig. 5.20.

At the edge of the NoC, flits are injected by the network interfaces such that they never use the same link at the same time (see Fig. 5.20). As a result, link-level flow control and scheduling can be omitted, and only one-flit buffer per link is necessary. This scheme assumes the propagation speed of individual flits through the NoC is fixed and known in advance. Flits wait in the network interfaces until they are injected in the NoC according to the TDM schedule. Note that network interfaces require some kind of end-to-end flow control for lossless operation, probably with virtual-circuit buffering (Section 5.7.2).

The TDM is used to globally schedule all flows, and hence end-to-end bandwidth and latency guarantees are easily given ([37]; Section 5.7.3).

The concept of globally scheduling all virtual circuits is used by NuMesh [95] (for parallel processing), Nostrum's looped containers [74], adaptive SoCs (aSoCs) global scheduling [64, 65], and Æthereal's contention-free routing [44, 89].

**Circuit management**

Set up and tear down of virtual circuits can take place as described at the start of this section. Options include static or dynamic reservations, backtracking, and multicast. Care must be taken that deadlock does not occur during set up. This can be achieved by retracting a circuit [44, 105], or by dropping data [31]. Alternatively, a programming can be centralized in a single entity such as a programmable processor [44, 62, 106].

Set-up, (negative) acknowledgment, and tear-down can also be encoded as messages (usually just of one flit), like in the ATM [3] and pipelined circuit switching [33]. Set-up and tear-down messages program routers, using a message-based or memory-mapped protocol. Æthereal [44] and Mango [6] are examples of this approach. Control messages can use a different switching scheme (e.g., wormhole (WH) packet switching) and different QoS class (e.g., BE). Using messages allows circuit management to use the NoC itself, eliminating a separate control interconnect to configure the NoC [49, 106].

## 5.4.2 Packet Switching

In (virtual) circuit switching a complete (shared) path is reserved before data is sent. In contrast, in packet switching, no link reservations are made, and the packets constituting a message make their way independently from sender to receiver, perhaps along different routes, and with different delays. Omitting the set-up phase removes the start-up time (set-up until acknowledgment), but without link reservations, packets of different flows may attempt to use a link at the same time. This is called *contention*, and requires that all but one packet must wait until the link becomes available again. The start-up waiting time of circuit switching (followed by a fixed minimal latency in the routers) is replaced by a zero start-up time and a variable delay due to contention in the each router along the packet's path. For this reason, QoS guarantees are harder to deliver in packet-switched NoCs than in circuit-switched NoCs.

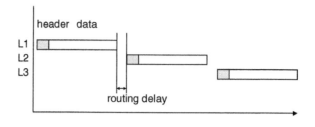

**■ FIGURE 5.6**

Store-and-forward switching.

The lack of resource reservations means that there is no limit to the number of flows (or connections) that a link or the NoC as a whole can support, in contrast to (virtual) circuit switching.

In packet switching, packets of different flows are automatically distributed over different links through the network (if dynamic routing is used), and interleaved on links. As a result, links can be dimensioned for (at least) the average amount of traffic. Virtual-channel or virtual-circuit buffering also interleave packets of different virtual circuits on a link, but all packets of a virtual circuit always follow the same static path, which must be dimensioned for (at least) the sum of the average traffic on the virtual circuits. Instead, pure circuit switching requires that each link supports the worst-case requirements of each circuit using it.

There are three basic packet-switching schemes: store and forward (SAF), virtual cut through (VCT), and WH switching. They are discussed next.

### SAF switching

SAF switching is the simplest form of packet switching. A packet is sent from one router to the next only when the receiving router has buffer space for the entire packet (Fig. 5.6). Hence, packet transmission cannot stall and the notion of flit is not required (flits are equal to packets). Routers forward a packet only when it has been received in its entirety. As a result, the latency per router and the buffer size are at least equal to the size of the packet. Given that minimizing buffer size is critical in NoCs [88], few NoCs have used this basic technique. In particular, Nostrum [74] uses hot-potato routing [35] where the notions of phit, flit, and packet are conflated (to their link width of 128 bits).

### VCT switching

VCT switching reduces the per router latency by forwarding the first flit of a packet as soon as space for the entire packet is available in the next router (middle transmission in Fig. 5.7). The other flits then follow without delay. If no space is available, the whole packet is buffered (last transmission in

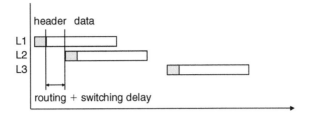

VCT switching.

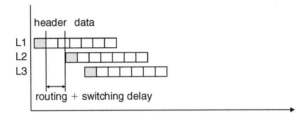

Wormhole switching.

Fig. 5.7). Buffering requirements of SAF switching and VCT switching are the same, and no NoC has used this basic technique, therefore.

### WH switching

WH switching improves on VCT switching by reducing the buffer requirements to one flit. This is achieved by forwarding each flit of a packet when there is space for that flit in the receiving router (Fig. 5.8). If no space is available in the next router for the entire packet (as required by VCT switching but not by WH switching), the packet is left strung out over two or more routers. This blocks the link, which results in higher congestion than with SAF and VCT switching. Link blocking can be alleviated by multiplexing virtual links (or virtual channels) on one physical link (cf. Section 5.4.1) [26]. WH switching is also more susceptible to deadlock than SAF and VCT switching due to the newly introduced usage dependencies between links. Virtual channels and/or routing schemes can be used to avoid deadlock (see Section 5.6.2).

Almost all NoCs use WH switching without virtual channels [13, 59] use restricted topologies (usually partial mesh, with some form of dimension-ordered routing) to avoid deadlock. SPIN [1] uses a fat tree topology with deflection routing and packet reordering at the receiving network interface, which also avoids routing deadlock. Æthereal's [88] BE service class uses WH switching class and avoids deadlock for any topology through a combination of turn-prohibition routing and end-to-end flow control [55]. Other approaches [8, 62], use WH switching with virtual channels.

### 5.4.3 Combinations of Different Switching Techniques

Circuit switching and packet switching have different characteristics that may be useful to combine in a single NoC. With circuit switching it is relatively easy to guarantee bandwidth and latency guarantees for a given set of (virtual) circuits (see Section 5.4.1). Packet switching, on the other hand, has no notion of (virtual) circuits and can therefore support any number of concurrent flows. However, it is hard to give hard bandwidth and latency guarantees.

Virtual circuit switching and packet switching can be combined by allocating (at least one) virtual channel to each class. Guaranteed services are mapped to virtual circuits and BE services to packets. Æthereal [88], Nostrum [74], and Mango [8] use this approach. The first two use TDM and Mango uses "asynchronous latency guarantee" scheduling. Note that although WH switching with very long (infinite) packet lengths also creates virtual circuits [49, 62], but to offer end-to-end performance guarantees the right scheduling discipline is essential.

Basic packet switching is appropriate for very dynamic applications where flows are short lived, change frequently, or very variable in their demands, due to the absence of flow reservations, and set-up and tear-down phases. Relatively short messages (synchronization traffic, cache lines, etc.) to many different modules (e.g., distributed memories and caches) makes connection-oriented QoS inappropriate. Connection-less packet switching with BE service class is therefore natural for CMP NoCs. However, to offer assured (guaranteed) service classes, connection-oriented packet switching, based on (TDM) virtual circuits, has been used successfully for ASICs and ASSPs [98].

## 5.5 NoC ADDRESSING AND ROUTING

While the *switching technique*, as discussed in Section 5.4, controls how data is buffered and transported between routers, the routing layer determines the paths over which data follows through the network. Before we do this we briefly describe how the start and end points of the route are indicated through various addressing schemes.

## 5.6 NoC ADDRESSING

In order to route information within the chip unique identifications or addresses must be assigned to each reachable destination. It is important to notice that such destinations may have a hierarchical relationships or layers. For example, a certain modules (e.g., subsystems or tiles) in the chip may have multiple submodules, such as processors and local shared memories. Each processor may execute multiple programs and each program may be composed of multiple tasks and threads. Consequently such an identity can be described as thread 3 of program 7 of processor 1 at module 6.

To hide implementation details logical addresses are often used instead of physical addresses. A single physical address, such as a memory (location) may be known as different logical addresses or vice versa. The former is useful when processors with different memory maps (e.g., 24 or 32 bit addresses, or with different layout) share a single memory to communicate. The latter can be used, for example, for security or virtualization purposes, or to allow different processors to boot from the same fixed address, but with different boot code. The mapping (or translation or resolution) of logical to physical address may be done in software or hardware, distributed or centralized, fixed at design time or configurable at run time [90].

The physical address space in each NoC implementation may be assigned according to the number of different modules (e.g., we use 4 bits for an NoC that has less than 16 modules), the relative location of the module in the NoC (e.g., XY coordinate in a grid or a node name in a fat tree). Logical addresses may relate to functional names (e.g., the external memory interface, a coprocessor unit) or to a global address space (e.g., 16, 24, or 32 bits).

Finally, different flows may belong to different service classes, introduced in the previous Section 5.2. Identifiers may be needed to distinguish flows for QoS purposes, therefore. (Because different flows between the same addresses may below to different flows, separating control and data, for example.) These include identifiers for individual transactions (for reordering), for communication threads (as used e.g., in AXI [2]), and flows/connections [91].

Addressing is discussed in more detail in Chapter 6 on network interfaces (e.g., address translations) and Chapter 7 on NoC programming models (e.g., message passing versus shared memory).

## 5.6.1 NoC Routing

In the following we focus on routing data in NoCs and particularly emphasize the planar mesh topology, which is popular for NoCs. The NoC routing mechanism is responsible for correct and efficient routing of packets (or circuits) that are traversing the network from sources to destinations. The routing protocol deals with resolution of the routing decision made at every router. Unlike traditional communication or interconnection networks, NoCs need not follow rigid networking standards, therefore, a multiple routing schemes can be evaluated and compared for each NoC implementation. There are several potentially conflicting metrics that need to be balanced:

- *Power*: minimize the power required to route packets. This means that packets may follow the minimal power path likely to be identical to the traditional shortest distance routing [13, 86]. In certain cases, for example, when DVS is applied in a non-uniform way, each router and link may offer a different packet switching power consumption [94].

- *Area and VLSI resources*: the routing mechanism itself consumes hardware resources like finite state machines, and addresses tables. It potentially also uses NoC bandwidth if routers exchange information.

- *Performance*: reduce the delay and maximize the traffic utilization of the network.

- *Robustness to traffic changes*: certain routing schemes (e.g., static routing schemes) may perform very well to an expected traffic pattern but poorly to changing traffic patterns. Other schemes (e.g., dynamic routing) may behave better to a larger spectrum of traffic conditions.

Routing schemes can be classified into several different categories. In particular routing can be static or dynamic, distributed or performed at the source, and minimal or non-minimal. We describe each in turn.

### Static and dynamic routing

The routing decision at every router can be *static* (also called oblivious or deterministic) or *dynamic* (or adaptive).

In a static routing scheme permanent paths from a given source to a given destination are defined and are used regardless of the current state of the network. This routing scheme does not take into account the current load of network links and routers when making routing decisions. Note that static routing may use single path or split the traffic in a predefined way among multiple paths between a source and a destination.

In a dynamic routing scheme, routing decisions are made according to the current state of the network (load, available links). Consequently, the traffic between a source destination changes its route with time.

Static routing is simpler to implement in terms of router logic and interaction between routers. A major advantage of a single path static routing is that all packets with the same source and destination are routed over the same path and can be kept in order. In this case, there is no need to number and reorder the packets at the network interface. Static routing is clearly more appropriate when traffic requirements are steady and known ahead of time, and therefore is preferable for NoCs for ASICs or ASSPs. Dynamic or adaptive routing may utilize alternative paths when certain directions become congested and therefore have the potential of supporting more traffic using the same network topology. Consequently it may be preferable in irregular and unpredictable traffic conditions, which are more common to the CMP NoCs.

**Distributed and source routing**

Routing techniques (both static and dynamic) can be further classified according to where the routing information is held and where routing decisions are made.

In *distributed* routing, each packet carries the destination address, for example the XY coordinates of the destination router or network interface, or a module number. The routing decision is implemented in each router either by looking up the destination addresses in a routing table or by executing a hardware routing function [14]. Using this method, each network router contains a predefined routing function whose input is the destination address of the packet and its output is the routing decision. When the packet arrives at the input port of the router, its output port is looked up in the table or calculated by the routing logic according to the destination address carried by the packet. Note that in order to reduce routing table space a specific distributed routing technique may restrict its supported network topology, the type of routes that can be defined and the naming convention of network destination. A very common example in NoCs is the XY routing (also termed dimension routing [24]) for mesh networks where destinations are named after their geographical (XY) coordinates and the intermediate routing function is limited to the comparison between the router address and the destination address. Interval routing [11] suggests a similar reduced table space methodology for general topology networks.

In *source routing* the pre-computed routing tables are stored in the network interface at the system modules. When a source router (network interface) transmits a packet, it looks up the source routing information according to the destination address at the source routing table and includes it in the header of the packet. Each packet carries in its header the routing choices for each hop along its path (typically the output link identity). When the packet arrives at a network router, its next routing output port is extracted from the header routing field. The routing field is then shifted in order to expose the relevant routing choice for the next router on its path. In comparison to distributed routing, source routing does not need any intermediate routing tables or functions, it may also eliminate the destination address field required in distributed. On the other hand, it requires a source route header in the packet header (whose size increases with the path length) and requires additional routing tables with specific entries for each source (these can be located in the source network interface or in a centralized pool). Æthereal uses source routing [44].

While distributed routing and source routing describe the way packets and router interact to perform intermediate routing decisions, both schemes leave a full degree of freedom in deciding on the selection of the (static) packet routes. Most distributed routing designs make a conscious limitation on the selected routes. In order to reduce the amount of logic required, routes are based only on the destination address rather than on the source and destination address pair.

### Minimal and non-minimal routing

A final classification criterion distinguishes between *minimal* and *non-minimal* distance routing. In contrast to traditional interconnection networks, the additional power consumption introduced by non-minimal routes may be prohibitively expensive in an NoC [57]. Note that in heterogeneous NoCs where links between routers are of different speed and length, minimal power routing is not equivalent to a minimal link hop routing [12].

## 5.6.2  Deadlock

The major constraint for any routing algorithm is assuring the freedom from *deadlock*. In packet switching networks whenever a packet or flit is transferred between neighboring routers, it releases a buffer at the transmitter and occupies a previously free buffer at the receiver. Consequently, such a transfer requires the availability of a free buffer at the receiver, or the flit is held (in a lossless network), due to link-level flow control, until such a buffer is freed at the receiver. Deadlock occurs when one or more packets in the network become blocked and stay blocked for an indefinite time, waiting for an event that cannot happen. A typical example, as depicted in Fig. 5.9 [80], is a situation where four packets are routed in a circular manner between the routers in a square mesh. The packet occupying channel $c_1$ is waiting for $c_2$ and that channel is allocated to a packet that wants to use $c_3$. That channel in turn is held by a packet that is requesting $c_4$ and the packet on this channel completes the circle by waiting for $c_1$. No packet can advance since the required resource, in this case the channel, is already held by another packet and will never be released.

While WH switching is the prevalent switching technique for NoCs (due to its relatively small buffering requirements), it is prone to suffer from deadlocks. Since only the header flit carries routing information all flits belonging to the same packet must be contiguous and cannot be interleaved with flits belonging to other packets [29]. If a flit is blocked due to busy resources all the trailing flits of that packet will also be stopped and keep blocking the resources they occupy in terms of channels and buffers. This can result in chained blocking [26] where the resources of a blocked packet again causes other packets to block, a property that makes WH routing very susceptible to deadlock [29, 36].

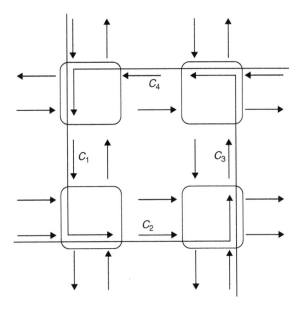

■ **FIGURE 5.9**

Deadlocked scenario with packets waiting for channels in a cyclic manner.

### Deadlock avoidance

The prominent strategy for dealing with deadlock is avoidance and most deadlock-free routing algorithms are deduced by the strategy of [29, 36]: (1) choose a particular routing algorithm, (2) check whether this algorithm is deadlock free, and (3) if needed, add hardware resources or restrict routing to make the algorithm deadlock free.

Deadlock freedom is analyzed by building a dependency graph of the shared network resources. Whenever a packet is holding a resource while requesting another there is a *dependency* between them and if this dependency graph is cyclic then we have a circular wait. Analysis of the graph can be done statically, with all possible dependencies represented in the graph, or dynamically, with the graph reflecting the current state of the system. Static analysis is unnecessarily pessimistic since dependencies that are mutually exclusive can be part of a cycle that never occurs at run time. However, the big advantage with static analysis is that it can be done beforehand and thus detach this computationally intensive problem from the actual system.[1] Thus, deadlock is usually avoided by employing a restrictive routing algorithm.

---

[1] Much research has been focused on distributed cycle detection mechanisms that allow for progressive deadlock recovery based on only local information while keeping recovery overhead to a minimum [19, 20, 71].

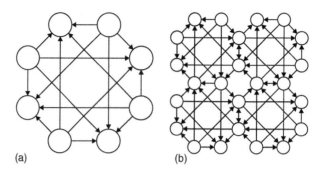

                        (a)                        (b)

■ **FIGURE 5.10**

Channel dependency graph of a mesh in which the routers use edge buffers. (a) For one router and (b) for the entire mesh.

Deadlock freedom in SAF and VCT networks, which are identical from a deadlock perspective [29], is proved using a buffer-ordering technique of Ref. [51]. In a WH-switched network blocked messages remain in the network and hence continue to occupy the links until contention is resolved. Therefore, we cannot use restrictions in buffer allocation to prevent deadlock, but must instead restrict routing over the communication channels, be they physical or virtual.

The dependencies of a WH-switched network are captured in a *channel dependency graph* [29]. Figure 5.10 illustrates the channel dependencies of a minuscule network. With a fully dynamic routing function every input channel has the possibility of forwarding a packet to any output channel and the channel dependency graph of a single router thus looks as shown in Fig. 5.10(a). The graph is obviously acyclic since all the input and output ports of the router are dangling. In Fig. 5.10(b) we see the corresponding graph for a simple two-by-two mesh and this figure contains numerous cycles. There is thus a risk of channel deadlock in the network.

Dally et al. [29] propose a necessary and sufficient condition for deadlock freedom in the case of a static routing function. This proof technique is extended in Refs [27, 69] to also cover dynamic routing.

Introducing virtual channels increases the degrees of freedom in alleviating restrictions in the choice of channels (see Section 5.4.1) [40] and can aid both in avoiding deadlock as well as increasing network throughput by reducing the effects of chained blocking [26]. By dissociating the buffers (associated with channels) from the actual physical channels, a blocked packet in one virtual channel does not preclude packets residing on other virtual channels [76] as depicted in Fig. 5.11.

The problem of deadlocks in a WH NoC can be alleviated by the use of multiple virtual channels. The idea is that more buffers can be allocated in the receiver size to different flows (e.g., a Virtual Channel per destination). However, the number of Virtual Channels that are required

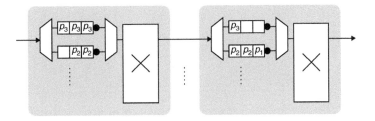

■ **FIGURE 5.11**

A packet $p_2$ is blocked by a previous packet $p_1$ and is occupying the physical channel between the routers. Packet $p_3$ that follows $p_2$ is allowed to proceed and pass $p_2$ by using the buffers associated with another virtual channel.

to completely solve the deadlock problem for any routing scheme, is large [28, 33] and therefore costly in terms of VLSI resources. There are general mechanisms for avoiding deadlocks by restricting the routing algorithms. Glass and Ni [40] call a specific pair of input–output links around a router as a *turn* and introduce several deadlock-free routing schemes based on *turn prohibition*. Enough turns are prohibited to break all dependency cycles while still maintaining connectivity between every pair of routers. The Æthereal BE service class uses the turn model for routing on arbitrary topologies [54]. The most simplistic approach to turn prohibition is the well-known dimension-ordered routing. It is a static and minimal distance routing algorithm where a packet is routed in one dimension at a time, finding the correct coordinate in each dimension before continuing with the next. In a two-dimensional mesh this method is known as XY routing and is very popular due to its simplicity. When a packet is sent it will be forwarded in the X dimension until it reaches the X coordinate of the destination; only then is it forwarded in the Y dimension until it reaches its goal.

An illustration of the turn model in a two-dimensional mesh is shown in Fig. 5.12. There are eight possible turns that form two cycles if no turns are prohibited. With the static XY routing four of the turns are prohibited and it is clear that no cycles can be formed from the remaining turns. It is not necessary to prohibit this many turns to guarantee deadlock freedom. The *west-first* routing algorithm [41], illustrated in Fig. 5.12(c), prohibits only two of the eight turns and manages to break all cycles thereby constituting a deadlock-free routing algorithm.

Many contemporary NoC proposals based on a regular mesh topology have implemented the above dimension-order (XY) based routing. It is shown in Refs [13, 52] that this may cause a significant imbalance in the traffic utilization of the mesh links, even when traffic requirements are symmetric. When traffic patterns are known ahead of time (for ASICs), this can be dealt with using NoCs with asymmetric link capacities such as QNoC [13, 52], where each link capacity is planned according to its expected load. This eliminates the need for an expensive implementation

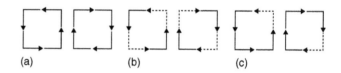

(a)            (b)            (c)

■ FIGURE 5.12

Illustration of turn prohibition in two-dimensional mesh. (a) Unprohibited mesh with two cycles, (b) f[
prohibited turns of XY routing, and (c) two prohibited turns of west-first routing.

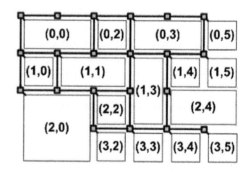

■ FIGURE 5.13

SoC modules interconnected by an irregular mesh NoC.

of an dynamic load balance routing that requires the reordering of packe
at the network interfaces. A simple alternative to link capacity planni
is to use a combination of alternatively (or randomly) choosing betwe
XY and YX routing as described in Refs [58, 93] that is close to optimal
terms of maximum utilization in a homogenous mesh. Therefore, the "to

each table, which affects NoC performance, depends on its size and thus on the network size.

Reference [14] develops hardware-efficient routing techniques that reduce the VLSI cost of routing in irregular mesh topology NoCs. The techniques are based on a combination of a fixed routing function (such as "route XY" or "don't turn") and reduced routing tables for both distributed and source routing approaches. The entries in the reduced routing tables are created only for destinations whose routing decisions differ from the output of the default routing function. Random simulations of different topologies and flow scenarios are used for comparing and estimating the VLSI cost savings obtained by different algorithms. This mechanism is found to be superior (around five times lower cost) to the traditional source routing and routing table-based techniques.

### 5.6.3  NoC Dynamic Routing Schemes[2]

Dynamic routing is an efficient alternative to balance the traffic load over a given NoC where the NoC traffic is unpredictable or changes with time (e.g., the CMP NoC type). Note that when a source destination traffic is split over multiple paths, packets may arrive out of order and re-sequencing buffers at the destination may be required. The simplest method of dynamic routing is termed *deflection routing* or *hot-potato* routing. It was suggested for metropolitan networks [72], optical burst networks [15], and interconnection networks [35]. In this scheme, when a packet enters a router it will be sent toward a preferred output port according to a routing table or a routing function as described above. However, if the preferred port is busy (blocked by a backpressure from a neighboring router, or captured by another packet) an alternative port will be selected. Here the router has no additional buffers in which to store the packets before they are moved, and each packet is constantly transferred until it reaches its final destination. The packet is bounced around like a "hot potato," sometimes moving further away from its destination because it has to keep moving through the network. This is in contrast to SAF switching where the network allows temporary storage at intermediate locations (Section 5.4.2). Deadlocks cannot happen in deflection routing when the number of input and output ports of a switch are identical and new local packets are not allowed in when all inputs are busy. It is guaranteed that any packet in a router will be transferred in the next cycle to any of the output ports and therefore the router can receive a new packet over all its inputs from neighboring routers (so no backpressure is sent among routers). *Livelocks* may happen in deflection routing and needs to be resolved. A livelock situation happens when a packet is sent

---

[2] We acknowledge contributions from Andreas Hansson [53, 54] for this section.

over and over and never reaches its final destination. Simple priority rules can resolve it [17].

While there is a broad literature on the use of deflection routing in various networks, one may question the impact of this scheme on the NoC power consumption due to the long routes packets may take in the network. However, the benefits of an dynamic routing scheme for the CMP model may overcome the disadvantages. First, deflection routing automatically spreads traffic to alternate routes when primary routes are in demand. Second, routes can be selected according to profitability [107] where routes that move packets closer to the destination are favorable over paths that lead packets away from it. Finally, in the NoC environment, the backpressure from neighboring routers may be more effective than in off-chip networks due to the relative proximity of routers to each other. Nilsson et al. [82] suggest avoiding excessive oscillations by exchanging "stress values" among neighboring routers. Packets are routed away from "stressed areas." Ye et al. [107] leverage the previous idea and suggests a contention-look-ahead routing based on flits in a WH network.

Several more dynamic routing technologies have been suggested to NoCs. An interesting technique that switches between the XY and the YX in an dynamic way is described in Ref. [58]. Gossip-based routing which is based on a broadcast of the information to all destinations is suggested in Refs [17, 34].

SPIN [49] uses dynamic routing in a fat tree topology. In routing from one module to another, any path to a common ancestor in the tree may be taken. A unique path then exists from that ancestor to the destination module. Although all paths between a pair of modules have the same length, delays may be different on different paths due to congestion and re-sequencing buffers are necessary at the receiver network interface.

An interesting balance between static and dynamic routing may be very suitable for the FPGA or CMP NoCs. In some NoCs paths can be configured at power up [39] or reconfigured at run time [7, 83, 90]. The latter usually happens based on a change in the application, for example, starting a new functionality [78]. It could, however, also be applied to deal with variation within a single application. The paths that are newly loaded can either be computed at run time or precomputed at design time. In this way, traffic can be distributed over the NoC in a way that is tailored to the application (or mode) at hand. ASICs, ASSPs, and FPGAs often have relatively static application modes (Fig. 5.1), and NoCs for these systems will benefit from this approach.

## 5.7  CONGESTION CONTROL AND FLOW CONTROL

In previous sections we have shown how data is transported through the network (Section 5.4), along which routes (Section 5.5), to offer the

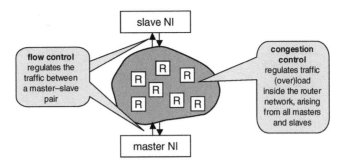

Scope of congestion control and flow control.

required QoS properties (Section 5.2). In this section we describe two fundamental phenomena that have to be addressed in order to deliver the various QoS levels.

In particular, packets travelling through the router network may want to use the same resources (buffers, links) at the same time. This is called *contention*. Like an oil stain, contention can propagate, leading to *congestion* and reduced network performance. In Section 5.7.1, we illustrate the problem in more detail, and describe a number of *congestion control* techniques.

A separate but related problem arises when two communicating parties (a master and a slave) are not balanced in terms of data injection and consumption rates. A slow (or non-responding) slave can cause a backlog of packets inside the router network. In Section 5.7.2 we describe a number of *flow control* techniques that address this problem.

Figure 5.14 shows the scope of congestion control and flow control. In a nutshell, congestion control keeps the router network free of traffic jams, while flow control ensures that no sender overwhelms any of its receivers.

In Section 5.7.3 we show that to offer desirable communication services (QoS) such as guaranteed minimum bandwidth, guaranteed maximum latency and jitter, both congestion control and flow control are required.

## 5.7.1 Congestion Control

Figure 5.15 illustrates the basic problem. To send a message from a producer to a consumer, it must be accepted by a buffer in the master network interface (master NI in the figure). Then the packets constituting the message traverse links and routers (L and B). At some point in the future the message has to be accepted by a buffer in the slave network interface (Slave NI in the figure). Therefore, a series of resources (buffers, links) is used at different points in time. Managing resources spatially (which links and/or buffers) and temporally (at which points in time) can prevent overloading the resources individually (contention) as well as collectively (congestion).

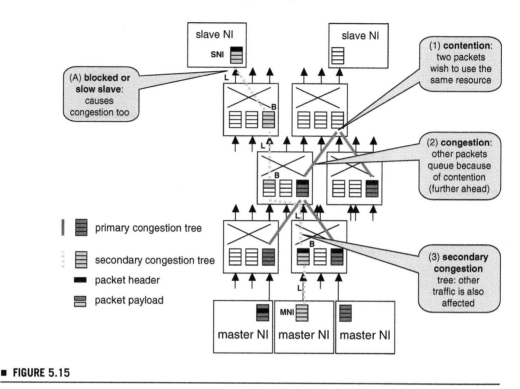

■ **FIGURE 5.15**

Contention and congestion.

Figure 5.15 shows that congestion is the result of contention: when multiple packets wish to use the same link, the waiting packet (on the left) causes queues behind it (comment (1) in the figure). The contention is caused inside the network by two packets contending for a single output link or buffer. This occurs, for example, in the center of a mesh when dimension-ordered routing is used. A congestion tree results, which affects not only packets destined for the shared link or slave (2, solid lines, and the dark-shaded packets) but also packets that only share a link in the congestion tree (3, dashed lines, and the light-shaded packets). Congestion lowers the effective utilization of the network (because packets are waiting or deflected and take a longer route), and thus congestion should be avoided for performance reasons. In addition, congestion also makes end-to-end bandwidth and latency harder to compute (because dynamic interactions between shared resources must be modeled), and must be circumscribed if performance guarantees are to be given.

We describe several techniques to limit congestion, starting with reactive techniques that do not require explicit resource reservations. Then we detail techniques that use resource reservations that reduce or even eliminate contention entirely.

Congestion control methods are also classified as being closed loop (based on feedback) or open loop (preventive) in the computer network

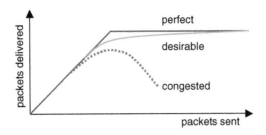

Dropping packets reduces the effective utilization of a network [100].

domain [100]. Many of the methods that exist for computer networks can be reused on chip (such as traffic shaping, token buckets, and leaky buckets). However, because NoCs have until now, not dropped packets many methods are also not applicable (such as load shedding).

### Congestion control without resource reservations

- A localized solution to contention is deleting (dropping) packets. When two or more packets wish to simultaneously use the same resource, all but one are deleted. In the short term, this solves the contention and also avoids any congestion. However, on a larger time scale, dropped packets must be re-sent. Dropping therefore leads to *more* traffic in the long run, reducing the effective utilization of the NoC, as shown in Fig. 5.16. Until now, no NoC has used packet dropping for congestion control.

- Alternatively, dynamic routing schemes (cf. Section 5.6.3) can be used to send packets around the contention. SPIN [50] is an example of this approach, and uses dynamic routing with WH switching. Nostrum [74] uses another approach, namely deflection routing with SAF switching. If multiple packets should be routed to the same output port, then one is picked and the remaining packets are routed to other outputs. By always routing all incoming packets to non-conflicting outputs, only one packet needs to be stored for each input. To reduce the SAF latency, short packets are preferred, at the cost of wider inter-router links.

  In all dynamic routing schemes, packets can arrive out of order, which requires packet numbering and reordering buffers in addition to depacketization buffers. Although these can be combined with buffers to hide variations in packet arrival (jitter) [49], random-access memories must be used to store packets, instead of potentially much cheaper FIFOs.

- Local methods can be extended to take information of a router neighborhood into account. In case of dynamic routing, routers

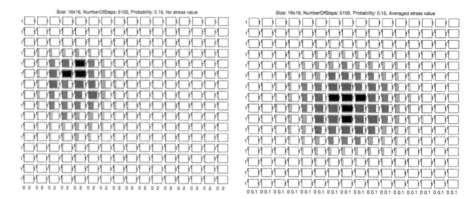

■ **FIGURE 5.17**

Dynamic routing around mesh center hot spot [82].

can exchange congestion information, and in this way route more efficiently around hot spots [82, 107]. Figure 5.17 shows how congestion is centered on a small area in the left graph. When congestion control is added, the load spreads over a larger area and is reduced by a factor 20 (maximum load changes from 0.2 to 0.01), as shown in the right graph.

■   Reducing the injection rates of packets into the network tackles congestion at a higher level. Rather than routing packets around congested areas, the number of packets in the network is reduced. Packet injection rates at network interfaces can be regulated either based on preset/precomputed values or based on measurements in the NoC (e.g., links [101] or network interfaces [84]). A (set of) rate controllers can regulate the packet injection rate, based on statistical information such as the offered load versus average latency distribution. For example, by measuring the utilization of a critical shared link, the injection rate (offered load) can be kept under limits that correspond to desired average latency. Van den Brand [101] shows how latency can be bounded by using NoC monitors [22] and multi-input multi-output *model-predictive controllers* (MPCs) (Fig. 5.18). An example is given in Fig. 5.19, where the monitor P observes the utilization of link L, and reports this information to the MPC. The controller regulates the packet injection rates of the producers to bound the link utilization (to 75% in the example). The graphs show the difference between link utilization of link L without and with use of the controller, respectively. The link utilization translates directly to latency experienced by BE traffic.

Note that all methods described above are reactive, and reduce the effects after contention or congestion has been detected. For this reason,

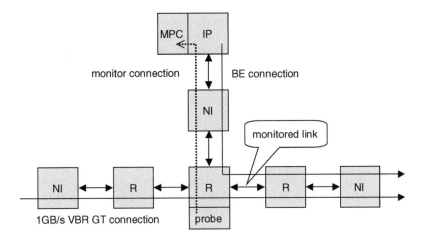

A small NoC with a monitoring probe and MPC [101].

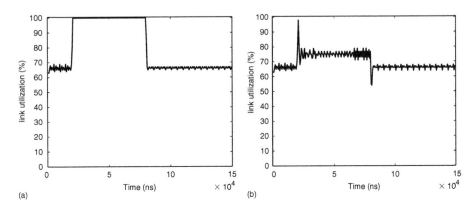

■ **FIGURE 5.19**

Link utilization with (a) and without (b) NoC congestion control [101].

it is hard to provide hard end-to-end guarantees on bandwidth or latency. This is much easier when resources are reserved, and congestion is constrained to be within required bounds as described below.

### Congestion control with resource reservations

As shown in Fig. 5.15, network resources such as buffers and links, are used in time and in space. Essentially, contention occurs when two packets use the same resource at the same time. It can therefore be avoided entirely, or bounded to a desired level by scheduling the time and location of all packets in the network. This is the aim of the category of congestion control methods described here, which reserve resources before they are used.

Producer and consumer network interfaces and the routers along the path(s) between them contain buffers and are interconnected by links (Fig. 5.15). Buffers and/or links are shared. Here we are concerned with resources that pertain to congestion, that is, in the router network only (Fig. 5.14). The resources in the network interfaces are managed by flow control, and will be described in Section 5.7.2.

A network user (usually a master–slave pair) must specify its requirements to the network before the network can decide what network resources are necessary for that flow. User requirements are stated using the service classes defined in Section 5.2. The network can then compute if sufficient resources are available to support the requested flow. If there are, the flow is accepted, else it is denied. This *admission control* (or setup phase) is necessary to prevent invalidating the guaranteed QoS of flows already present in the network by overloading the network. Furthermore, the traffic that the communicating parties (e.g., master and slave) inject in the network must conform to the service class that was reserved for their flow. For example, if a CBR flow with a maximum of 50 MByte/s was requested, the master should not produce more data than promised. To guarantee the services of other flows, they should be enforced (*traffic policing*). That is, if a user does not obey the agreed traffic class, the network will enforce the contract, for example, by not accepting data. Finally, when a flow finishes, the network must be notified, after which the resources that were reserved can be freed for reuse by other flows (tear-down phase). The set-up and tear-down phases take time, and it is advisable therefore to use a flow for a longer period of time. (Although SOCBus has used flow for single transactions [105].) The aforementioned flows are often called connections [91], which are communication pipes between a master and (multiple) a slaves.

Below we describe two different methods to eliminate or reduce contention to within known bounds. They guarantee that packets that are injected in the network at the initiator network interface arrive at the target network interface within a predetermined interval. We highlight the strong relation of scheduling and buffering schemes (alluded to in Section 5.4.1):

- In the most extreme scheme, contention can be eliminated entirely by scheduling the time and location of all packets globally, using TDM. Packets are injected at the network's edge in such a way that they never collide at any link. This assumes that propagation speed of individual flits through the NoC is fixed and known in advance. This corresponds to a global notion of time (although not necessarily implemented through a synchronous clock). Flits wait in the network interfaces until they are injected in the NoC according to the TDM schedule. For lossless operation, some kind of end-to-end flow control is also necessary (described below in Section 5.7.2).

  The concept of globally scheduling all virtual circuits is used by NuMesh [95] (for parallel processing), Nostrum's looped

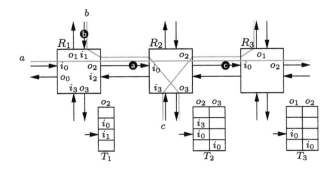

■ **FIGURE 5.20**

Eliminating contention through TDM [44].

containers [74], aSoCs global scheduling [64, 65], and Æthereal's contention-free routing [44, 89].

We illustrate the concept with contention-free routing, as used in Æthereal [44], but the other methods are essentially the same. Every TDM slot table $T$ has $S$ time slots (rows), and $N$ router outputs (columns). There is a logical notion of synchronicity: all routers in the network are in the same fixed-duration slot. In a slot $s$ at most one flit can be read/written per input/output port. In the next slot, $(s+1)$ module $S$, the packets are written to their appropriate output ports. The entries of the slot table map outputs to inputs for every slot: $T(s, o) = i$, meaning that flits from input $i$ (if present) are passed to output $o$ at times $s + kS, k \in \mathbb{N}$. An entry is empty when there is no reservation for that output in that slot. There is no contention, by construction, because there is at most one input per output for each slot.

Figure 5.20 illustrates the operation of contention-free routing. It shows a snapshot of a router network with three routers $R_1$, $R_2$, and $R_3$ at slot $s = 2$, indicated by the arrows pointing to the third entry

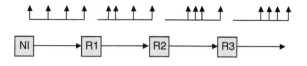

■ **FIGURE 5.21**

Characterizing congestion in rate-controlled methods [108].

they have to wait for their slots in the network interfaces, even if there are no other flits in the network that they could clash with. However, the worst-case latency is the same for TDM and rate-control described below. Note that with TDM bandwidth and latency, guarantees are (inversely) coupled, that is, a lower latency requires a larger slice of the bandwidth. Low-bandwidth high-priority flow (e.g., control traffic) may therefore be relatively expensive to offer (in terms of bandwidth overprovisioning).

■ The essence of *rate-control schemes* [108] is that contention is allowed, but within known bounds. By regulating the amount of traffic that is injected at the network edge, the maximum contention that can arise at any point in the network can be computed. End-to-end latency follows from this. Each router is seen as an independent server, which means that routers must be non-blocking, that is, packets must not interfere with each other in switch or buffer usage. Each router accepts incoming traffic with a certain pattern, and produces outgoing traffic with a potentially modified pattern. Different schedulers can be used in the routers, with different results, in terms of average and worst-case latency, jitter, buffering requirements, etc. Starting with periodic traffic, two things can be observed in Fig. 5.21 [108]: (1) the incoming traffic pattern of a flow can be distorted due to network load fluctuations, (2) the distortion may make the traffic burstier and cause instantaneously higher rates. In the worst case, the distortion can be accumulated, and downstream routers potentially face burstier traffic than upstream routers. Therefore, the source network interface traffic characterization may not be applicable inside the network.

Figure 5.22 shows an example of how different flows can use a router network. The encircled links are the bottlenecks, determining the maximum allowed bandwidth ratios for the flows. At most three solid flows (C1, C2, C3, C6) use the same link, and they are allocated one-third of the link bandwidth each. The dashed flows (C3, C4) each share a link with at most one other flow, and can therefore use up to half of the link bandwidth. Essentially, any allocation of rates to flows is allowed, as long as the sum of rates allocated at each link does not surpass the link's capacity. At the network interfaces, the flows are constrained to not inject more than their allocated bandwidth ratio, for a fixed periodic time interval.

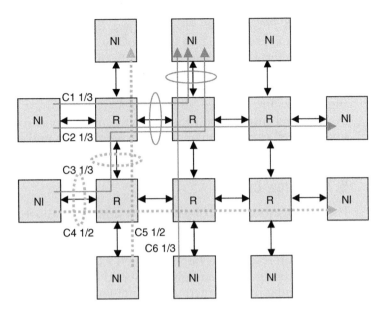

■ **FIGURE 5.22**

Flows and rates.

Buffers are required to deal with local bursts. The exact maximum burst size depends on the scheduling scheme used, but also on the time interval over which the flow rates are enforced (described above). This, together with the non-blocking requirement, means that virtual-circuit buffering with output queueing, or SAF switching with output queueing can be used. The latter results in large buffers, and has not been used for NoCs. Mango uses the former, with both round-robin arbitration [10] and priority-based ALG [9]. Both scheduling methods are work-conserving, which can result in lower average latencies in case the NoC is not fully utilized, unlike TDM as described before. Moreover, the ALG scheduler also decouples bandwidth and latency guarantees [9, 109]. However, ALG cannot schedule 100% of the available bandwidth for guaranteed services, unlike TDM and some other schedulers.

Reservation of resources can eliminate or bound congestion, but it comes at a price. Network users must negotiate their requirements with the network, which requires insight into their communication behavior, and which takes time (set-up and tear-down phases). Reservations may have to be for the worst-case traffic instead of the average case traffic, which can lead to overdimensioning of the NoC.[3] This is possible for ASIC,

---

[3] This depends on the exact service classes that are offered. But routers implementing guaranteed services may be smaller and faster than those implementing BE services [44], counteracting the impact of worst-case dimensioning.

ASSP, and FPGA NoCs. However, just like for connection-oriented circuit or packet switching, in CMP NoCs with dynamic applications where flows are short lived, change frequently, or very variable in their demands, resource reservations may be difficult to achieve.

Positive points are the possibility to offer hard guarantees regarding bandwidth, latency, jitter, etc. This has a host of advantages for system-level integration [45, 47], such as compositional system design and reduced verification effort, because modules and applications do not interfere with one another.

## 5.7.2 Flow Control

Using congestion control techniques described above, we can eliminate or reduce the congestion that packets encounter in the router network. However, they do not guarantee that there is space in buffers at the target network interface.[4] This happens when a target does not keep up with the incoming traffic, or even fails to respond entirely (e.g., through incorrect programming or malfunction). Its network interface buffers will fill up, and incoming packets must wait in the network. For example, the root of the dashed congestion tree in Fig. 5.15 could be caused by a slow slave, marked A. This may congest and even completely block the network, no matter what congestion method we use. Even when using congestion-free TDM, a full buffer still inhibits progress of other packets because their reserved link or buffer in the router network is, in fact, not available. Essentially, to avoid this phenomenon, packets from a master to a slave that cannot be accepted by a slave network interface must not inhibit the progress of packets on different flows (connections). Flow control, described below, ensures this.

Congestion control in the router network and flow control for network interfaces have different but related aims. Congestion control by itself is useful, for example, to increase performance, but not essential. However, flow control is often necessary even without congestion control, for example, to avoid deadlock [55]. (For example, the Æthereal BE traffic class uses no congestion control but does use flow control.)

Flow control deals with individual flows, that is, flows between a master and a slave (in its simplest form). Flow control ensures that the master does not send more data to the router network than can be accepted by the slave and its network interface. We describe a number of flow control techniques. First, three methods that do not require resource reservations:

1. The simplest form of flow control is to ensure that packets always find space in the slave network interface buffer. If the buffer is

---

[4] One may argue that the network interface buffers are no different than router buffers, but then the problem we will describe just shifts one "hop," into the slave. For clarity, we will omit any discussion of the behavior and internals (buffering, pipelining, etc.) of the slave, and concentrate on the network interfaces.

full packets are deleted (dropped), according to wine (drop new packets) or milk (drop old packets) policies. As discussed before, dropping increases congestion in the long run. For this reason, dropping packets as a flow control measure has not been used in NoCs. Note that no resource reservations are required for this method, and it fits quite well with dropping as a congestion control technique.

2.  Returning packets that do not fit in buffers at the target network interface to the sender. To avoid introducing deadlock, the sender *must* accept the rejected packets. Although this technique has been used in computer networks, it is not (yet) used by any NoC [97].

3.  Deflection routing for end-to-end flow control is used by SPIN [49] (and could be a natural option for Nostrum [74]). In other words, when a packet does not fit in the slave network interface buffer, it is sent back in the router network. Note that this is not the same as sending back the packet to the sender described above. There the slave does not intend accepting the packet at all, while here the slave temporarily refuses a packet with a view to accepting it later. This method combines well with using deflection routing for congestion control.

Methods that require resource reservations:

■  Another method to ensure that packets always find space in the slave network interface buffer is to ensure that packets are only sent by the master network interface when there is guaranteed to be space in the slave network interface buffer. This can be achieved by *end-to-end flow control* based on credits or sliding windows [38, 96]. (Note that ACK/NACK, Go-Back-N, etc. are flow control schemes most commonly used for the link-layer and are described in Chapter 4.) These methods can be used with any buffering scheme. However, when virtual-circuit buffering (Section 5.4.1) is used a better approach exists, as described in the next point.

Figure 5.23 shows how credit-based end-to-end flow control operates. A master sends requests to a slave, who optionally returns responses. End-to-end flow control is required independently for both the request and the response flows. The figure shows routers that use shared buffers for all request and response traffic for all flows in the NoC. The routers have degree 3 with input buffers are shown. However, the buffer organization is not relevant here. Packets are allowed to leave the master network interface request buffer (1) only when there is space for them in the slave network interface request buffer (2). This is accomplished by tracking the free space in a credit counter (a). The counter is initialized to the capacity of the slave network interface request buffer (2), and is decremented

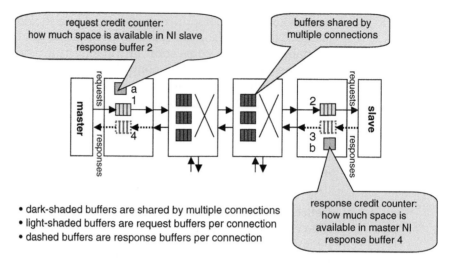

Credit-based end-to-end flow control.

whenever a packet is sent. Whenever a slave removes a packet from its request buffer (2), it sends a credit (in a packet) back to the master, who adds it to its credit counter (a). The slave network interface request buffer must be large enough to hide the delay introduced by sending back credits, otherwise the master will stall as long as the credits are in transit. Responses from slave to master are handled analogously.

Credit traffic can consume a substantial amount of the bandwidth (31% is quoted in Ref. [50]). As an optimization, the requested credit packets can be combined with response packets (and vice versa). This is called piggy backing [100], and can save substantial bandwidth (Fig. 5.24). As the burst size grows, the relative overhead introduced by credit packets decreases. With larger burst sizes, more credits are reported in a credit packet. Consequently, the number of credit packets (i.e., overhead introduced by credits) decreases. The drawback of increasing burst sizes is that larger buffers are required to accommodate the bigger bursts.

Æthereal [90] uses standard credit-based end-to-end flow control as described as above. Nostrum has been extended [42] to use the same flow control scheme.

SPIN [50] uses the same basic scheme, except that the credit counter is initialized with a value higher than the receiving buffer capacity. This can be interpreted as using the router buffers as an extension of the receiving buffer. However, this optimization introduces a dependency of flow control on congestion control. In SPIN this is solved by using deflection routing for both.

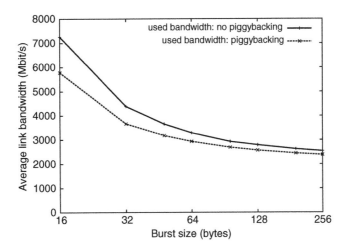

■ **FIGURE 5.24**

Piggy backing can save up to 20% on the average link bandwidth [90].

QNoC [105] uses end-to-end credits for both congestion control and flow control. Credits are not handed out on a flow (single master–single slave) basis as above. Instead a slave manages the requests of all masters that communicate with it in a single queue, and hence a single set of credits. A two-phase protocol is now used: a master requests credits from the slave using a high-priority traffic class, then it can send its data using the standard priority traffic class. This method is useful for heavily used slaves (i.e., those that cause congestion) with many masters (i.e., end-to-end flow control per master–slave pair is too expensive).

■ The end-to-end flow control methods ensure that when packets enter the router network then they will after some time (depending on congestion) arrive at the receiving network interface, and find space in the appropriate buffer. Queueing in the router network must be avoided because packets of other flows *that use the same router buffers* may be blocked. In other words, by giving all flows their own buffers in every router along their paths (virtual circuit buffering, Section 5.4.1), a packet that is blocked causes no harm to other flows.

As briefly mentioned in Section 5.4, routers synchronize the transfer of flits to ensure that the sending buffer is not empty, and that the receiving buffer does not overflow. This is performed by link-level flow control, as described in Chapter 4. Note that link-level flow occurs per virtual circuit, and if one virtual circuit is blocked, another may still progress. Now, end-to-end flow control is automatically obtained by a chain of link-level flow controls, as shown

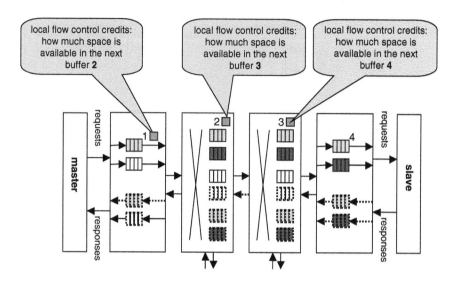

- master–slave uses light-shaded solid and dashed circuits for requests and responses, resp.
- the master communicates to another slave using the white-shaded circuits
- the slave communicates with another master using the dark-shaded circuits

▪ **FIGURE 5.25**

Virtual-circuit buffering with credit-based local flow control.

in Fig. 5.25. Although the figure shows credit-based link-level flow control, other techniques are also possible (see Chapter 4).

Mango [8] and [104] use this approach. Although currently not the case, Nostrum [74] could use this approach for its containers (under the restriction that containers transport data from initiator to target only).

We have described congestion control and flow control, and why each is desirable or necessary. In the next section we see that both together are required if NoC offers QoS performance guarantees.

### 5.7.3 Congestion Control and Flow Control for QoS

We have described how congestion control and flow control regulate the interaction or interference between different packets and flows in the network. In either case, resources can be (explicitly) reserved or not, as summarized in Table 5.2.

As mentioned a number of times in this section, without resource reservations it is in general impossible to give hard (100%) guarantees. If either congestion control or flow control does not use resource reservations, no end-to-end guarantees such as minimum bandwidth and maximum

**TABLE 5.2** ■ Overview of congestion control and flow control.

|  | Without reservations | With reservations |
|---|---|---|
| Congestion control | Cropping<br>Dynamic routing<br>Adaptive injection | Contention-free routing<br>Rate control |
| Flow control | Dropping<br>Refusing<br>Deflection | End-to-end flow control<br>Virtual-circuit flow control |

latency service classes can be offered. Congestion control is required to eliminate or characterize the interference between different flows in the router network. Flow control characterizes the interaction between the sender and the receiver. Without it a single slow or non-responding master or slave can cause loss of data (if dropping is used), or congestion or even complete blockage of the NoC. No bandwidth or latency guarantees can be given in all these cases.

When no hard performance guarantees can be given, the NoC behavior can be described statistically. Under the assumption of particular traffic patterns, distributions of for example, packet delays can be computed, as well as statistical bounds (for a given confidence level). For example, Guerrier [49] shows that in SPIN network latencies occur with an exponentially decreasing probability (Fig. 5.26). Note that the number of packets (vertical axis) is exponential. The lowest network latencies are obtained at the lowest (27%) offered load, where over 90% of packets arrive within 50 cycles. The almost horizontal curve corresponds to a saturated NoC at 47% offered load. Beyond this load the module-to-module latency rapidly becomes unbounded.

Figure 5.26 shows that very long latencies will occur, although infrequently. However, note that "improbable" failures (e.g., $10^{-14}$ chance of being "late") occur often, with high NoC operating frequencies (e.g., 500 MHz), concurrent transactions (e.g., 20 active packets): every $20 \times 5 \times 10^{12} \times 10^{-14} =$ every second. In an NoC with BE service classes only, reducing the failure rate to an acceptable level may mean drastically reducing the offered load, or equivalently, the utilization of the NoC.

However, there are advantages when resources do not need to be reserved. The set-up and tear-down phases are not required, and at any point in time resources are automatically distributed over flows (although perhaps unfairly [103]). No resource re-allocation (reconfiguration) is necessary, when NoC usage changes.

Few NoCs have implemented both congestion control and flow control with a view to give hard QoS guarantees. In particular, Mango [6], Æthereal [44], and SonicsMX [104] offer complete solutions. However, a number of NoCs can be extended with techniques described above (usually end-to-end or chained link-level flow control), and they too will be

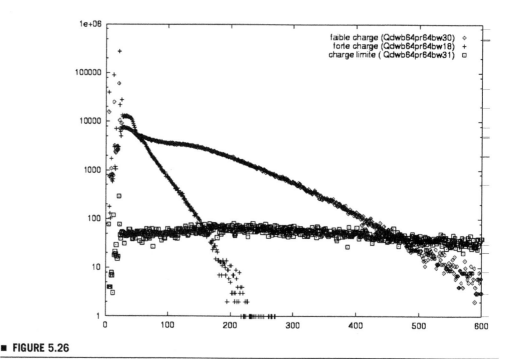

**■ FIGURE 5.26**

Network latency distributions for 27%, 45%, and 47% offered load [49].

able to offer the same level of QoS guarantees. This category contains Nostrum [74], aSoC [64], and Refs [62, 106].

NoCs differ in the granularity of resource reservations they make. There are two ends of the spectrum. The first precisely maps traffic classes to a larger number of tailored service classes, where the other more coarsely maps traffic classes to fewer, more generic service classes (Section 5.2). The former enables accurate traffic scheduling for high effective NoC utilization, at the cost of precise traffic characterizations with tailored service classes. The latter reduces the burden of precise traffic characterization, at the cost of lower effective NoC utilization. Hence, the respective approaches trade off accuracy of mapping of traffic classes to service classes versus effective NoC utilization. The QoS approaches, introduced in Section 5.2.1, of QNoC [13] (four service classes) and SPIN [49] (a single BE service class) may be interpreted as examples of these respective approaches, with Nostrum [74] and Æthereal [44] being in between (with guaranteed-bandwidth and BE service classes).

Which approach is most suitable depends on the application requirements and NoC type (ASIC, ASSP, FPGA, or CMP). For example, real-time streaming applications (such as video) tend to be long lived, and minimum bandwidth and low jitter must be guaranteed, but low latency is less important. Resource reservations to optimally schedule link and buffer utilization reduces the cost of the system, at the cost of explicit flow set-up.

This is acceptable in ASIC and ASSP NoCs, which are the natural implementation for this sort of applications for reasons of performance and cost. On the other hand, cache traffic of embedded processors of CMPs, is unpredictable in terms of (instantaneous) bandwidth usage and the average and worst-case bandwidth are very different. Average low latency is essential for good performance. Hence, precise characterization of cache traffic is difficult, and overdimensioning would be very expensive. As a result, most traffic classes may be mapped to the BE service class, which needs no or few reservations.

## 5.8  SUMMARY

In this chapter we described the main principles in designing the networking and transport layers of NoCs. These layers include the specification of the following NoC characteristics: switching technique, topology, addressing and routing, and end-to-end congestion and flow control schemes. We presented this complex problem as a constrained optimization process. The NoC designer should minimize the cost of the NoC which is expressed in terms of VLSI area and power consumption added to his overall chip design. The constraints are the level of services (service classes) to be provided for the traffic patterns of the SoC under consideration. Since NoCs can be designed anew for each SoC implementation, this optimization process is expected to be repeated for each new SoC design, using NoC CAD tools.

Following this definition, it is clear that the SoC traffic characteristics strongly impact the NoC characteristics. We therefore classified SoCs according to how much is known in advance about their functionality, and consequently about their expected module-to-module communication patterns (Fig. 5.1).

We reviewed topical state-of-the-art solutions for each of the important NoC characteristics: switching mechanisms, QoS implementations, topological design, routing mechanisms, congestion and flow control techniques. We showed how the NoC model impacts the choices available, for each of these, as well as the relationships and trade-offs between them.

## REFERENCES

[1]  A. Adriahantenaina, H. Charlery, A. Greiner, L. Mortiez and C.A. Zeferino, "SPIN: A Scalable, Packet Switched, On-Chip Micro-Network," *Proceedings of the Design, Automation and Test in Europe Conference and Exhibition (DATE)*, 2003.

[2]  *ARM, AMBA AXI Protocol Specification*, June 2003. Available at http://www.arm.com/products/solutions/axi_spec.html

[3]   ATM Forum, *ATM User-Network Interface Specification*, Prentice Hall, Upper Saddle River, NJ, July 1994, Version 3.1.

[4]   C. Aurrecoechea, A.T. Campbell and L. Hauw, "A Survey of QoS Architectures," *Multimedia Systems*, Vol. 6, No. 3, 1998, pp. 138–151.

[5]   D. Bertozzi and L. Benini, "Xpipes: A Network-on-Chip Architecture for Gigascale Systems-on-Chip," *IEEE Circuits and Systems Magazine*, Vol. 4, No. 2, 2004, pp. 18–31.

[6]   T. Bjerregaard, *The MANGO Clockless Network-on-Chip: Concepts and Implementation*, Ph.D. thesis, Informatics and Mathematical Modelling, Technical University of Denmark, DTU, 2006.

[7]   T. Bjerregaard, S. Mahadevan, R. Grøndahl and J. Sparsø, "An OCP Compliant Adapter for GALS-Based SoC Design Using the MANGO Network-on-Chip," *Proceedings of the International Symposium on Systems on Chip (SoC)*, 2005.

[8]   T. Bjerregaard and J. Sparsø, "A Router Architecture for Connection-Oriented Service Guarantees in the MANGO Clockless Network-on-Chip," *Proceedings of the Design, Automation and Test in Europe Conference and Exhibition (DATE)*, March 2005, pp. 1226–1231.

[9]   T. Bjerregaard and J. Sparsø, Scheduling Discipline for Latency and Bandwidth Guarantees in Asynchronous Network-on-Chip," *Proceedings of the International Symposium on Asynchronous Circuits and Systems (ASYNC)*, March 2005, pp. 34–43.

[10]  T. Bjerregaard and J. Sparsø, "Implementation of Guaranteed Services in the MANGO Clockless Network-on-Chip," *IEE Proceedings: Computers and Digital Techniques*, 2006.

[11]  H.L. Bodlaender, J. van Leeuwen, R. Tan and D. Thilikos, "On Interval Routing Schemes and Treewidth," *Information and Computation*, Vol. 139, No. 1, November 25, 1997, pp. 92–109.

[12]  E. Bolotin, I. Cidon, R. Ginosar and A. Kolodny, "Cost considerations in Network on Chip," *Integration, the VLSI Journal*, Vol. 38, No. 1, October 2004, pp. 19–42.

[13]  E. Bolotin, I. Cidon, R. Ginosar and A. Kolodny, "QNoC: QoS Architecture and Design Process for Network on Chip," *Journal of Systems Architecture*, Vol. 50, No. 2–3, February 2004, pp. 105–128. Special issue on Networks on Chip.

[14]  E. Bolotin, I. Cidon, R. Ginosar and A. Kolodny, "Efficient Routing in Irregular Topology Nocs," *Technion, CCIT Report*, Vol. 554, No. 5, September 2005.

[15]  F. Borgonovo, L. Fratta and J.A. Bannister, "On the Design of Optical Deflection-Routing Networks," *Proceedings of the Annual Joint Conference of the IEEE Computer and Communications Societies (INFOCOM)*, 1994, pp. 120–129.

[16]  R. Braden, D. Clark and S. Shenker, "Integrated Services in the Internet Architecture: An Overview," *Internet Drafts*, RFC 1633, June 1996. Available at http://www.rfc_archive.org/getrfc.php?rfc=1633

[17]  J. Brassil and R.L. Cruz, "Bounds on Maximum Delay in Networks with Deflection Routing," *IEEE Transactions on Parallel and Distributed Systems*, Vol. 6, No. 7, 1995, pp. 724–732.

[18]  I. Cidon, R. Guérin and A. Khamisy, "On Protective Buffer Policies, *IEEE/ACM Transactions on Networking*, Vol. 2, No. 3, 1994, pp. 240–246.

[19] I. Cidon, F.M. Jaffe and M. Sidi, "Local Distributed Deadlock Detection by Cycle Detection and Clustering," *IEEE Transactions on Software Engineering*, Vol. 13, No. 1, 1987, pp. 3–14.

[20] I. Cidon, J.M. Jaffe and M. Sidi, "Distributed Store-and-Forward Deadlock Detection and Resolution Algorithms, "*IEEE Transactions on Communication*, Vol. 35, No. 11, 1987, pp. 1139–1145.

[21] I. Cidon and I. Keidar, "Zooming in on Network on Chip Architectures," *Technion, CCIT Report*, Vol. 565, No. 5, December 2005.

[22] C. Ciordaş, T. Basten, A. Rădulescu, K. Goossens and J. van Meerbergen, An Event-Based Network-on-Chip Monitoring Service," *ACM Transactions on Design Automation of Electronic Systems*, Vol. 10, No. 4, October 2005, pp. 702–723. HLDVT'04 Special Issue on Validation of Large Systems.

[23] D.J. Culler, J.P. Singh and A. Gupta, *Parallel Computer Architecture: A Hardware/Software Approach*, Morgan Kaufmann Publishers, San Francisco, CA, 1999.

[24] W. Dally and B. Towles, *Principles and Practices of Interconnection Networks*, Morgan Kaufmann Publishers Inc., San Francisco, CA, USA, 2003.

[25] W.J. Dally, "Express Cubes: Improving the Performance of $k$-ary $n$-cube Interconnection Networks," *IEEE Transactions on Computers*, Vol. 40, No. 9, September 1991, pp. 1016–1023.

[26] W.J. Dally, "Virtual-Channel Flow Control," *IEEE Transactions on Parallel and Distributed Systems*, Vol. 3, No. 2, March 1992, pp. 194–205.

[27] W.J. Dally and H. Aoki, Adaptive routing using virtual channels, Technical Report, Laboratory for Computer Science, MIT, Cambridge, Massachusetts, September 1990.

[28] W.J. Dally and H. Aoki, "Deadlock-Free Adaptive Routing in Multicomputer Networks Using Virtual Channels," *IEEE Transactions on Parallel and Distributed Systems*, Vol. 4, No. 4, April 1993, pp. 466–475.

[29] W.J. Dally and C.L. Seitz, "Deadlock-Free Message Routing in Multiprocessor Interconnection Networks," *IEEE Transactions on Computers*, Vol. 36, No. 5, May 1987, pp. 547–553.

[30] W.J. Dally and B. Towles, "Route Packets, Not Wires: On-Chip Interconnection Networks," *Proceedings of the Design Automation Conference (DAC)*, 2001, pp. 684–689.

[31] A. DeHon, Robust, high-speed network design for large-scale multiprocessing, A.I. Technical report 1445, Massachusetts Institute of Technology, Artificial Intelligence Laboratory, Cambridge, MA, September 1993.

[32] J. Dielissen, A. Rădulescu, K. Goossens and E. Rijpkema, "Concepts and Implementation of the Philips Network-on-Chip," In *Workshop on IP-Based System-on-Chip Design*, Grenoble, France, November 2003.

[33] J. Duato, S. Yalamanchili and L. Ni, "*Interconnection Networks – An Engineering Approach*, Morgan Kaufmann, San Francisco, CA, 2003.

[34] T. Dumitras and R. Mărculescu, "On-Chip Stochastic Communication," *Proceedings of the Design, Automation and Test in Europe Conference and Exhibition (DATE)*, 2003.

[35] U. Feige and P. Raghavan, "Exact Analysis of Hot-Potato Routing," *Proceedings of the Annual Symposium on Foundations of Computer Science*, October 1992, pp. 553–562.

[36]  E. Fleury and P. Fraigniaud, "A General Theory for Deadlock Avoidance in Wormhole-Routed Networks," *IEEE Transactions on Parallel and Distributed Systems*, Vol. 9, No. 7, July 1998, pp. 626–638.

[37]  O.P. Gangwal, A. Rădulescu, K. Goossens, S. González Pestana and E. Rijpkema, "Building Predictable Systems on Chip: An Analysis of Guaranteed Communication in the Æthereal Network on Chip," in P. van der Stok (editor), *Dynamic and Robust Streaming in and Between Connected Consumer-Electronics Devices*, Vol. 3. *Philips Research Book Series*, Springer, Berlin, Germany, 2005, Chapter 1, pp. 1–36.

[38]  M. Gerla and L. Kleinrock, "Flow Control: A Comparative Survey," *IEEE Transactions on Communications*, Vol. COM-28, No. 4, 553–574, April 1980.

[39]  R. Gindin, I. Cidon and I. Keidar, NoC architecture for future fpgas, Department of EE, CCIT Report 579, Technion, March 2006.

[40]  C.J. Glass and L.M. Ni, "The Turn Model for Adaptive Routing," *International Symposium on Computer Architecture*, May 1992, pp. 278–287.

[41]  C.J. Glass and L.M. Ni, "The Turn Model for Adaptive Routing," *Journal of the ACM*, Vol. 41, No. 5, September 1994, pp. 874–902.

[42]  S. González Pestana, K. Goossens, A. Rădulescu and R. Thid. Framework and performance metric definitions: A first step towards network-on-chip benchmarking, Technical Note 2006/00003, Philips Research, January 2006.

[43]  K. Goossens, J. Dielissen, O.P. Gangwal, S. González Pestana, A. Rădulescu and E. Rijpkema, "A Design Flow for Application-Specific Networks on Chip with Guaranteed Performance to Accelerate SOC Design and Verification," *Proceedings of the Design, Automation and Test in Europe Conference and Exhibition (DATE)*, March 2005, pp. 1182–1187.

[44]  K. Goossens, J. Dielissen and A. Rădulescu, "The Æthereal Network on Chip: Concepts, Architectures, and Implementations," *IEEE Design and Test of Computers*, Vol. 22, No. 5, September–October 2005, pp. 21–31.

[45]  K. Goossens, J. Dielissen, J. van Meerbergen, P. Poplavko, A. Rădulescu, E. Rijpkema, E. Waterlander and P. Wielage, "Guaranteeing the quality of Services in Networks on Chip," In A. Jantsch and H. Tenhunen (editors), *Networks on Chip*, Kluwer, Norwell, MA, 2003, Chapter 4, pp. 61–82.

[46]  K. Goossens, O.P. Gangwal, J. Röver and A.P. Niranjan, "Interconnect and Memory Organization in SOCs for Advanced Set-Top Boxes and TV – Evolution, Analysis, and Trends," In J. Nurmi, H. Tenhunen, J. Isoaho and A. Jantsch (editors), *Interconnect-Centric Design for Advanced SoC and NoC*, Kluwer, Norwell, MA, 2004, Chapter 15, pp. 399–423.

[47]  K. Goossens, J. van Meerbergen, A. Peeters and P. Wielage, "Networks on Silicon: Combining Best-Effort and Guaranteed Services, in *Proceedings of the Design, Automation and Test in Europe Conference and Exhibition (DATE)*, March 2002, pp. 423–425.

[48]  A.G. Greenberg and N. Madras, "How Fair Is Fair Queueing," *Journal of the ACM*, 1992, pp. 568–598.

[49]  P. Guerrier, *Un Réseau D'Interconnexion pour Systémes Intégrés*, Ph.D. thesis, Université Paris VI, Paris, France, March 2000.

[50]  P. Guerrier and A. Greiner, "A Generic Architecture for On-Chip Packet-Switched Interconnections," *Proceedings of the Design, Automation and Test in Europe Conference and Exhibition (DATE)*, 2000, pp. 250–256.

[51] K.D. Günther. "Prevention of Deadlocks in Packet-Switched Data Transport System," *IEEE Transactions on Communications*, Vol. 29, April 1981, pp. 512–524.

[52] Z. Guz, I. Walter, E. Bolotin, I. Cidon, R. Ginosar and A. Kolodny, "Efficient Link Capacity and QoS Design for Wormhole Network-on-Chip," *Proceedings of the Design, Automation and Test in Europe Conference and Exhibition (DATE)*, March 2005, pp. 9–14.

[53] A. Hansson, K. Goossens and A. Rădulescu, UMARS: A unified approach to mapping and routing on a combined guaranteed service and best-effort network-on-chip architecture, Technical Report 2005/00340, Philips Research, April 2005.

[54] A. Hansson, K. Goossens and A. Rădulescu, "A Unified Approach to Constrained Mapping and Routing on Network-on-Chip Architectures," *International Conference on Hardware/Software Codesign and System Synthesis (CODES+ISSS)*, September 2005, pp. 75–80.

[55] A. Hansson, K. Goossens and A. Rădulescu, Analysis of message-dependent deadlock in network-based systems on chip, Technical Report 2006/00230, Philips Research, Eindhoven, The Netherlands, March 2006.

[56] M. Harmanci, N. Escudero, Y. Leblebici and P. Ienne, "Quantitative Modelling and Comparison of Communication Schemes to Guarantee Quality-of-Service in Networks-on-Chip," *Proceedings of the International Symposium on Circuits and Systems (ISCAS)*, 2005, pp. 1782–1785.

[57] J. Hu and R. Marculescu, "Exploiting the Routing Flexibility for Energy/Performance Aware Mapping of Regular NoC Architectures," *Proceedings of the Design, Automation and Test in Europe Conference and Exhibition (DATE)*, 2003, pp. 688–693.

[58] J. Hu and R. Marculescu, "DyAD – Smart Routing for Networks on Chip," *Proceedings of the Design Automation Conference (DAC)*, June 2004.

[59] J. Hu and R. Marculescu, "Energy- and Performance-Aware Mapping for Regular NoC Architectures, *IEEE Transactions on CAD of Integrated Circuits and Systems*, Vol. 24, No. 4, April 2005, pp. 551–562.

[60] P. Humblet, A. Bhargava and M.G. Hluchyj, "Ballot Theorems Applied to the Transient analysis of nD/D/1 Queues," *IEEE/ACM Transactions on Networking*, 1993, Vol. 1, No. 1, pp. 81–95.

[61] A. Jalabert, S. Murali, L. Benini and G. De Micheli, "XpipesCompiler: A Tool for Instantiating Application Specific Networks on Chip," *Proceedings of the Design, Automation and Test in Europe Conference and Exhibition (DATE)*, 2004.

[62] N. Kavaldjiev, G.J.M. Smit and P.G. Jansen, "A Virtual Channel Router for On-Chip Networks," *Proceedings of the International SOC Conference (SoCC)*, September 2004.

[63] H.T. Kung, T. Blackwell and A. Chapman, "Credit-Based Flow Control for ATM Networks: Credit Update Protocol, Adaptive Credit Allocation and Statistical Multiplexing." *SIGCOMM*, 1994, pp. 101–114.

[64] A. Laffely, J. Liang, P. Jain, N. Weng, W. Burleson and R. Tessier, "Adaptive Systems on a Chip (aSoC) for Low-Power Signal Processing, *Proceedings of the Asilomar Conference on Signals, Systems, and Computers*, 2001.

[65] A.J. Laffely, *An Interconnect-Centric Approach for Adapting Voltage and Frequency in Heterogeneous System-on-a-Chip*. Ph.D. thesis, University of Massachusetts, Amherst, MA, 2003.

[66] K. Lee, S.-J. Lee and H.-J. Yoo, "A Distributed Crossbar Switch Scheduler for On-Chip Networks," *Proceedings of the Custom Integrated Circuits Conference*, 2003.

[67] K. Lee, S.-J. Lee and H.-J. Yoo, "A High-Speed and Lightweight On-Chip Crossbar Switch Scheduler for On-Chip Interconnection Networks," *Proceedings of the International Conference on European Solid-State Circuits*, 2003.

[68] C. Leiserson, "Fat-Trees: Universal Networks for Hardware-efficient super-computing," *IEEE Transactions on Computers*, October 1985, Vol. C-34, No. 10, pp. 892–901.

[69] D.H. Linder and J.C. Harden, "An adaptive and Fault Tolerant Wormhole Routing Strategy for k-ary n-cubes," *IEEE Transactions on Computers*, Vol. 40, No. 1, January 1991, pp. 2–12.

[70] A. Litman and S. Moran-Schein, "Fast, Minimal, and Oblivious Routing Algorithms on the mesh with bounded Queues," *Journal of Interconnection Networks*, 2001, Vol. 2, No. 4, pp. 445–469.

[71] P. López, J.M. Martínez and J. Duato, "A Very Efficient Distributed Deadlock Detection Mechanism for Wormhole Networks," *Proceedings of the International Symposium on High-Performance Computer Architecture (HPCA)*, February 1998, pp. 57–66.

[72] N.F. Maxemchuk, "Routing in the Manhattan Street Network," *IEEE Transactions on Communication*, Vol. COM-35, No. 2–3, May 1987, pp. 503–512.

[73] D.E. McDysan and D.L. Spohn, *ATM: Theory and Application*, McGraw-Hill, Inc., New York, USA, 1994.

[74] M. Millberg, E. Nilsson, R. Thid and A. Jantsch, "Guaranteed Bandwidth Using Looped Containers in Temporally Disjoint Networks within the Nostrum Network on Chip," *Proceedings of the Design, Automation and Test in Europe Conference and Exhibition (DATE)*, 2004.

[75] M. Millberg, E. Nilsson, R. Thid, S. Kumar and A. Jantsch, "The Nostrum Backbone – a Communication Protocol Stack for Networks on Chip," *Proceedings of the International Conference on VLSI Design*, 2004, pp. 693–696.

[76] P. Mohapatra, "Wormhole Routing Techniques for Directly Connected Multicomputer Systems." *ACM Computing Surveys*, Vol. 30, No, 3, 1998, pp. 374–410.

[77] S. Murali, L. Benini and G. de Micheli, "Mapping and Physical Planning of Networks on Chip Architectures with Quality of Service Guarantees," *Proceedings of the Design Automation Conference, Asia and South Pacific (ASPDAC)*, 2005.

[78] S. Murali, M. Coenen, A. Rădulescu, K. Goossens and G. De Micheli, "A Methodology for Mapping Multiple Use-Cases on to Networks on Chip. *Proceedings of the Design, Automation and Test in Europe Conference and Exhibition (DATE)*, March 2006, pp. 118–123.

[79] S. Murali and G. De Micheli, "SUNMAP: A Tool for Automatic Topology Selection and Generation for NOCs," *Proceedings of the Design Automation Conference (DAC)*, June 2003.

[80] L.M. Ni and P.K. McKinley, "A Survey of Wormhole Routing Techniques in Direct Networks," *IEEE Computer*, February 1993, Vol. 26, No. 2, pp. 62–76.

[81] K. Nichols, S. Blake, F. Baker and D. Black. "Definition of the Differentiated Services Field (DS Field) in the IPv4 and IPv6 Headers," The

RFC archive – RFC 2474. December 1998, pp. 1126–1127. Available at http://www.rfc_archive.org/getrfc.php?rfc=2474

[82] E. Nilsson, M. Millberg, J. Öberg and A. Jantsch, "Load Distribution with the Proximity Congestion Awareness in a Network on Chip," *Proceedings of the Design, Automation and Test in Europe Conference and Exhibition (DATE)*, 2003.

[83] V. Nollet, T. Marescaux, P. Avasare, D. Verkest and J.-Y. Mignolet, Centralized Run-Time Resource Management in a Network-on-Chip Containing Reconfigurable Hardware Tiles," *Proceedings of the Design, Automation and Test in Europe Conference and Exhibition (DATE)*, March 2005, pp. 234–239.

[84] V. Nollet, T. Marescaux and D. Verkest, "Operating-System Controlled Network on Chip. *Proceedings of the Design Automation Conference (DAC)*, June 2005, pp. 256–259.

[85] OCP International Partnership, *Open Core Protocol Specification*, 2001, pp. 1342–1347. Available at www.ocpip.org

[86] U.Y. Ogras, J. Hu and R. Marculescu, "Key Research Problems in NoC Design: A Holistic Perspective," *International Conference on Hardware/Software Codesign and System Synthesis (CODES+ISSS)*, September 2005.

[87] P. Pande, C. Grecu, Ivanov and R. Saleh, "Design of a Switch for Network on Chip Applications," *Proceedings of the International Symposium on Circuits and Systems (ISCAS)*, 2003.

[88] E. Rijpkema, K. Goossens, A. Rădulescu, J. Dielissen, J. van Meerbergen, P. Wielage and E. Waterlander, "Trade Offs in the Design of a Router with Both Guaranteed and Best-Effort Services for Networks on Chip," *IEE Proceedings: Computers and Digital Technique*, September 2003, Vol. 150, No. 5, pp. 294–302.

[89] E. Rijpkema, K. Goossens and P. Wielage, "A Router Architecture for Networks on Silicon," *Proceedings of Progress 2001, 2nd Workshop on Embedded Systems*, Veldhoven, the Netherlands, October 2001.

[90] A. Rădulescu, J. Dielissen, S. González Pestana, O.P. Gangwal, E. Rijpkema, P. Wielage and K. Goossens, "An Efficient On-Chip Network Interface Offering Guaranteed Services, Shared-Memory Abstraction, and Flexible Network Programming," *IEEE Transactions on CAD of Integrated Circuits and Systems*, January 2005, Vol. 24, No. 1, pp. 4–17.

[91] A. Rădulescu and K. Goossens, "Communication Services for Networks on Chip," In S.S. Bhattacharyya, E.F. Deprettere and J. Teich (editors), *Domain-Specific Processors: Systems, Architectures, Modeling, and Simulation*, Marcel Dekker, New York, 2004, pp. 193–213.

[92] D. Signenza-Tortosa, J. Nurmi, "Proteo: A New Approach to Network-on-Chip," in *Proceedings of the IAS TED International Conference on Communication Systems and Networks (CSN '02)*, Malaga, Spain, September 9–12, 2002.

[93] D. Seo, A. Ali, W.-T. Lim, N. Rafique and M. Thottethodi, "Near-Optimal Worst-Case Throughput Routing for Two-Dimensional Mesh Networks, *International Symposium on Computer Architecture*, 2005, pp. 432–443.

[94] L. Shang, L.-S. Peh and N.K. Jha, "Dynamic Voltage Scaling with Links for Power Optimization of Interconnection Networks," IEEE Computer Society, *HPCA '03: Proceedings of the 9th International Symposium on*

*High-Performance Computer Architecture*, Washington, DC, USA, 2003, pp. 91.

[95] D. Shoemaker, *An Optimized Hardware Architecture and Communication Protocol for Scheduled Communication*, Ph.D. thesis, Electrical Engineering and Computer Science Department, Massachusetts Institute of Technology, Cambridge, MA, May 1997.

[96] V. Shurbanov, D. Avresky, P. Mehra and W. Watson, "Flow Control in Server-net Clusters," *The Journal of Supercomputing*, June 2002, Vol. 22, No. 2, pp. 161–173.

[97] Y.H. Song and T.M. Pinkston, "A Progressive Approach to Handling Message-Dependent Deadlock in Parallel Computer Systems," *IEEE Transactions on Parallel and Distributed Systems*, 2003, Vol. 14, pp. 259–275.

[98] F. Steenhof, H. Duque, B. Nilsson, K. Goossens and R. Peset Llopis, "Networks on Chips for High-End Consumer-Electronics TV System Architectures," *Proceedings of the Design, Automation and Test in Europe Conference and Exhibition (DATE)*, March 2006, pp. 148–153.

[99] P. Stravers and J. Hoogerbrugge, "Homogeneous Multiprocessing and the Future of Silicon Design Paradigms," *International Symposium on Technology, Systems, and Applications Proceedings* 2001, 184–187.

[100] A.S. Tanenbaum, *Computer Networks*, Prentice-Hall, Upper Saddle River, NJ, 1996.

[101] J.W. van den Brand, C. Ciordaş and T. Basten, Runtime Networks-on-Chip Performance Monitoring, Technical Report 2006/00218, Philips Research, Eindhoven, The Nertherlands, March 2006.

[102] E. Waingold, M. Taylor, D. Srikrishna, V. Sarkar, W. Lee, V. Lee, J. Kim, M. Frank, P. Finch, R. Barua, J. Babb, S. Amarasinghe and A. Agarwal, "Baring It All to Software: Raw Machines," *IEEE Computer*, September 1997, Vol. 30, No. 9. pp. 86–93.

[103] I.Z. Walter, Quality of Service in Network-on-Chip, Master's thesis, Technion, Israel Institute of Technology, Haifa, Israel, August 2005.

[104] W.-D. Weber, J. Chou, I. Swarbrick and D. Wingard, "A Quality-of-Service Mechanism for Interconnection Networks in System-on-Chips. *Proceedings of the Design, Automation and Test in Europe Conference and Exhibition (DATE)*, March 2005.

[105] D. Wiklund, *Development and Performance Evaluation of Networks on Chip*, Ph.D. thesis, Department of Electrical Engineering, Linköping University, Linköping, Sweden, 2005.

[106] P.T. Wolkotte, G.J. Smit, G.K. Rauwerda and L.T. Smit, "An Energy-Efficient Reconfigurable Circuit Switched Network-on-Chip," *Proceedings of the International Parallel and Distributed Processing Symposium (IPDPS)*, April 2005.

[107] T.T. Ye, L. Benini and G. De Micheli, "Packetization and Routing Analysis of On-Chip Multiprocessor Networks," *Journal of Systems Architecture*, February 2004, Vol. 50, No. 2–3, pp. 81–104. Special issue on Networks on Chip.

[108] H. Zhang, "Service Disciplines for Guaranteed Performance Service in Packet-Switching Networks," *Proceedings of the IEEE*, October 1995, Vol. 83, No. 10, pp. 1374–1396.

[109] H. Zhang and D. Ferrari, "Rate-Controlled Static-Priority Queueing. *Proceedings of the Annual Joint Conference of the IEEE Computer and Communications Societies (INFOCOM)*, 1993, pp. 227–236.

# CHAPTER 6

# NETWORK INTERFACE ARCHITECTURE AND DESIGN ISSUES*

Network interfaces (NIs) are usually denoted as the glue logic necessary to adapt communicating cores to the on-chip network. Historically, embedded system processors have been natively designed with bus-specific interfaces [3–9]. As a consequence, hardware sub-modules of communication architectures have been directly exposed to core interfaces. The request-grant signaling between AMBA (*Advanced Microcontroller Bus Architecture*) AHB (*Advanced High-performance Bus*) master interfaces and the bus arbiter is an example thereof, and causes the arbiter design to be instance specific, thus not reusable without modifications across a number of different hardware platforms.

As the level of system integration in *Systems on Chip* (SoCs) began to rise, system designers faced the need to reuse pre-designed and pre-verified computation units across different platforms and with different communication architectures. Therefore, the need for effective plug-and-play design styles pushed the development of standard interface sockets, allowing to decouple the development of computational units from that of communication architectures [10, 11]. The *Open Core Protocol* (OCP [12]) was devised as an effective means of simplifying the integration task through the standardization of the core interface protocol. It also paves the way for more cost-effective system implementations, since it can be custom-tailored based on the complexity and on the communication requirements of the connected cores. AMBA AXI (*Advanced eXtensible Interface*) introduced a similar paradigm shift in the AMBA family of communication protocol specifications: in fact, it defines a point-to-point protocol between a master and a slave interface with advanced communication semantics.

---

* This chapter was provided by Davide Bertozzi of University of Ferrara, Italy.

The common feature of such advanced state-of-the-art communication protocols is that they are core-centric transaction-based protocols which abstract away the implementation details of the system interconnect. Transactions are specified as though communication initiator and target were directly connected, taking the well-known layered approach to peer-to-peer communication that proved extremely successful in the domain of *local- and wide-area networks* (LANs and WANs, respectively). Referring to the reference ISO/OSI model, the NI front-end can be viewed as the standardized core interface at session layer or transport layer, where bus-agnostic high-level communication services are provided.

With the advent of *networks on chip* (NoCs), the same standard interface sockets can be reused for interfacing communicating cores to the network, but the semantic gap between point-to-point protocols and network protocols increases, thus making the design of NIs potentially more complex. This gap mostly depends on the switching technique used in the network.

Circuit-switching networks require the least level of adaptation, in that data are simply propagated across the pre-established circuits (reserved buffering resources and channel bandwidth). The cost incurred for circuit set-up and tear-down, namely latency overhead and waste of bandwidth, is compensated by a more predictable performance and more relaxed buffering requirements.

Although there are many common issues, integrating cores through packet-switched networks makes the NI design process much more complicated and subject to a number of trade-offs. First, data has to be structured as packets or even flits (smaller units of flow control), and additional control information has to be transmitted in addition to the payload: at least flit type, destination address, routing path, and source identifier. The packetization process is critical with respect to overall network latency, and several implementation trade-offs have been explored to determine the most efficient design points or to account for different network architectures: software versus hardware implementations [13] and synchronous versus asynchronous packetization [14, 15].

An important issue when creating packets or when dividing them up into flits is selecting packet length and flit size. Almost all NIs try to tailor the packet length to the specific network transaction to be carried out. For instance, short request packets are commonly used for burst read transactions, and in some networks single-flit packets are effectively deployed [16]. Another issue concerns the trade-off between packet length and initial transmission latency. Larger packets can be built waiting for the initiator core to deliver more data, while at the same time increasing packet delivery latency. In this scenario, starvation at the user/application level should be carefully avoided, and a flush signal may be implemented for this purpose, which temporarily overrides the packet accumulation threshold. The flush signal is controlled by processor cores and is a standard technique in modern communication protocols, such as *device transaction level* (DTL) [17], which has a similar flush signal, or AXI, which forces transmission of potentially buffered write data with an unbuffered write command. The flushing mechanism can also be applied to packets carrying flow control information. This is the case of cumulative credit packets which limit the amount of upstream signaling associated with end-to-end credit-based flow control. Eventually, credits can also be piggybacked on data packets. In general, since communication latency reduction is one of the biggest challenges in NoC design, such cumulative transmission mechanisms are usually deployed with extreme care.

Another degree of freedom when designing packet-switched NoCs is flit width selection. Smaller flits can be used to reduce buffering requirements at the switches, wiring congestion or data-path complexity. Less wires are also used in presence of link serialization techniques exploiting the high communication frequencies made available by optimized interconnects operating in the multi-GHz domain [18]. In contrast, larger flit widths inherently provide more bandwidth at the cost of larger building blocks and of place-and-route problems.

In all NoC prototypes, buffering resources account for a large fraction of NI (but also switch) area and power consumption. It can be interestingly observed that reducing the amount of sequential logic results either in performance drops (like in *best-effort* (BE) packet-switched networks [19]) and unpredictability (e.g., starvation-critical routing algorithms for buffer-less switches [20]) or in increased network protocol complexity (like in circuit-switching for circuit management or for end-to-end flow control [21–23]). Moreover, sequential logic represents the ultimate load of the clock distribution network, which therefore turns out to be the most power-hungry component of NoCs [24]. For this reason, NIs synchronizing clocked interface sockets with asynchronous NoCs have been proposed, trying to leverage the advantages of both design styles [15]. The NI role with respect to synchronization issues is critical even in presence of fully synchronous systems, since the advent of the NoC paradigm makes it possible to run the network at a higher speed with respect to the attached cores. Frequency decoupling is therefore a key NI task which significantly impacts its architecture.

This chapter covers the above NI design issues and architectural trade-offs in a structured way. After describing the main services offered by the NI, the main components of its architecture will be illustrated, namely the standard interface socket, the packetization stage, the buffering and flow control stage. For each component, besides addressing implementation issues and the trade-offs spanned by the configuration space, its impact on the overall system complexity and performance will also be captured. In fact, design choices made at the NI affect switching element design strategies and even system-level performance and power dissipation, thus making the point for communication-centric design [26]. Finally, a few representative NI architectures among those proposed in the open litera-ture will be illustrated, trying to capture the distinctive features of each of them. In particular, case studies of NIs for *quality of service* (QoS) oriented, BE, and asynchronous NoCs will be presented. More interestingly, for each solution a system-level perspective will be taken, pointing out the contri-bution of NIs to overall NoC area and power in real-life MPSoC platforms where state-of-the-art interconnects have been replaced by NoCs.

## 6.1  NI SERVICES

In order to have a comprehensive overview of the tasks carried out by NIs in NoC architectures, this section presents a classification of the services offered by NIs in categories that have different design challenges [27].

Implementation or not of a service and careful customization of chosen ones depends on the application needs, on the core and NoC protocol fea-tures, and is a platform-instance-dependent decision. Services span from session and transport-level services decoupling computation from commu-nication, up to lower-level ones such as packetization and clock-domain crossing.

### 6.1.1  Adaptation Services

These are the basic core wrapping services. Their role is to adapt the com-munication protocol of the component to the communication protocol of the network. Of course, the challenge here is to minimize performance loss from a latency viewpoint.

#### Core interfacing

This task is extremely critical in the context of highly integrated multipro-cessor SoCs. In fact, beyond implementing a high-performance physical connection between the NI and the corresponding core, the *layering* con-cept is traditionally applied [11, 28]. Layering naturally decouples system-processing elements from the system they reside in, and therefore enables design teams to partition a design effort into numerous activities that can proceed concurrently since they are minimally dependent. This can dramatically accelerate final product delivery schedules. Layering also nat-urally enables core reuse in different systems. Since an individual core's interface is independent of (decoupled from) the system, with the right core interface design, the core can remain unchanged as it is reused in subsequent system designs that support that interface. By selecting an industry standard interface, there is no added time for this reuse approach since all cores require such an interface.

**Packetization**

This is the very basic service the wrapper should offer: taking the incoming signals specifying processor core transactions, and building packets respective to the NoC communication protocol. This service should carefully optimize packet and flit size in order to achieve high-performance network operation and reduce implementation complexity of network building blocks. The amount of memory needed by the packetization service depends on the complexity of the packet size decision, on the packet structure, on the routing algorithm, and on the number of signals sampled at the interface with attached cores (e.g., high- or low-complexity interfaces with advanced or simple processor cores, respectively, bitwidth of transaction fields such as width of data and address links).

## 6.1.2  Clock Adaptation

Clock distribution is to become a hard issue in future SoCs. That is why future SoCs will probably be *locally synchronous and globally asynchronous* (GALS). Even if the clock frequency is the same over the chip, phase adaptation will be needed for communications, as in Refs [29, 30]. Also, NoCs are composed of rather simple elements and rely on path segmentation, and thus they can potentially run at higher frequencies in order to decrease the latency seen by the networked computation units. For example, SoCBus micronetwork designers expect it to run at 1.2 GHz [31], which is 4 or 5 times the running frequency of average cores. xpipes NoC has been showed to work at 885 MHz (post-layout characterization), and impact of this operating speed on system performance has been described in Ref. [25]. In general, the reason behind the high clock speed of NoC architectures is twofold. First, path segmentation by means of retiming stages at switch inputs and/or outputs. Latching switch outputs and/or sampling switch inputs ensures high operating speeds. Moreover, link pipelining can be used to change (increasingly worse at each technology node) link delay into latency [32–34]. Placing repeater stages at pre-defined distance in the link takes the effect of not bounding the link data introduction rate to the link delay. Second, a separate design and optimization of the data-path and of the control path can potentially lead to higher clock speed implementations. In some NoC synthesis examples, it has been showed that the data-path can run at a higher speed than the control path [22, 35] and that the critical path of a switch still lies in the arbitration logic and not in the interconnects [36]. One architecture exploiting this fact is the one in Ref. [35], where a flit is composed of multiple physical words (phits). Since the control path is slower than the data-path, a multi-phit flit allows the data-path to be kept busy while the switching control logic is taking a new decision. Alternatively, circuit switching also takes advantage of this decoupled design strategy [22].

### Bandwidth and latency guarantees

If the NoC implements dynamic latency and/or bandwidth guarantees, then the wrapper has to manage this service. This means for example to build and send virtual circuit set-up or tear-down packets, or to allocate multiple buffering resources and design complex packet schedulers in the wrapper in order to handle traffic with different QoS requirements or maintain connections. Processor cores should not directly handle this, but they should be able to require, through the interface communication protocol with the NoC wrapper, transactions with different associated QoS levels, unless the association of processor core transactions with service classes has been statically performed.

## 6.1.3    Network Services

They are the classical services that a transport layer is expected to implement in computer networks. Implementing them or not strongly depends on the features of the underlying network.

### Transactions ordering

In packet-switched networks implementing dynamic routing schemes, packets can potentially arrive unordered. This is most of the time not acceptable for on-chip communications because it raises a memory consistency issue. Wrappers must in this case reorder the transactions before forwarding them to the component. This typically requires a large amount of buffering resources. Moreover, source NIs can force a sequence of subsequent transactions to complete in-order or out-of-order. This can be accomplished by means of transaction identifiers grouping those transactions that should be completed in-order. If the transactions have a response phase and the network cannot guarantee in-order delivery, the NI has to reconstruct the right sequence of responses as they are passed to the connected cores.

### Reliable transactions

The on-chip communication medium has been considered until now as a reliable (error-free) medium. According to some studies [37] this is expected to be no longer true in future deep sub-micron technologies. The wrapper could be involved in providing reliable network transactions by inserting parity check bits in packet tails at the packetization stage or by implementing end-to-end error control. Even in case link-level error control is implemented, the NI-to-switch link should be made robust with respect to communication errors. This might require special features of the flow control mechanism. Further details in Chapter 4.

**Flow control**

When a given buffering resource in the network is full, there needs to be a mechanism to stall the packet propagation and to propagate the stalling condition upstream. The flow control mechanism is in-charge of regulating the flow of packets through the network, and of dealing with localized congestion. ACK/NACK, credit-based or on–off flow control are the most widely used strategies in NoC prototypes. The NI is involved with the generation of flow control signals exchanged with the attached switch. Moreover, packet flow control has to be carried out also with respect to the connected cores. In fact, when the network cannot accept new packets any more because of congestion, new transactions can be accepted from the connected core just at the same, provided the necessary amount of decoupling buffering resources is implemented in the NI. This mechanism allows to decouple (to a certain extent) core computation with its requests for non-blocking communication services. However, when buffers are full, the core behavior is impacted by the congestion in the network, since it has to be stalled if it requires further communication services.

## 6.1.4 Functional Services

These services add new functionality to the system. In principle, many alternative implementations do exist for these functions (software, dedicated hardware, and processor core modifications), but in some cases their implementation in the NI achieves the best performance and allows design reuse.

**Cache coherence**

This is a new issue brought by the introduction of NoCs. Indeed cache coherency can be achieved on a bus at very low cost by means of snoop devices, leveraging the shared nature of the communication medium. Cache coherence on a network is no longer an easy task because snooping is not possible. New protocols are therefore needed to allow the use of multiprocessor systems at low cost. This service may require cache modifications to allow access to their directories.

**Security**

In some sensitive SoCs (e.g., manipulating private information), the security of transactions is an issue. To deal with this need, the wrapper could offer to some cores an encryption service in order to prevent *Tempest*-like pirating (electromagnetic emanation analysis of the chip). Another security system would be to filter the target addresses, allowing only some communications. This could be useful to prevent a sensitive part of the chip from communicating with another part.

**Low power**

Many SoCs are battery-operated, and therefore exhibits tight power budgets. The power concern becomes even more critical in presence of network-centric systems. In fact, early NoC prototypes are showing a significant contribution of the NoC to the system power dissipation. Beyond electrical- and gate-level low-power design techniques (e.g., deployment of low-swing signaling, low-power flip-flops, clock gating), higher-level techniques are likely to achieve larger savings. For instance, switching off some components and waking them up is one well-known system-level power management technique that could be applied also to network building blocks (especially NI). The operating system could be in charge of managing component deactivation and reactivation. Power characterization and modeling of NoC components is still in its early stage, and will be devoted increasing efforts in the short term by the NoC community. In the next sections, architecture and design issues for the main sub-modules of a NI will be described in detail.

## 6.2   NI STRUCTURE

A generic template for the NI architecture is reported in Fig. 6.1. The structural view of this hardware block includes a front-end and a back-end sub-module. The network front-end implements a standardized point-to-point protocol allowing core reuse across several platforms. This solution allows core developers to focus on developing core functions, thus avoiding the need for advance knowledge regarding potential end-systems. The interface assumes the attributes of a *socket*, that is, an industry-wide well-understood attachment interface which should capture all signaling between the core and the system (such as data-flow signaling, errors, interrupts, flags, software flow control, and testing). A distinctive requirement for this standard socket is to enable the configuration of specific interface instantiations along a number of dimensions (bus width, data handshaking, etc.).

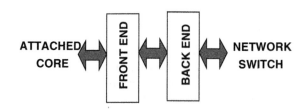

■ **FIGURE 6.1**

Generic NI architecture.

It is common practice to implement the front-end interface protocol so to keep backward compatibility with existing protocols, such as AMBA AXI [38], OCP [12], VCI [1, 2], and DTL [17]. This objective is achieved by using a transaction-based communication model [21], which assumes communicating cores of two different types: masters and slaves. Masters initiate transactions by issuing requests, which can be further split in commands and write data (corresponding to the address and write signal groups in AXI). Examples of commands are read and write. One or more slaves receive and execute each transaction. Optionally, a transaction can also involve a response issued by the slave to the master to return data or an acknowledgement of transaction completion (corresponding to the read data and write response groups in AXI). This request – response transaction model directly matches the bus-oriented communication abstractions typically implemented at core interfaces, thus reducing the complexity of core wrapping logic.

As a consequence, the NI front-end can be viewed as an implementation of the session layer in the ISO/OSI reference model. Traditionally, the session layer represents the user's interface to the network. It determines when the session is opened, how long it will be used and when to close it. Moreover, it controls the transmission of data during the session, supports security and name lookup enabling computers to locate each other. Flow control strategies and QoS negotiations can be also implemented at this layer. Similarly, in the NoC domain, the point-to-point communication protocol between processor core and NI front-end can be used to specify:

- The kind of transaction(s) that the core requests the targets to carry out.

- Session management. In the simplest solution, sessions are opened and closed. However, in case of a connection loss, the session protocol might try to recover it. If a connection is not used for a long period of time, it might be closed down and reopened for the next use. This happens transparently to the higher layers (i.e., to the application code the processor core is executing). The session layer might also provide synchronization points in the stream of exchanged packets.

- Requirements for end-to-end communication services that the lower layers should expose. They might concern communication reliability, predictability, ordering, and parallelism. For instance, a read transaction could be required to the memory, associated with the requirement of *guaranteed throughput* (GT) or best effort data delivery across the communication medium.

High-level communication services made available to processor cores at the session layer must be implemented by the transport layer, which is still unaware of the implementation details of the system interconnect. The transport layer relieves the upper layers from any concern with providing reliable, sequenced, and QoS-oriented data transfers. It is the first true and basic end-to-end layer [40]. For instance, the transport layer is in charge of establishing connection-oriented end-to-end communications [21, 41], as opposed to datagram-like communications [42], thus providing end-to-end control and information transfers with the QoS needed by the application program.

The transport layer provides transparent transfer of data between end nodes using the services of the network layer. These services, together with those offered by the link layer and the physical layer, are implemented in the NI back-end (Fig. 6.1). Data packetization and routing-related functions can be viewed as essential tasks performed by the network layer, and are tightly interrelated. In fact, source routing implementation requires routing table look-ups at the sender NI, and integration of routing information in packet headers. In contrast, when adaptive routing is used, the *look-up tables* (LUTs) are distributed at the network switches, and the packet just needs to indicate the destination node. Clearly, network layer implementation is strongly affected by the features of the adopted NoC. The NI also provides, in its back-end, data-link layer services. Primarily, communication reliability has to be ensured by means of proper error-detecting strategies and effective error recovery techniques. Moreover, flow control is handled at this layer, by means of upstream (downstream) signaling regulating data arrival from the processor core (data propagation to the first switch in the route), but also of piggybacking mechanisms and buffer/credit flushing techniques. Finally, the physical channel interface to the network has to be properly designed. NoCs have distinctive challenges in this domain, consisting of clock-domain crossing, high-frequency link operation, low-swing signaling, and noise-tolerant communication schemes.

## 6.3    EVOLUTION OF COMMUNICATION PROTOCOLS

This section focuses on the point-to-point communication protocol between an NI and the connected processor cores. This kind of protocol specifications is quite recent, and represents the latest evolution stage of industry-wide communication protocols for multiprocessor SoCs. Interestingly, this stage has introduced a paradigm shift in protocol specifications that can potentially speed up the development and industrial adoption of NoCs.

Let us follow the evolution of system interconnect protocols by observing the features introduced by successive releases of AMBA protocol specifications [43]. The family of AMBA protocols represents the *de facto* standard for the design of communication architectures for high-performance embedded microcontrollers.

AMBA AHB is one of the first protocol specifications released by ARM, and supports the efficient connection of processors, on-chip memories, and off-chip external memory interfaces with low-power peripheral macrocell functions. AMBA AHB assumes that the system backbone consists of a shared communication resource connecting multiple system cores. The structure of an AHB-compliant bus is reported in Fig. 6.2(a).

This bus specification defines two split and unidirectional data links (one for reads and one for writes), but only one of them can be active at any time. Only one bus master can own the data wires at any time, preventing the multiplexing of requests and responses on the interconnect signals. Transaction pipelining (i.e., split ownership of data and address lines) is supported to provide for higher throughput. In practice, each transaction consists of an address and a data phase. While the data phase of the $i$th transfer is in progress, the address for the $i + 1$th transfer can be anticipated on the address lines. However, address sampling is only allowed when the data phase completes successfully, and if this latter takes a few cycles to complete, the configuration of the address lines is frozen and their sampling postponed as well. As a consequence, the AHB-compliant transaction pipelining cannot be considered as a means of allowing multiple outstanding transactions.

Bursts are supported by AHB masters and arbiter as a way to amortize arbitration time, but AHB slaves do not have a native burst notion. Incrementing and wrapping bursts can be carried out, and the non-posted paradigm for write transactions is implicitly assumed (the "ready" signal driven by the slave and sampled by the master indicates successful completion of the write transfer). An interesting feature is represented by the native support for bursts of undefined length.

Based on the AHB specification, it is not possible to directly connect a master to a slave interface. In fact, both kinds of interfaces include signals that are connected to internal bus building blocks, such as the request-grant handshaking signals which connect master interfaces to the arbiter. Therefore, the internal bus architecture is directly exposed to the core interface. As a consequence, the bus architecture is instance-specific, since the bus components have to be modified and re-instantiated if the number (or the characteristics) of communicating actors is changed.

As the level of system integration keeps increasing, new protocol specifications and system-level interconnection schemes have been defined within the AMBA framework. The objective is twofold. On one hand, topology evolutions are required to make a larger bandwidth available to communicating cores than that offered by simple shared busses. On the other hand, more advanced and efficient protocols are needed in order to better exploit the bandwidth offered by the communication medium. In fact, the fraction of physical maximum bus bandwidth that can be effectively deployed by the initiators depends on protocol efficiency, measured

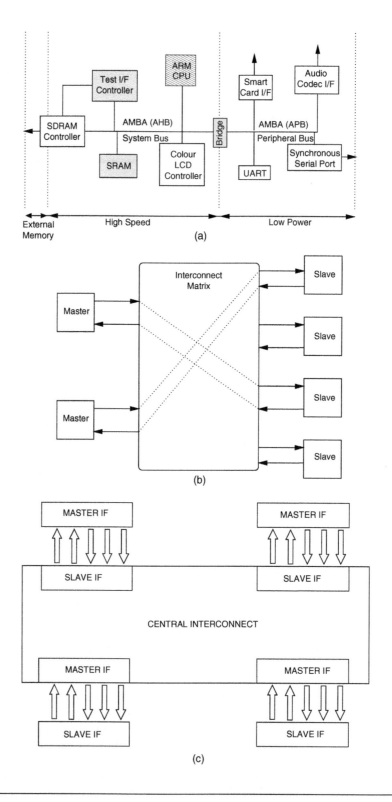

■ **FIGURE 6.2**

Evolution of AMBA family of communication architectures.

by the percentage of useful data transfers over the total number of bus busy cycles. Protocol efficiency is impacted by metrics such as rearbitration time, contention-related waiting time, recovery time from transfer errors, or inefficient bus occupancy due to slow slaves.

Multi-layer AHB is the specification of an interconnection scheme that overcomes the limitations of shared busses, and therefore represents a topology evolution. It enables parallel access paths between multiple masters and slaves in a system by using a more complex interconnection matrix and gives the benefit of increased overall bus bandwidth and a flexible system architecture (see Fig. 6.2(b)). Interestingly, master and slave modules compliant with the standard AHB protocol can be used without the need for modifications within the multi-layer approach. Further, each AHB layer which is connected at the inputs of the interconnection matrix can be simplified when it features only one master. In this case, no arbitration nor split/retry transaction support is required, and a lightweight AHB-Lite protocol can be used.

The only hardware that has to be added to the standard AHB transport infrastructure is the multiplexer block to connect the multiple masters to the peripherals. The key point about multi-layer AHB architecture is that more physical communication resources are made available to allow for parallel communication flows. Therefore, shared communication resources (where competing bus access requests need to be serialized by means of a centralized arbitration mechanism) tend to be replaced by partial or full crossbars and, in the long run, by networks of interconnections to provide for more scalability.

AMBA 3.0 (AXI) is the latest generation of AMBA interfaces, and represents an advanced interconnect protocol. AMBA AXI builds upon the concept of point-to-point connection, as depicted in Fig. 6.2(c). It exhibits complex features, such as the support for multiple outstanding transactions (with out-of-order or in-order delivery selectable by means of transaction IDs), for burst transactions with only the first address issued, for register insertion for timing closure transparent to the protocol, and for time interleaving of traffic toward different masters on internal data lanes. Five different logical monodirectional channels are provided in AXI interfaces, and activity on them is largely asynchronous and independent. The main differences between AMBA 2.0 and AMBA 3.0 are illustrated in Fig. 6.3, as reported in Ref. [58].

Since AMBA AXI defines a point-to-point master–slave interface, it can be used to connect (i) a communication initiator to a bus, (ii) a communication target to a bus, or (iii) directly an initiator to a target. In fact, the connection between any two devices translates into the instantiation of a master interface and of the symmetrical slave interface. The interface definition enables a variety of different interconnect implementations, which are not addressed by the AXI specification. System designers can therefore opt for the following approaches to central interconnect design. Single shared address and data lanes could be used, or alternatively multiple

| AMBA AHB | AMBA AXI |
|---|---|
| FIXED PIPELINE FOR ADDRESS AND DATA TRANSFERS | FIVE INDEPENDENT CHANNELS FOR ADDR/DATA AND RESPONSE |
| BIDIRECTIONAL LINK WITH COMPLEX TIMING RELATIONSHIPS | EACH CHANNEL IS UNIDIRECTIONAL EXCEPT FOR SINGLE HANDSHAKE FOR RETURN PATH |
| HARD TO ISOLATE TIMING | REGISTER SLICES ISOLATE TIMING |
| LIMITS FREQUENCY OF OPERATION | FREQUENCY SCALES WITH PIPELINING |
| INEFFICIENT ASYNCHRONOUS BRIDGES | HIGH-PERFORMANCE ASYNCHRONOUS BRIDGING |
| SEPARATE ADDRESS FOR EVERY DATA ITEM | BURST BASED – ONE ADDRESS PER BURST |
| ONLY ONE TRANSACTION AT A TIME | MULTIPLE OUTSTANDING TRANSACTIONS |
| FIXED PIPELINE FOR ADDRESS AND DATA | OUT-OF-ORDER DATA |
| ONLY ONE TRANSACTION AT A TIME | SIMULTANEOUS READS AND WRITES |
| DOES NOT NATIVELY SUPPORT THE ARM V6 ARCHITECTURE | NATIVE SUPPORT FOR UNALIGNED AND EXCLUSIVE ACCESSES |
| NO SUPPORT FOR SECURITY | NATIVE SECURITY SUPPORT |

■ **FIGURE 6.3**

Feature comparison: AMBA 2 AHB protocol versus AMBA 3 AXI protocol.

data lanes could be implemented. This stems from the consideration that in most systems, the address channel bandwidth requirement is significantly less than the data channel bandwidth requirement. Such systems can achieve a good balance between system performance and interconnect complexity by using a shared address bus with multiple data busses to enable parallel data transfers. To the limit, designers could opt for a multi-layer approach, with multiple address and data busses [58].

## 6.3.1   Considerations

The mixing of various traffic types in the same SoC design (application convergence), the increasing integration density, the interconnect delay bottleneck, and time-to-market pressures (driving most designs to make heavy use of synthesizable RTL rather than manual layout, and thus restricting the choice of available bus implementation solutions) have driven the evolution of bus architectures. New *de facto* standards at first introduced split and retry techniques, removal of tri-state buffers and multi-phase clocks, transaction pipelining, and various attempts to define standard communication sockets.

However, in many cases processing units reuse strategies and associated compatibility requirements have slowed down the introduction of bus evolutions driven by technology changes. On the other hand, the support for processing units with growing complexity has required many changes in the bus implementation, but more importantly in the bus interface, with major impact on computation units' reusability and new designs.

The real issue is that busses do not decouple activities at different levels of abstraction, such as transaction (or communication service) specification, transport protocol definition, arbitration algorithms execution, and physical layer implementation. This is the key reason for bus architectures cannot closely follow process evolution, nor system architecture evolution. Changes to bus physical implementation might have serious ripple effects upon the implementation of higher-level bus behaviors, although this has to be assessed case by case. Historically, the replacement of tri-state techniques with multiplexers has had little effect upon the transaction levels. Conversely, the introduction of flexible pipelining to ease timing closure has had massive effects on all bus architectures up through the transaction level. Similarly, system architecture changes may require new transaction types or transaction characteristics. Recently, such new transaction types as exclusive accesses have been introduced near simultaneously within OCP 2.0 and AMBA AXI socket standards. Out-of-order response capability is another example. Unfortunately, such evolutions typically impact the intended bus architectures down to the physical layer, if only by addition of new wires or op-codes. Thus, the bus implementation must be redesigned. In practice, bus architects must always make compromises between the various driving forces, and resist change as much as possible.

In the data communications space, LANs and WANs have successfully dealt with similar problems by employing a layered architecture. By relying on the OSI model, upper and lower layer protocols have independently evolved in response to advancing transmission technology and transaction-level services. The decoupling of communication layers using the OSI model has successfully driven commercial network architectures, and enabled networks to follow very closely both physical layer evolutions (from the Ethernet multi-master coaxial cable to twisted pairs, ADSL, fiber optics, wireless, etc.) and transaction-level evolutions (TCP, UDP, streaming voice/video data). This has produced incredible flexibility at the application level (web browsing, peer-to-peer, secure web commerce, instant messaging, etc.), while maintaining upward compatibility (old-style 10 Mb/s or even 1 Mb/s Ethernet devices are still commonly connected to LANs).

Following the same trends, communication protocol developers for SoCs have started to provide specifications only for point-to-point interfaces. This has been the paradigm change introduced by AMBA AXI with respect to AMBA AHB. Signals connecting a master interface to a slave one are not specific for a particular bus implementation, and this allows to directly connect a master to a slave interface. Of course, the slave interface can be implemented by a bus, and can therefore represent a socket for the plug-and-play connection of communication initiators to the bus, provided they implement the symmetric master interface. Similarly, communication targets can be connected to the bus as well, by implementing slave interfaces matching a corresponding master interface on the bus.

A bus-independent standard interface specification has two relevant benefits. First, there are no constraints limiting the evolution of bus topology. The definition of the multi-layer architecture by ARM was an early example thereof: a shared bus can be replaced by a crossbar-like interconnection matrix without the need to change the AHB protocol. The protocol can be sometimes even simplified, as showed by the AHB-Lite solution. In the long run, this paves the way for the industrial adoption of NoCs, since they represent the forward-looking infrastructure for scalable on-chip communication that could replace pre-existing system interconnects while keeping the same interface toward connected hardware blocks. As an example, many NoC prototypes, such as Aethereal [21] or xpipes [44], implement the OCP or the AXI (or even proprietary protocols such as DTL [17]) interface at their boundaries.

Second, since the master–slave standard interface abstracts from bus/network internals, it just specifies the transactions and the high-level communication services that an initiator requires. The associated information are not related to the transport, arbitration, or physical-level mechanisms implemented in the system interconnect, but are those required by the communicating peer to carry out the specific transaction (e.g., address, data, and burst control signals for memory accesses). Therefore, such high-level protocols are positioned at the session or transport layer in the ISO/OSI reference model, since the transport layer is the first layer to be independent of network implementation.

Unfortunately, most point-to-point interface specifications assume bus semantics. For instance, they assume that two consecutive transactions issued by the same master and addressing the same slave device will be delivered in order. For NoCs, this might not be the case, for instance in those cases where dynamic adaptive routing is implemented. This example points out the need, in some cases, to implement an adapter with the task of bridging the gap between point-to-point and network protocol mismatches.

More in general, the communication abstractions modeled in the interface protocol are extremely critical for effective decoupling. On one hand, they have to match the semantics of high-level communication services required by communicating cores. On the other hand, if the semantic gap between point-to-point protocol and bus/network protocol increases a lot, the interfacing logic becomes overly complex and the semantic translation of transactions of one protocol into the transactions of the other one might become inefficient or even impossible.

In other cases, the interface logic is required to fully exploit distinctive NoC features. For example, Aethereal NI [21] introduces connection-based communication which does not only relax ordering constraints (as for busses), but also enables new communication properties, such as end-to-end flow control based on credits, or GT. All these properties can be set for each connection individually.

The following section illustrates the features of AMBA AXI and OCP protocols more in detail, then the main mismatches that may need to be addressed in NIs between interface and network protocols are described. The interface protocols are hereafter covered taking the designer viewpoint, with many practical details to better understand the applicability to real-life systems. For a more comprehensive description, the interested reader should refer to the open specification of each protocol.

## 6.4 POINT-TO-POINT COMMUNICATION PROTOCOLS

### 6.4.1 AXI Overview

The AMBA AXI protocol family defines a new set of on-chip interface protocols for SoC designs, and is interoperable with existing bus technology defined in the AMBA 2 specification. The AXI protocol is an advanced microprocessor system bus interface which aims at meeting the interface requirements of a wide range of components and targets low-latency and high-bandwidth designs.

The distinctive features of the AXI protocol include [58]:

- Separate address/control and data phases.

- Support for unaligned data transfers using byte strobes.

- Burst-based transactions with only start address issued.

- Separate read and write data channels to enable low-cost *direct memory access* (DMA).

- Ability to issue multiple outstanding addresses.

- Out-of-order transaction completion.

- Easy addition of register stages to provide timing closure.

- Protocol includes optional extensions that cover signaling for low-power operation.

In practice, the combination of these features brings the following advantages to system interconnects based on AMBA AXI:

- Independently acknowledged address and data channels.
- Out-of-order completion of bursts.
- Exclusive access (atomic transaction).
- System-level cache support.
- Access security support.
- Unaligned address and byte strobe.
- Static burst, which allows bursts to FIFO memories.
- Low-power mode.

The AXI protocol is burst based. Bursts are managed in quite a different way with respect to AMBA AHB. For instance, address information can be issued ahead of the actual data transfer. Combining this feature with the multiple outstanding transaction support and the out-of-order completion option, we can get the waveform behavior of Fig. 6.4, where the enhanced

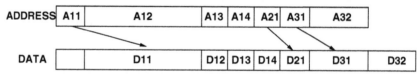

MULTIPLE ADDRESSES FOR BURST
FIXED ADDRESS TO DATA PIPELINE
DATA IN THE SAME ORDER AS ADDRESS

(a)

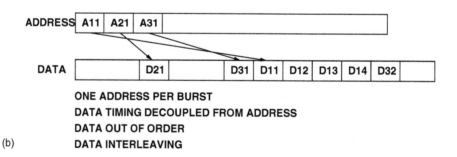

ONE ADDRESS PER BURST
DATA TIMING DECOUPLED FROM ADDRESS
DATA OUT OF ORDER
DATA INTERLEAVING

(b)

■ FIGURE 6.4

Feature comparison: burst transaction completion with AMBA protocols. Burst transactions is the (a) AMBA AHB and (b) AMBA AXI protocol.

flexibility of AXI can be observed. The AXI protocol includes ID signals allowing out-of-order transactions: all transactions with a given ID must be ordered, but there is no restriction on the ordering of transactions with different IDs.

AMBA AXI differs significantly from previous AMBA protocols also by the introduction of channels. Each of the five independent channels consists of a set of information signals and uses a two-way VALID and READY handshake mechanism. The information source uses the VALID signal to show when valid data or control information is available on the channel. The destination uses the READY signal to show when it can accept the data. Both the read data channel and the write data channel also include a LAST signal to indicate when the transfer of the final data item within a transaction takes place.

Read and write transactions each have their own address channel. The appropriate address channel carries all of the required address and control information for a transaction. The Read data channel conveys both the read data and any read response information from the slave back to the master. The Read data channel includes the data bus, which can be 8, 16, 32, 64, 128, 256, 512, or 1024 bits wide and a read response indicating the completion status of the read transaction. The Write data channel conveys the write data from the master to the slave. The Write data channel includes the data bus, which can be 8, 16, 32, 64, 128, 256, 512, or 1024 bits wide, and 1-byte lane strobe for every eight data bits, which indicates which bytes of the data bus are valid. Finally, a write response channel provides a way for the slave to respond to write transactions. All write transactions use completion signaling. The completion signal occurs once for each burst, not for each individual data transfer within the burst.

Each AXI channel transfers information in only one direction, and there is no requirement for a fixed relationship between the various channels. In practice, activity on the different channels is mostly asynchronous (e.g., data for a write can be pushed to the write channel before or after the write address is issued to the address channel), and can be parallelized, allowing multiple outstanding read and write requests. All these features are very important, because they enable the insertion of a register slice in any channel, at the cost of an additional cycle of latency, without violating protocol specification. Register slices can therefore be used to pipeline a connection for higher speeds, or to enable the use of multiple clock domains for low power.

The main innovation introduced by AMBA AXI lies certainly in the fact that it is a protocol definition for the description of interfaces. Therefore, it can be used to connect (i) a master with an interconnect, (ii) a slave with an interconnect, and (iii) a master directly with a slave. The design of the central interconnect can then be tailored to the specific system needs. In most systems, the address channel bandwidth requirement is significantly less than for the data channel. Such systems can achieve a good balance between system performance and interconnect complexity by using a shared address bus with multiple data busses to enable parallel data transfers. More in general, most systems use one of three interconnect approaches [58]:

1. *Shared Address and Single Data busses* (SASD)

2. *Shared Address buses and Multiple Data busses* (SAMD)

3. *Multi-layer with Multiple Address busses and Multiple Data busses* (MAMD)

With SASD, one master and slave can be active per channel. When one master sends an address to one slave, no other master can use the address bus. This is similar to the structure of the AMBA AHB specification. With SAMD, one master and slave can be active per address channel. Multiple masters and slave pairs can be active on other channels (read/write data and response channel). For example, if master1 sends write data to slave1, master2 can send write data to slave2 at the same time and does not have to wait for the bus being freed. The number of active pairs is design dependent. Obviously the parallel nature of this operation significantly improves the performance of the subsystem. With MAMD, multiple pairs can be active on the address channels as well. This provides the maximum of interconnect flexibility and yields the highest-performance subsystem interconnect architecture. This topology is also the most complex scenario to verify as multiple masters and slaves can be active at any time. Verifying the interaction between all ports will be critical to the successful operation of the resulting system.

### Advanced options

AXI provides support for system-level caches and other performance enhancing components by means of cache information signals. These signals provide additional information about how the transaction can be processed. In particular, ARCACHE and AWCACHE signal provides the bufferable, cacheable, and allocate attributes of the transaction. For instance, when ARCACHE[1] (the cacheable bit in the read address channel) is high, it means that the transaction at the final destination does not have to match the characteristics of the original transaction. In other words, a location can be pre-fetched or can be fetched just once for multiple read transactions. To determine if a transaction should be cached, this bit should be used in conjunction with the *read allocate* (RA) bit ARCACHE[2]. When RA is high, it means that if the transfer is a read and it misses in the cache, then it should be allocated.

The AXI protocol does not determine the mechanism by which buffered or cached data reaches its destination. For example, a system-level cache might have a controller to manage cleaning, flushing, and invalidation of cache entries. Another example is a bridge containing a write buffer, which might have control logic to drain the buffer if it receives a non-bufferable write with a matching transaction ID.

AXI also supports complex system designs, where it can be necessary to both the interconnect and other devices in the system to provide protection against illegal transactions.

To enable the implementation of atomic access primitives, the ARLOCK[1:0] or AWLOCK[1:0] signals provides exclusive access and locked access. The exclusive access mechanism enables the implementation of semaphore-type operations without requiring the bus to remain locked to a particular master for the duration of the operation. The advantage of exclusive access is that semaphore-type operations do not impact either the critical bus access latency or the maximum achievable bandwidth. The slave must have additional logic to support exclusive access. The AXI protocol provides a fail-safe mechanism to indicate when a master attempts an exclusive access to a slave that does not support it.

When the ARLOCK[1:0] or AWLOCK[1:0] signals for a transaction show that it is a locked transfer then the interconnect must ensure that only that master is allowed access to the slave region until an unlocked transfer from the same master completes. The arbiter within the interconnect is used to enforce this restriction. Locked accesses require that the interconnect prevents any other transactions occurring while the locked sequence is in progress and can therefore have an impact on the interconnect performance. It is recommended that locked accesses are only used to support legacy devices.

Finally, the low-power clock control interface consists of the following signals: a signal from the peripheral indicating when its clocks can be enabled or disabled, and two handshake signals for the system clock controller to request exit or entry into a low-power state.

### 6.4.2  OCP Overview

OCP is a freely available, bus-independent protocol that meets the core-centric considerations discussed above. As a highly configurable interface, it comprises a continuum of protocols that share a common definition. OCP explicitly supports side-band signals via optional extensions to the basic OCP data set. These side-band signals include, for example, reset, interrupt, error, control/status information, etc. In addition, a generic flag bus accommodates any unique core signaling needs. An optional OCP test interface extension supports scan, JTAG, and clock control, enabling core debug and manufacturing test. System designers can therefore tailor a specific OCP configuration to match core requirements exactly. Through configuration procedures, OCP can support simple, low-performance cores with very simple and frugal OCP interfaces, as well as complex, high-performance cores with more complex interfaces. The system integrator

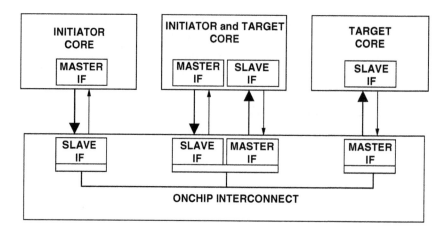

■ **FIGURE 6.5**

OCP core interfaces.

is free to choose the on-chip interconnect that best suits the system requirements of the application, then effectively *wraps* that interconnect to present OCP interfaces to the cores.

The signal names are part of the OCP protocol. In short, an OCP connection has a Master entity and a Slave entity. Some simple notes on OCP nomenclature and the protocol:

- The Master drives all signals having a name starting with the letter M.

- The Slave drives all signals having a name starting with the letter S.

- There are simple handshakes for the protocol.

- The master and slave can both assert flow control.

- All transfers and signals are synchronous to the rising edge of OCP clock.

Let us try to capture the features of the OCP protocol by means of three implementation case studies. They are reported in Fig. 6.5, and are now described into the details [11].

### Bus bridge

A bus bridge might interconnect a PCI, USB, or other bus standard to OCP. The controller would have an external (to the SoC) PCI or USB interface and the internal SoC interface would be OCP. Bus bridges usually act as both a master and a slave on the internal SoC interconnect. The master sends the bus traffic to the desired location and the slave writes or reads the bus bridge internal control or status registers.

The slave consists of a simple OCP interface and most likely needs a few side-band signals. A likely slave core OCP signal set for this example is:

- **MCmd**   Master Command (e.g., read/write)

- **MAddr**   Master Address (up to 32 bits)

- **MData**   Master Data (write data; 8, 16, 32, 64, 128 bits wide)

- **SCmdAccept**   Slave Command Accept

- **SResp**   Slave Response

- **SData**   Slave Data (read data, must be the same size as MData)

- **SError**   Slave Error, error from the bridge

- **SInterrupt**   Slave Interrupt, interrupt from the bridge

- **Control**   Control bits for the bus bridge

- **Clk**   The Clock signal

- **Reset_N**   The Reset signal

If more than one interrupt signal were needed, SFlags could provide up to eight additional interrupts. The bus bridge master might have the following signals:

- **MCmd**   Master Command

- **MAddr**   Master Address (up to 32 bits)

- **MData**   Master Data (write data; 8, 16, 32, 64, 128 bits wide)

- **MBurst**   Master Burst

- **SCmdAccept**   Slave Command Accept

- **SResp**   Slave Response

- **SData**   Slave Data (read data, must be same size as MData)

- **Clk**   The Clock signal

- **Reset_N**   The Reset signal

With PCI, optional OCP thread signals might enhance the interface. This would allow the interface to support concurrency and out-of-order processing of transfers. The optional OCP complex extension signals support multiple threads. Transactions within different threads have no ordering requirements, and can be processed out of order. Within a single thread of data flow, all OCP transfers must remain ordered. Threads for a PCI interface might be useful to access different memory and I/O regions.

- **MThreadID**  Master Thread Identifier (up to 16 different threads)
- **SThreadID**  Slave Thread Identifier (up to 16 different threads)

### Processor interface

A processor interface usually only requires an OCP master. The signals would be similar to those of the master interface of the bridge but would normally include byte enable signals for less than single-word transfers. A likely core OCP signal set for this example is:

- **MCmd**  Master Command
- **MAddr**  Master Address (up to 32 bits)
- **MData**  Master Data (write data; 8, 16, 32, 64, 128 bits wide)
- **MBurst**  Master Burst
- **MByteEn**  Master Byte Enable
- **SCmdAccept**  Slave Command Accept
- **SResp**  Slave Response
- **SData**  Slave Data (read data, must be same size as MData)
- **SError**  Slave Error, input to the processor
- **SInterrupt**  Slave Interrupt, usually the NMI pin
- **SFlag**  Slave Flags, other interrupts to the processor (up to eight flags)
- **Clk**  The Clock signal
- **Reset_N**  The Reset signal

Some newly available processors support concurrent instruction and data cache-miss fetches. OCP threads directly support this. Hence, the following signals could be needed to enhance concurrent memory operations for such processors:

- **MThreadID**  Master Thread Identifier (up to 16 different threads)
- **SThreadID**  Slave Thread Identifier (up to 16 different threads)
- **MThreadBusy**  Master Thread Busy
- **SThreadBusy**  Slave Thread Busy

The master and slave thread busy signals permit each thread flow control. Processors might have several fetches outstanding, but not the resources to handle all of them simultaneously. The thread busy signals therefore enable an OCP master processor or the target slave to control the transfer flows as necessary.

### Memory subsystem

A memory subsystem may interface to DRAM, DDR, SRAM, or FLASH. The OCP signals require some OCP complex extensions to maintain high-bandwidth utilization. The memory subsystem is probably multi-threaded, to service multiple memory banks. The memory controller will also have OCP simple extensions such as burst and byte enable to service requests efficiently. A likely core OCP signal set for this example is:

- **MCmd**   Master Command (e.g. read/write)
- **MAddr**   Master Address (up to 32 bits)
- **MData**   Master Data (write data) (8, 16, 32, 64, or 128 bits wide)
- **MBurst**   Master Burst
- **MByteEn**   Master Byte Enable
- **SCmdAccept**   Slave Command Accept
- **SResp**   Slave Response
- **SData**   Slave Data (read data, must be the same size as MData)
- **MThreadID**   Master Thread Identifier (up to 16 different threads)
- **SThreadID**   Slave Thread Identifier (up to 16 different threads)
- **MThreadBusy**   Master Thread Busy
- **SThreadBusy**   Slave Thread Busy
- **Clk**   The Clock signal
- **Reset_N**   The Reset signal

The memory subsystem might also utilize a larger bit width on MData and SData than the memory to which it interfaces. This makes each system interconnect transfer maximally efficient.

### Signals common to each example

Each of the preceding examples can have scan and JTAG signals. These test and scan signals might be common to each core, but the number of scan chains might differ. OCP allows test structures as part of the interface, not as a separate entity. Having scan and JTAG access to each of the example cores might require the following signals:

- **ScanCtl**   Scan Control
- **ScanIn**   Scan In (up to 256 scan-in ports or chains)
- **ScanOut**   Scan Out (up to 256 scan-out ports or chains)
- **ClkByp**   Clock Bypass, use Test Clock instead of normal clock
- **TestClk**   Test Clock
- **TCK JTAG**   Test Clock, uses IEEE 1149.1 definitions for all signals
- **TDI JTAG**   Test In
- **TDO JTAG**   Test Out
- **TMS JTAG**   Test Mode Select
- **TRST_N**   JTAG Reset

### 6.4.3   Bus versus Network Semantics

Introducing networks as on-chip interconnects radically changes the communication as compared to direct interconnects such as busses or switches. This is because of the multi-hop nature of a network, where communication modules are not directly connected, but separated by one or more network nodes. This is in contrast with the prevalent existing interconnects (i.e., busses) where modules are directly connected. The implications of this change reside in the arbitration (which must change from centralized to distributed), and in the communication properties (e.g., ordering or flow control).

This subsection outlines the main differences between NoCs and busses [45] that might generate semantic translation problems of the transactions at the NI. Addressing this gap is key for an efficient decoupling of computation and communication in NoC-based MPSoCs. The complexity and performance of an NI depends on how the following differences are handled.

**Programming model**

The programming model of a bus typically consists of load and store operations which are implemented as a sequence of primitive bus transactions. Bus interfaces typically have dedicated groups of wires for command, address, write data, and read data [43].

A bus is a resource shared by multiple cores. Therefore, before using it, cores must go through an arbitration phase, where they request access to the bus, and block until the bus is granted to them. A bus transaction involves a request and possibly a response. Modules issuing requests are called masters, and those serving requests are called slaves. If there is a single arbitration for a request–response pair, the bus is called non-split. In this case, the bus remains allocated to the master of the transaction until the response is delivered, even when this takes a long time. Alternatively, in a split bus, the bus is released after the request to allow transactions from different masters to be initiated. However, a new arbitration must be performed for the response such that the slave can access the bus.

For both split and non-split busses, both communication parties have direct and immediate access to the status of the transaction. In contrast, network transactions are one-way transfers from an output buffer at the source to an input buffer at the destination that causes some action at the destination, the occurrence of which is not visible at the source. The effects of a network transaction are observable only through additional transactions. A request–response type of operation is still possible, but requires at least two distinct network transactions. Thus, a bus-like transaction in an NoC will essentially be a split transaction.

### Transaction ordering

Traditionally, all transactions on a bus are ordered (Peripheral VCI [2], AMBA [39], DTL [17], or CoreConnect PLB and OPB [4]). This is possible at a low cost, because the interconnect, being a direct link between the communicating cores, does not reorder data. However, on a split bus, a total ordering of transactions on a single master may still cause performance penalties, when slaves respond at different speeds. To solve this problem, recent extensions to bus protocols allow transactions to be performed on connections.

Ordering of transactions within a connection is still preserved, but between connections there are no ordering constraints (e.g., OCP [12] or Basic VCI [2]). A few of the bus protocols allow out-of-order responses per connection in their advanced modes (e.g., Advanced VCI [2]), but both requests and responses arrive at the destination in the same order as they were sent. In an NoC, ordering becomes weaker. Global ordering can only be provided at a very high cost due to the conflict between the distributed nature of the networks, and the requirement of a centralized arbitration necessary for global ordering. Even local ordering, between a source–destination pair, may be costly. Data may arrive out of order if it is transported over multiple routes. In such cases, to still achieve an in-order delivery, data must be labeled with sequence numbers and reordered at the destination before being delivered.

### Atomic chains of transactions

An atomic chain of transactions is a sequence of transactions initiated by a single master that is executed on a single slave exclusively. That is, other masters are denied access to that slave, once the first transaction in the chain claimed it. This mechanism is widely used to implement synchronization mechanisms between master modules (e.g., semaphores).

On a bus, atomic operations can easily be implemented, as the central arbiter will either (a) lock the bus for exclusive use by the master requesting the atomic chain or (b) not granting access to a locked slave. In the former case, the time resources are locked is shorter because once a master has been granted access to a bus, it can quickly perform all the transactions in the chain (no arbitration delay is required for the subsequent transactions in the chain). Consequently, the locked slave and the bus can be opened up again in a short time. This approach is used in AMBA and CoreConnect. In the latter case, the bus is not locked, and can still be used by other modules, however, at the price of a longer locking duration of the slave. This approach is used in VCI and OCP.

In an NoC, where the arbitration is distributed, masters do not know that a slave is locked. Therefore, transactions to a locked slave may still be initiated, even though the locked slave cannot accept them. Consequently, to prevent deadlock, these other transactions must be either dropped or transactions in the atomic chain must be able to bypass them to be served. Moreover, the duration a module is kept locked is much longer in case of NoCs, because of the higher latency per transaction.

### Deadlock

In busses, deadlock is generally not an issue. Deadlock can still occur at the application level (e.g., an atomic chain of transactions that locks the bus, but never unlocks it), but this is not caused by the interconnect itself. In a network, deadlock becomes a more important issue, and special care has to be taken in the network design to avoid deadlock. Deadlock is mainly caused by cycles in the buffers. To avoid deadlock, either network nodes must drop packets when their buffers are filled or routing must be cycle-free.

A second cause of deadlock is atomic chains of transactions. The reason is that while a module is locked, the queues storing transactions may get filled with transactions outside the atomic transaction chain, blocking the access of the transaction in the chain to reach the locked module. If atomic transaction chains must be implemented (to be compatible with processors allowing this, such as MIPS), the network nodes should be able to filter the transactions in the atomic chain, or be allowed to drop those blocking them.

### Media arbitration

An important difference between busses and NoCs is in the medium arbitration scheme. In a bus, master modules request access to the interconnect, and the arbiter grants the access for the whole interconnect at once. Arbitration is centralized, as there is only one arbiter component. It is also global as all the requests as well as the state of the interconnect are visible to the arbiter. Moreover, when a grant is given, the complete path from the source to the destination is exclusively reserved.

In a non-split bus, arbitration takes place once when a transaction is initiated. As a result, the bus is granted for both request and response. In a split bus, requests and responses are arbitrated separately.

In an NoC arbitration is also necessary, as it is a shared interconnect. However, in contrast to busses, the arbitration is distributed, because it is performed at every router, and is based only on local information. Arbitration of the communication resources (links, buffers) is performed incrementally as the request or response advances.


### Destination name and routing

For a bus, the command, address, and data are broadcasted on the interconnect. They arrive at every destination, only one of which activates based on the broadcasted address, and executes the requested command. This is possible because all modules are directly connected to the same bus.

In an NoC, it is not feasible to broadcast information to all destinations, because it must be copied to all routers and NIs. This floods the network with data. The address is better decoded at the source to find a route to the destination module. A transaction address has, therefore, two parts:

1.  a destination identifier,
2.  an internal address at the destination.


### Latency

Transaction latency is caused by two factors: the access time to the bus, which is the time until the bus is granted, and the latency introduced by the interconnect to transfer the data. For a bus, where the arbitration is centralized, the access time is proportional to the number of masters connected to the bus. The transfer latency itself typically is constant and relatively low, because the modules are linked directly. However, the speed of transfer is limited by the bus speed, which is relatively low.

In an NoC, arbitration is performed at each router for the following link. The access time per router is small. Both end-to-end access time and transport time increase proportionally to the number of hops between master and slave. However, network links are unidirectional and point to point and, hence, can run at higher frequencies than busses, thus lowering the latency. From a latency prospective, using a bus or a network is a trade-off between the number of modules connected to the interconnect (which affects access time), the speed of the interconnect, and the network topology.

### Data format

In most modern bus interfaces the data format is defined by separate wire groups for the transaction type, address, write data, read data, and return acknowledgments/errors (e.g., VCI, OCP, AMBA, DTL, or CoreConnect). This is used to pipeline transactions. For example, concurrently with sending the address of a read transaction, the data of a previous write transaction can be sent, and the data from an even earlier read transaction can be received. Moreover, having dedicated wire groups simplifies the transaction decoding; there is no need for a mechanism to select between different kinds of data sent over a common set of wires.

Inside a network, there is typically no distinction between different kinds of data. Data is treated uniformly and passed from one router to another. This is done to minimize the control overhead and buffering in routers. If separate wires would be used for each of the above-mentioned groups, separate routing, scheduling and queuing would be needed, and the cost of routers would increase proportionally.

In addition, in a network at each layer in the protocol stack, control information must be supplied together with the data (e.g., packet type, network address, or packet size). This control information is organized as an envelope around the data. That is, first a header is sent, followed by the actual data (payload), followed possibly by a trailer. Multiple such envelopes may be provided for the same data, each carrying the corresponding control information for each layer in the network protocol stack.

### Buffering and flow control

Buffering data of a master (output buffering) is used both for busses and NoCs to decouple computation from communication. However, for NoCs output buffering is also needed to marshal data, which consists of (a) (optionally) splitting the outgoing data in smaller packets which are transported by the network, and (b) adding control information for the network around the data (packet header). To avoid output buffer overflow the master must not initiate transactions that generate more data than the currently available space.

Similarly to output buffering, input buffering is also used to decouple computation from communication. In an NoC, input buffering is also required to unmarshal data. In addition, flow control for input buffers differs for busses and NoCs. For busses, the source and destination are directly linked, and destination can therefore signal directly to a source that it cannot accept data. This information can even be available to the arbiter such that the bus is not granted to a transaction trying to write to a full buffer. In an NoC, however, the destination of a transaction cannot signal directly to a source that its input buffer is full. Consequently, transactions to a destination can be started, possibly from multiple sources, after the destination's input buffer has filled up.

Several policies can be adopted when an input buffer is full, and only some representative ones are listed here. One policy is not to accept additional incoming transitions and to store them in the network. However, this approach can easily lead to network congestion, as the data could be eventually stored all the way to the sources, blocking the links in between. Another policy is to accept incoming transactions at a full destination, and drop some data in the input buffer. Congestion is avoided but data is lost, and this is undesirable. To avoid input buffer overflow connections can be used, together with end-to-end flow control. At connection set up between a master and one or more slaves, buffer space is allocated at the NIs of the slaves, and the NI of the master is assigned credits reflecting the amount of buffer space at the slaves. The master can only send data when it has enough credits for the destination slave(s). The slaves grant credits to the master when they consume data.

## 6.5 LATEST ADVANCES IN PROCESSOR INTERFACES

Recently a new trend of processor architecture for SoC design has emerged: customizable processors [47]. These processor cores are highly configurable (e.g., registers and memories) and allow SoC system designers to define new instructions and registers to match target application performance requirements. Usually, the customization process is tool assisted. For instance, the Tensilica's Xtensa processor extension synthesis compiler (XPRES [46]) uses three techniques to create optimized Xtensa processor configurations: operator fusion, multiple data (SIMD) vectorization, and flexible-length instruction extensions (FLIX) [47].

At the same time, the need to provide high-performance computation is pushing the development of multiprocessing architectures based on the instantiation of multiple customizable processor cores and hardware accelerators. Therefore, high-bandwidth inter-processor communication is becoming a critical issue for state-of-the-art SoCs. In order to overcome the traditional bus bottleneck, designers of configurable cores are introducing multiport access to the processor's internal execution units. This trend is leading to significant changes in processor interfaces.

Let us consider the relevant example of hardware extensions for messaging. Message-passing software communications naturally correspond to data queues, but can also be implemented through a global shared memory, where messages are written and read by accessing the bus. In this case, the synchronization overhead is also significant, and translates into more bus traffic as well. Direct processor-to-processor connections reduce communication cost and latency by allowing data to move directly from one processor's registers to the registers and execution units of another.

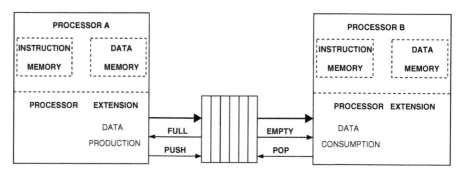

■ **FIGURE 6.6**

Processor interface extension for queue-assisted data communication.

The highest-bandwidth mechanism for task-to-task communication is hardware implementation of data queues. One data queue can sustain data rates as high as one transfer every cycle, or more than 10 Gbytes per second for wide operands (tens of bytes per operand at a clock rate of hundreds of MHz) because queue widths need not be tied to a processor's bus width or general register width. The handshake between data producer and consumer is implicit in the interfaces between the processors and the queue's head and tail (see Fig. 6.6).

The data producer pushes the data into the tail of the queue, assuming the queue is not full; if the queue is full, the producer stalls. When ready, the data consumer pops data from the head of the queue, assuming the queue is not empty; if the queue is empty, the consumer stalls. The SoC designer can also create non-blocking push and pop operations for queues. Such queue operations in the data producer explicitly check for a full queue before attempting a push. The data consumer can explicitly check for an empty queue before attempting a pop. These mechanisms let the data producer or consumer move to other work in lieu of stalling.

Application-specific processors' instruction-set extensions allow direct implementation of queues. An instruction can specify a queue as one of the destinations for result values or use an incoming queue value as one operand source. Such operations can create a new data value or use an incoming data value during each cycle on each queue interface.

A complex processor extension can perform multiple queue operations per cycle, combining input from two input queues with local data and sending values to two output queues. A queue's high aggregate bandwidth and low-control overhead enable using application-specific processors for applications with very high data rates, which processors with conventional bus or memory interfaces cannot handle.

Queue interfaces to processor execution units are an unusual feature of commercial microprocessor cores. They became part of an Xtensa LX processor through the *Tensilica instruction extension* (TIE) language

syntax [48]. One Xtensa LX processor can have more than 300 queue interfaces of variable width up to 1024 bits each. These limits are set beyond
the routing limits of current silicon technology so that the processor core's
architecture is not the limiting factor in a system design. The designer sets
the practical limit based on system requirements, computer-aided design
flow, and process technology selection. Using queues, designers can trade
off fast and narrow processor interfaces with slower and wider interfaces
to achieve bandwidth, performance, and power goals.

TIE queues serve directly as input and output operands of TIE instructions, just like a register operand, state, or memory interface. A key difference between queue interfaces and memory interfaces is that the system
designer can customize the width of each queue interface port to the
exact value desired, either wider or narrower than the processor's standard
memory interface ports.

Whereas memory accesses often exploit temporal locality, queue data is
naturally transient. Consequently, queue storage can typically be smaller
than a general-purpose memory buffer used for similar purposes. The
Xtensa LX processor includes two-entry buffering for every TIE queue
interface that the system designer defines. A queue interface's two-entry
buffer consumes a substantially smaller area than a memory load/store
unit, which can have large combinational blocks for alignment, rotation,
and sign extension of data as well as cache-line buffers, write buffers, and
complicated state machines. Thus, the processor area that TIE queue interface ports consume is under the designer's direct control and can be quite
small or as large as necessary.

The FIFO buffering incorporated into the Xtensa LX processor for TIE
queues serves three distinct purposes. First, it provides a registered and
synchronous interface to the external agent (the actual FIFO memory).
Second, for output queues, buffering provides two entries that prevent the
processor from stalling when the attached external FIFO memory is full.
Third, it hides the processor's speculative instruction executions from the
external FIFO memory. Further details in Refs [47, 48].

The availability of ports and queues tied directly to a configurable processor's execution units permits the use of processors in an application
domain previously reserved for hand-coded RTL logic blocks: flow-
through processing. Combining input and output queue interfaces with
designer-defined execution units makes it possible to create a firmware-
controlled processing block within a processor that can read values from
input queues, perform a computation on those values, and output the
results with a pipelined throughput of one complete input–compute–
output cycle per clock. Figure 6.7 illustrates a simple design of such a system with two 256-bit input queues, one 256-bit output queue, and a 256-
bit adder/multiplexer (mux) execution unit. Although this processor extension
runs under firmware control, its operation bypasses the processor's memory busses and load/store unit to achieve hardware-like processing speeds.

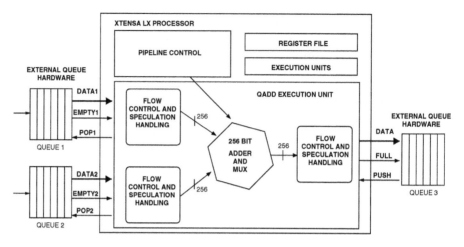

■ **FIGURE 6.7**

Flow-through processing.

This processor interface extension, as defined in the Xtensa LX processor, does not follow the general evolution trend represented by advanced interface protocols such as AMBA AXI and OCP. The underlying guiding principle is to *tightly integrate the processor interface in the processor architecture (ISA and pipeline)*. This is a new concept to SoC system engineering, which is likely to impact also the development of NIs for NoC-centric SoCs. In fact, taking this approach, network access operations could be directly exposed to the ISA, thus paving the way for a tighter coupling of computation and communication. This approach goes in the opposite direction with respect to traditional NIs, which rely on network-agnostic computation architectures. Each solution has its own trade-offs, as will be illustrated in the following section.

## 6.6  THE PACKETIZATION STAGE

### 6.6.1  Hardware–Software Implementation

The preparation of the packets is one of the key stages of NI architectures, and the associated latency is likely to significantly impact overall communication latency, as showed by early NoC prototypes [32].

Three possible implementations are feasible [13]:

1. ***Software-based library***: The operating system is extended in order to include primitives for implementing all the needed communication services.

2. ***On-core module***: The processor core itself is modified and parameterized, and the services are implemented in hardware.

3. ***Hardware based wrapper***: An independent hardware block (the NI) is located between the processor core and the interconnect.

The implementation of the packetization strategy varies depending on the reconfigurability and programmability of a specific core. In general, it is a performance/area/flexibility trade-off [13]. The packet communication process has essentially three stages: packet preparation, packet transmission, and packet handling at the receiver. In this chapter, we will primarily focus on the packet preparation stage for communication over the network. The packet handling stage at the destination is essentially complementary, and exhibits similar tendencies.

Without loss of generality, let us restrict our analysis to a simple distributed memory environment and let us study packetization implementation in this context. The system consists of a core that can access separate memory cores spread through the on-chip network. To the software executing on the processor core, the memory is one contiguous block present at a single location. The processor core is aware of the distributed nature of the memory space. When the software attempts to access a memory location, the destination core has to be identified and accessed. It will be demonstrated that this can be implemented in the three different schemes. The Xtensa Processor Core is used to explore the processor core customization solution for packetization.

We now need to briefly recall the generic structure of a packet. A packet can be tuned for a particular network, so as to reduce the packetization overhead. It essentially consists of three parts: the packet *header*, *payload*, and *tail*. The packet header contains the necessary routing and network control information, such as the destination and source addresses. When source routing is used, the destination address is ignored by the network, even if it can be used by the destination NI to generate responses (e.g., in read or non-posted write transactions). Instead, it is replaced by a routing field that specifies the route to the destination. A disadvantage of source routing is the added overhead of including the route field in the packet header. But the inclusion of such a field reduces the complexity of the routing logic on the network switches. It simplifies their routing decisions and their task will be to just look at the routing field and route the packet over the specified output port. This solution allows a topology-independent approach to routing implementation and is well suited only for deterministic routing strategies. When adaptive routing needs to be implemented for the purpose of balancing network load or recovering from malfunctions (of links or switches), then routing decisions have to be taken at the switches. In this case, the MSBs of the destination address are read from the packet by each switch, and the corresponding output port is selected based on an LUT. The LUT can also provide several options, which must be discriminated based on some feedback metric about congestion in the network (e.g., output buffer level, flow control information, etc.).

The packet data consists of essentially two types of information. The first is the control information that will be used by the receiving core to carry out the required transaction. A typical example is represented by the burst control information that each interface protocol makes available. OCP-compliant interfaces include burst-specific signals such as MAtomicLength, MBurstLength, MBurstPrecise, etc. [12]. The second will be the actual data, such as the memory address being accessed or data to be written to memory. The packet tail might contain redundancy bits for error detection or correction, but this part of the packet is optional. It depends on the error probability of the underlying network, on the complexity of error codecs, and on the availability of wiring resources that might lead to implement such redundant bits as parallel wires rather than serial bits. The packet structure has also to be tuned to the implementation of the packetization strategy.

### Software library for packetization

A software implementation of the packet preparation stage has to provide the user with a library of instructions that can be used to access a memory address in a distributed memory space. Some header files may be required by the library to specify (i) the address of the host or route information to the network elements; (ii) the packet structure (fields in the packet and corresponding bitwidth); and (iii) memory allocation information (e.g., the address space of the memory cores in the system).

When the user issues a memory access instruction, the corresponding packet is prepared according to the following steps:

- **Step 1**: Translate address by determining which memory core needs to be accessed, and determine the effective address at that memory core.

- **Step 2**: Prepare packet header by setting the source address and the route to the destination.

- **Step 3**: Examine program instruction requiring memory access, and set control flags in the packet data.

- **Step 4**: Set effective address in packet data.

- **Step 5**: If using error-detecting or-correcting codes, calculate the parity check bits and set them in the packet tail.

- **Step 6**: Assemble packet and deliver it to the network logic of the core.

A sample library code was developed in Ref. [13], and executed on the basic Xtensa Processor Core, configured with a 128-bit processor interface. The equivalent area result for this strategy can be expressed as the size of the software library code.

## On-core module for packetizing

An alternative implementation consists of exploiting the configurability of many processor cores for embedded systems. The Xtensa toolset has tools that allow the profiling of the executed instructions. From the profile one can obtain the required results such as the cycle count for the executed instructions. The TIE compiler also generates the required Verilog/VHDL files that can be then analyzed using Synopsys Design Analyzer, to obtain the timing and area costs.

Custom instructions can be therefore defined and tested for data packetization, as is done in Ref. [13]. By means of the TIE language, the same stages described in previous subsection were specified. The TIE code was successfully compiled with the TIE compiler and the execution of the custom instruction was tested on the Xtensa processor. The cycle count for this implementation was obtained using the *instruction set simulator* (ISS), provided with the Xtensa toolset. The ISS provides detailed information on the contents of the registers in use and the output available at the processor interface.

## Wrapper logic for packetizing

For cores that are neither programmable nor reconfigurable, the only option for interfacing with the networking logic of the tile is to utilize a wrapper, which would have the responsibility of packetizing and depacketizing the cores requests and responses. The wrappers have the responsibility of (i) receiving the contents from the core interface, preparing the packets, and dispatching them to the network logic of the tile and (ii) receiving the packets from the networking logic and presenting the contents to the core interface.

The work in Ref. [13] considers a packetizer module which is compliant with the VSI Alliance's VCI Standard Version 2 [2]. The basic interface (BVCI) is implemented. VCI interfaces are quite similar in philosophy to OCP. However, OCP can be considered as a superset of VCI, in that it handles control and test flows in addition to data flow. Moreover, it also provides industrial grade tools and services that ensure members can rapidly confirm compliance and maximize their productivity.

The packet structure in this implementation is dependent on the signals used in the interface protocol. The packetizer module maintains the address translation information, that is the mapping of memory addresses to destination core addresses. It analyses the content of the core request and tries to optimize on the amount of data being sent over the on-chip network, by filtering the redundant information from the packets.

**TABLE 6.1** ■ Latency and area results.

| Packetization strategy | Cycle count | Clock frequency | Latency | Area | Remark |
|---|---|---|---|---|---|
| Software library | 47 | 193 MHz | 243.5 ns | 118 kB | Code size |
| On-Core packetization | 2 | 185 MHz | 10.8 ns | 13 k | Gate count using 0.18 μm technology |
| Wrapper packetization | – | – | 3.02 ns | 4 k | Gate count using 0.35 μm technology |

### Comparison

An area and latency comparison of the different implementation alternatives for the packetization stage is reported in Table 6.1. The results summarize the hardware–software implementation effort carried out on the Xtensa Processor Core, as described in Ref. [13]. Latency in the case of the software library was determined using the cycle count obtained from the ISS and the clock frequency. This result will vary with different processors and implementations of the packetization library. In the case of the on-core packetization, the latency was determined in a similar way. However, the clock frequency was obtained through synthesis of the custom-processor specification. The lower clock frequency is due to the slow-down caused by the TIE logic that was incorporated into the processor core. This is an acceptable trade-off, in light of the performance improvement. This conservative result was obtained by using slow TSMC 0.18 μm technology library.

The result for the wrapper implementation is obtained using the 0.35 micron technology library. With better technology, there will be a further reduction in the latency. The latency result in this case provides the developer with the time taken through the *longest path in the wrapper*, and will enable him to decide on the clocking rate for the interface.

The area cost of the three implementations cannot be compared quantitatively. The results provide a measure of the cost that the system designer would incur using a particular strategy. The area for the software implementation was determined in terms of the code size of the software library. To determine the area of the TIE logic, for the on-core packetization, the TIE specifications were synthesized following the regular steps (i.e., compilation of the TIE specification and synthesis of the compiler output). The 13K gate count is an overly conservative estimate. The silicon area appeared to be under 0.2 mm$^2$. The area for the wrapper was determined using Synopsys Design Analyzer. It cannot be directly compared to the one obtained for the Xtensa Core, as the technologies used in both are considerably different.

**TABLE 6.2** ■ Comparison between packetization schemes.

| Type of implementation | Area | Expected latency | Complexity | Flexibility |
|---|---|---|---|---|
| Software library (on-core) | Low on HW area but increases code size | High | Increased code size | Requires programmable cores |
| RTL (HW) implementation (on-core) | Additional register and logic to packetize | Low | Additional registers and logic and an increase in instruction set | Requires programmable cores or development of modified cores |
| Wrapper RTL (HW) implementation (off-core) | Additional control, registers and logic to packetize | Low | Additional control, registers and logic. Ability to understand core operation | Can use existing cores. Modify wrappers for plug-and-play into different networks |

Although extracted from an early and non-homogeneous exploration framework, the above results (summarized in Table 6.2) give us an insight into the complexities and the intricacies that are involved when cores or their corresponding wrappers employ packet-switched networks for on-chip communication. Design decisions will be incumbent upon the latencies and the area overhead factors that have been pointed out in this analysis. Taking the network logic deeper into the processor core is a viable option only for extensible cores, while the hardware wrapper is a more general solution allowing clean reuse of attached units. In both cases, the latency is much lower compared to that incurred by a software implementation of data packetization. Most state-of-the-art NoC prototypes adopt the hardware wrapper solution. However, the queueing extensions introduced by recent configurable processor cores to their external interfaces, such as the Xtensa LX processor, pave the way for an in-built implementation of the packetization logic. The output queues, to which the processor can directly interface, can be viewed as the equivalent back-end stage of an NI. More research is needed to assess real applicability and effectiveness of this solution.

## 6.6.2  Packet Types

A first attempt to capture the composition of the packetized data flows that are exchanged between MPSoC nodes in network-centric systems was made in Ref. [49]. Packets are categorized in the following types:

- *Memory access request packets*: This packet is induced by cache misses that request data fetch from memories. The header of these packets contains the destination address of the target memory (node ID and memory address) as well as the type of memory operation requested (memory READ, for example). Because there is no data being transported, the payload is empty.

- *Cache coherence synchronization packets*: This kind of packet might be induced by the cache coherence mechanism in the system. This packet might come from the updated memory, and might be sent to all caches that have a copy of the updated data. The packet header would contain the memory tag and block address of the data. If the synchronization used *update* method, the packet would contain updated data as payload. If the synchronization used *invalidate* method, the packet header would contain the operation type (INVALIDATE, in this case), and the payload would be empty.

- *Data fetch packets*: It is the reply packet from memory, containing the requested data. The packet header contains the target address (the node ID of the cache requesting for the data). The data is contained in the packet payload.

- *Data update packets*: This packet contains the data that will be written back to the memory. It comes from a cache that requests the memory write operation. The header of the packet contains the destination memory address, and the payload contains the data.

- *IO and interrupt packets*: This packet type is used by IO operations or interrupt operations. The header contains the destination address or node ID. If data exchange is involved, the payload contains the data.

From the above analysis, we can see that most packets travel between memories and caches (assuming they are located on different network nodes), except those packets involved in I/O and interrupt operations. Although packets of different types originate from different sources, the length of the packets is determined by the size of the payload. In reality, there are two differently sized packets on the MPSoC network, *short packets* and *long packets*, as described below. Short packets are the packets with no payloads, such as the memory access request packets and cache coherence packets (invalidate approach). These packets can consist of only header and tail flits.

Long packets are the packets with payloads, such as the data fetch packets, the data update packets, and the cache coherence packets used in the update approach. These packets travel between caches and memories. The data contained in the payload are either from cache block or they are sent back to the node cache to update the cache block. Packets with payload size different from one cache block size will increase cache-miss penalty. The reason is twofold:

1. If each cache block is segmented into different packets, it is not guaranteed that all packets will arrive at the same time, and consequently the cache block cannot be updated at the same time.

2. If several cache blocks are to be packed into one packet payload, the packet needs to hold its transmission until all the cache blocks are updated. This will again increase the cache-miss delay penalty.

### 6.6.3 Impact of Packet Size

The packet payload size is an important parameter for system performance and energy, since it is not only related to the effectiveness of network operation and to the complexity and performance of NIs, but has a number of system-level implications. The early work in Ref. [49] points out the relationship between different design options at the architectural level and shows the impact of appropriate packet size on the performance-power trade-off.

Here we recall some of those results, which are based on a network energy model applied to the RSIM multiprocessor simulator [50]. The target platform consists of eight RISC processors connected by a 2D mesh interconnection network. Each processor has two levels of cache hierarchy, where both L1 and L2 caches use write-through methods for memory updates. The *invalidate* approach is used for cache coherence synchronization. Flits are 8 bytes long. The packet payload size is varied from 16Byte, 32Byte, 64Byte, 128Byte to 256Byte, obviously only for *long packets*, since small packets are assumed to be 2 flits long.

Changing the packet payload size will change the L2 cache block size that can be updated in one memory fetch. If we choose larger payload size, more cache contents can be updated. Therefore, the cache-miss rate will decrease.

On the other hand, whenever there is a L2 cache miss, the missed cache block needs to be fetched from the memories. The latency associated with this fetch operation is called a *miss penalty*. Some delays building up the miss penalty are likely to be quite sensitive to packet size, such as packetization delay and memory access delay. In fact, longer packets need longer time for packetization and memory access. Moreover, longer packets will actually cause more contention delay in the network due to pipelined packet propagation mechanism of wormhole switching. Combining all these factors, the overall cache penalty is likely to increase as the packet payload size increases. Overall, there exists an optimal payload size that can achieve the minimum execution time of running applications. This fact is showed in Fig. 6.8 for *sor*, *water*, *quicksort*, *lu* and *mp3d* benchmarks, ported from the Stanford SPLASH project.

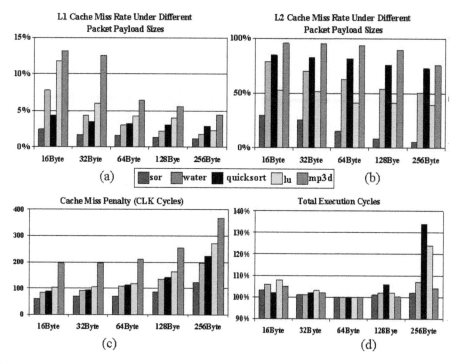

**■ FIGURE 6.8**

Impact of packet payload size on cache-miss rate and penalty.

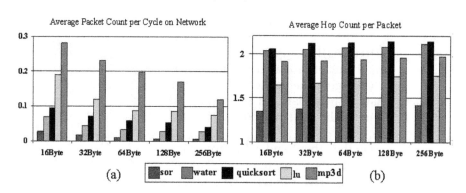

**■ FIGURE 6.9**

Average hop and packet count as a function of packet size.

Let us now try to capture the impact of packet size on network and system power. Packets with larger payload size will decrease the cache-miss rate and consequently decrease the number of packets on the network. This effect can be seen from Fig. 6.9(a). It shows the average number of packets on the network (traffic density) at one clock cycle. As the packet size

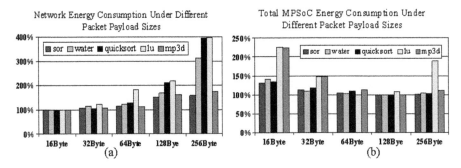

■ **FIGURE 6.10**

Network and total MPSoC energy consumption under different packet payload sizes.

increases, the number of packets decreases accordingly. Actually, with the same packet size, the traffic density of different benchmarks is consistent with the miss penalty.

In contrast, larger packet size will increase the energy consumed per packet, because there are more bits in the payload. Moreover, larger packets will occupy the intermediate node switches for a longer time, and cause other packets to be re-routed to non-shortest data-paths. This leads to more contention that will increase the total number of hops needed for packets traveling from source to destination. This effect is shown in Fig. 6.9(b). It shows the average number of hops a packet travels between source and destination. As packet size (payload size) increases, more hops are needed to transport the packets. Actually, increasing the cache block size will not decrease the cache-miss rate proportionally [51]. Therefore, the decrease of packet count cannot compensate the increase of energy consumed for packet transfer across the network. Figure 6.10(a) shows the combined effects of these factors under different packet sizes. The values are normalized to the measurement of 16 Bytes. As packet size increases, energy consumption on the interconnect network will increase.

Although increase of packet size will increase the energy dissipated on the network, it will decrease the energy consumption on cache and memory. Because larger packet sizes will decrease the cache-miss rate, both cache energy consumption and memory energy consumption will be reduced.

The total energy dissipated on MPSoC comes from non-cache instructions (instructions that do not involve cache access) of each node processors, caches, shared memories as well as the interconnect network. In order to see the packetization impact on the total system energy consumption, all MPSoC energy contributors should be put together to see how the energy changes under different packet sizes. The results are shown in Fig. 6.10(b), which indicates that MPSoC energy will decrease as packets size increases. However, when the packets are too large, as in the case of 256 Bytes in the figure, the total MPSoC energy will increase, since the increase of interconnect network energy will outgrow the decrease of energy on cache, memory, and non-cache instructions.

The added value of the analysis results reported above (taken from Ref. [49]) is to show how packet payload size impacts performance and energy of the interconnection network and of system memory components (caches and memories), so that a correlation can be identified between packet size and overall system metrics.

## 6.6.4  Impact of Flit Size

Another degree of freedom for system designers is represented by flit size, given a fixed packet payload size for each transaction. This parameter is in fact related to NoC bandwidth, and should be tuned to application requirements. However, contrarily to packet payload size, flit size largely determines the role of network buffering resources with respect to overall network power and area. A few recent studies have shed light on the trade-offs associated with flit sizing, and their major findings will be now briefly described. Contrarily to the analysis of the previous subsection, these results are more recent and therefore can leverage the latest developments in NoC synthesis back-end. They have been derived on the xpipes NoC architecture, as explained in Ref. [36].

The flit size parameter can be design time customizable, and primarily affects packetization logic at the NI. However, it significantly impacts the architecture of all network building blocks. As an example, let us assess the impact that flit size has on switch area and operating frequency. A baseline xpipes switch [44] was synthesized with a UMC 0.13 $\mu$m technology library with *16-* and *38-bit flits*. All instances have been placed and routed to get accurate reports. The reference switch configuration includes six input and output ports, output buffering with 3-flit buffer depth, fixed priority arbitration and ACK/NACK flow control.

Figure 6.11(a) and (b) depict performance when varying the flit width of packets. When moving from 16 to 38 bits, a huge area penalty of 64% can be observed. Such penalty is however less than linear, as flit width increased by 138%. This result is logical, since the area for the data-path (including buffers) has to scale linearly with flit width, but arbitration and control logic are unaffected. The maximum operating frequency is also almost unaffected by flit width, which suggests the worsening of wiring congestion not to be critical yet.

Let us now extend this kind of analysis to entire NoC-based MPSoC platforms, recalling the results of the work in Ref. [25]. The reference architecture is the one reported in Fig. 6.12. The NoC topology is regular, but it is not really a mesh since it features two communicating cores per node; this choice was taken to cut silicon requirements for the NoC in half. Thirty cores are attached to the fabrics, of which 15 masters and 15 slaves

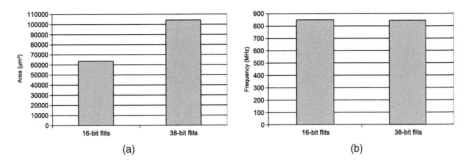

(a)                              (b)

■ **FIGURE 6.11**

Impact of flit size on switch area and performance.

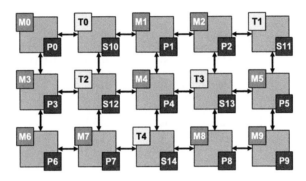

■ **FIGURE 6.12**

The platform under test: an xpipes "mesh-like" topology. M: ARM7 masters; T: traffic generators; P: privately accessed slaves; S: shared slaves.

(typically memory banks). Non-pipelined links were instantiated in the xpipes mesh.

The NoC was configured with two different flit sizes, namely 21 and 38 bits, to explore the dependency of power consumption and performance on this parameter. These numbers were chosen taking into account the length of each possible packet type and trying to optimize the resulting flit decomposition.

This platform was rendered into a cycle-accurate simulation environment and a synthesis flow was established to achieve a placed and routed chip layout. For the purpose of simulation, 10 ARM7 processor cores with caches were plugged to the master ports of the platform. Moreover, five custom traffic generators were inserted to better model the behavior of devices (such as *digital signal processors* (DSPs)) that perform transaction streams. System slaves (memories) were either accessed by a single master (private) or subject to contention due to requests by multiple masters (shared). A 0.13 $\mu$m, power characterized technology library was used

**TABLE 6.3** ■ Pre- and post-placement achievable frequencies.

| Max frequency | ×pipes (21-bit) | ×pipes (38-bits) |
|---|---|---|
| After synthesis | 910 MHz | 910 MHz |
| After place-and-route | 885 MHz | 885 MHz |
| Frequency drop | 2.7% | 2.7% |

**TABLE 6.4** ■ Power consumption of the fabrics.

| Power consumption | ×pipes (21-bits) | ×pipes (38-bits) |
|---|---|---|
| Global power | 377 mW | 501 mW |
| Sequential cell power | 296 mW | 416 mW |
| Sequential power ratio | 78.5% | 83.0% |

**TABLE 6.5** ■ Energy consumption of the fabrics.

| Energy consumption | ×pipes (21-bits) | ×pipes (38-bits) |
|---|---|---|
| Available bandwidth | 100 GB/s | 180 GB/s |
| Energy per injectable data | 3.77 mJ/GB | 2.78 mJ/GB |
| Length of benchmark run | 0.9 ms | 0.85 ms |
| Energy per benchmark run (fabric only) | 0.339 mJ | 0.426 mJ |
| Energy per benchmark run (1-W system) | 1.34 mJ | 1.37 mJ |
| Energy per benchmark run (5-W system) | 5.32 mJ | 5.13 mJ |

for synthesis, floorplanning, and place-and-route. For the place-and-route phase, cores and memories were represented by means of Hard blocks each obstructing a non-routable area of 1 mm². 

The resulting area for 15 switches and 30 NIs is 2.3 mm² for the 38-bit case, reducing to 1.7 mm² for the 21-bit case. The same operating frequency of 885 MHz was found, irregarding of flit size (see Table 6.3).

As a consequence, the NoC aggregate bandwidth amounts to 100 GB/s with 21-bit flits, growing to 180 GB/s with 38-bit flits. This comes at a power cost which is almost 25% higher for the network with 38-bit flits, due to the larger network buffers required. Power results are summarized in Table 6.4.

An interesting metric to consider is the energy required to complete a benchmark. The 38-bit flit interconnection network consumes 20.5% more energy. However, the impact on system-level energy is not that straightforward. In fact, we need to consider also power dissipated by other components. This is a very difficult task, since it depends on the specific component at hand. If we conservatively assume a power consumption of just 1 W at 400 MHz for all of the 15 cores, caches and memory blocks,

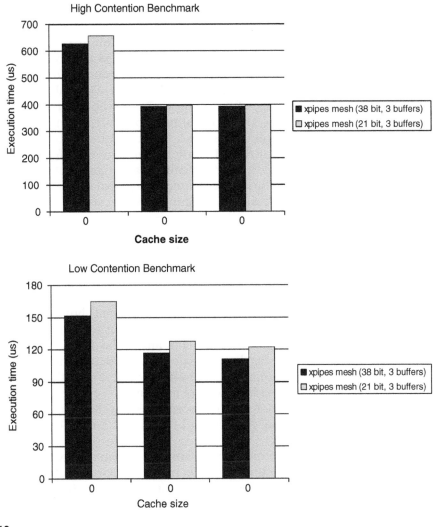

■ **FIGURE 6.13**

Execution time for two benchmarks with varying cache sizes: (a) high- and (b) low-contention benchmark.

the overall system power amounts to 1.487 W (21-bit NoC), as opposed to 1.611 W for the 38-bit configuration. The corresponding energy is reported in Table 6.5. Finally, with system components requiring 5 W, the 38-bit NoC turns out to be the most energy efficient.

Should the system designer opt for the 21-bit NoC solution (since it is less power-hungry), what would the performance implications be? The functional simulation results reported hereafter address this concern. Two multimedia loads were applied to the NoC, and global execution performance is illustrated in Fig. 6.13. ARM cores are supposed to run at

442.5 MHz and the NoC at 885 MHz, leveraging the dual clock support of xpipes. The test is repeated with different cache sizes for the ARM7 processors, since smaller caches translate into more cache misses and more congestion on the interconnect fabric. The 38-bit NoC configuration performs better than the 21-bit one, but not as much as one might think; this is because even the simpler NoC already provides abundant bandwidth. The main advantage of the 38-bit NoC is the ability to compress the same packets in less flits, therefore cutting down on packet latency.

This system-level analysis has confirmed the global impact that local design decisions at the NI can have on system performance and power consumption. Due to their criticality, packet parameters (packet size and flit width) should be selected based on a global decision framework, since their scope is not restricted within NI design issues.

## 6.7 END-TO-END FLOW CONTROL

The flow control mechanism in an NoC prevents buffers from overflowing and is an essential part of the NI back-end. In general, buffer overflow can occur both at the network switches and at the destination NIs. Therefore, in buffered switching, link-level flow control determines the way the downstream switch communicates buffer availability to the upstream switch. In turn, end-to-end flow control mechanisms ensure that a message producer node does not produce more messages than the consumer node can handle. Traditional flow control techniques are on–off, ACK/NACK, and credit-based [52].

Link-level flow control can be used to automatically prevent buffer overflow also at the destination node. In fact, the last switch in the routing path exchanges with the destination NI just those flits (or packets, depending on the switching technique) that can be stored in the downstream buffer. The NI is viewed just as another flow control stage in the packet delivery path. This technique ensures that even when the destination buffer is full, the remaining flits can be propagated across the network so to fill up the switch buffers. This distributed buffering strategy can significantly reduce flit delivery latency, since flits are stored as close to the destination as possible. Unfortunately, when the destination node is slow in absorbing flits, a macro-pipeline of outstanding flits ends up being stalled across the network, causing congestion and deadlock concerns. Virtual channels can be used as a workaround for this problem.

However, the latest NoC design experiences suggest that buffer minimization is the primary optimization goal for power-aware designs. Typically, at least two-slot buffering stages are needed at switch inputs and/or outputs for the purpose of flow control: one slot is used for normal flit propagation and the other one is a back-up slot needed to store

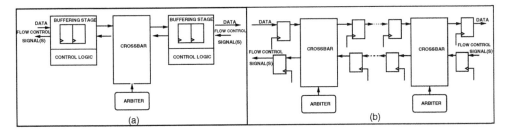

Reference architectures to assess the impact of flow control on switch performance.

incoming flits before the backpressure mechanism can stop the upstream switch propagation. Although simplified, this buffering architecture requires some control logic whose delay adds up to the data-path delay.

In order to better understand this, let us consider the switch architecture of Fig. 6.14(a), where the arbitration stage is decoupled from the data-path components (input/output flow control stage, input/output data sampling stage, and central crossbar). Typically, post-synthesis analysis of standard BE switches reveals that the critical path is in the arbitration logic, which has to execute the arbitration algorithm and has to drive the crossbar multiplexers [36]. Let us now assume to use this switch in a circuit-switching fashion: the upstream node (connected at a certain input port) asks the arbiter to get access to a certain output port by means of physical request and routing signals. The arbiter replies by driving the grant signal, which reserves the desired output port. This arbitration process results in an initial arbitration delay which is paid for circuit set-up across the switch. This architecture allows us to get the maximum speed from the data-path, and to assess the impact of flow control logic on data-path delay. The arbitration delay is translated into an equivalent number of data-path clock cycles. The critical path of the isolated data-path is affected by the control logic of the flow control stage, which limits maximum achievable speed.

Let us compare this solution with an equivalent architecture (depicted in Fig. 6.14(b)), where the flow control has been reduced to a pure retiming stage. With a UMC 0.13 $\mu$m technology library, adding up the propagation delay of the input sampling stage, a $5 \times 5$ crossbar delay and the final set-up time results in an operating frequency of more than 2 GHz [59]. However, the complexity has been moved to the NIs, since now an end-to-end flow control mechanism is required to prevent destination buffer overflow.

However, the implementation of end-to-end flow control is not straightforward. Let us continue our analysis on the baseline architecture of Fig. 6.14(b). When a source NI has to transmit one or more packets to a destination NI, a complete circuit has to be established through all the intermediate network switches along the routing path. One simple mechanism is to have the source NI sending request and routing signals to all

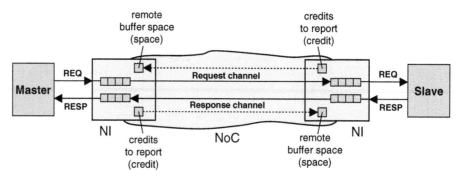

■ **FIGURE 6.15**

Connection implementation.

switch arbiters, and then waiting for all (backward-propagating) grant signals to be driven high by the arbiters. This mechanism, which is managed by the source NI, corresponds to the *media access control* (MAC) layer in the ISO/OSI layering specification. It is however not enough, since a higher (transport) layer end-to-end protocol (the flow control) is required to match the packet transmission rate at the source NI with the available storage locations at the destination NI. A credit-based flow control could be used for this purpose. Its implementation depends however on the buffering strategy at the NIs. If the destination NI has only one buffer for all incoming communications, as soon as the MAC layer connection has been established, the destination NI has to advertise the number of available buffering resources to the source NI, so to set its initial credit window.

The inherent limitations of this approach can be overcome by having a couple of dedicated buffers (one on each NI) for each peer-to-peer communication. The Aethereal NoC architecture implements this approach [21], as depicted in Fig. 6.15. Communication services are provided by means of connections, which consist of a request channel and of a response channel. For each connection, buffering resources are allocated in the corresponding NIs. Although the Aethereal NI architecture will be described in detail in Section 6.9, we now briefly isolate the credit-based end-to-end flow control implementation.

As shown in Fig. 6.15, for each channel, there is a counter (*space*) tracking the empty buffer space of the remote destination queue. This counter is initialized with the remote buffer size. When data is sent from the source queue, the counter is decremented. When data is consumed by the destination module at the other side, credits are produced in a counter in the remote NI (credit) to indicate that more empty space is available. These credits are sent to the producer of data (dashed line in Fig. 6.15) to be added to its space counter.

Credits can be piggybacked in the header of packets for data flowing in the other direction to improve NoC efficiency [21]. The cost for

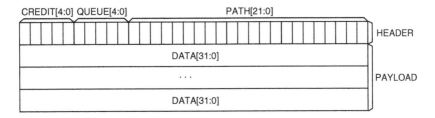

■ **FIGURE 6.16**

Aethereal packet format.

piggybacking credits is a change of the packet structure. As shown in Fig. 6.16, besides the routing information (path and queue identifier), five additional bits (27–31) are used in the Aethereal packet header for piggy-backing the credits. Thus, the maximum amount of credits that can be sent at a time is $2^5 = 32$. Note that at most *space* data items can be transmitted before credits must be received. This stresses the buffering needs for credit-based end-to-end flow control, reflected in an increased complexity and power of the destination NI of a communication channel. Such buffering requirements clearly depend on the round-trip latency, although the flit ejection rate from the network might play a masking role. In Ref. [21], the minimum between the data items in the queue and the value in the counter space is called the *sendable data*. Whenever a queue contains sendable data, it can be scheduled for transmission by the NI.

The circuit-switched NoC architecture in Ref. [23] employs a similar mechanism for end-to-end flow control. The only difference lies in the fact that credits are sent from the destination node by means of a backward-propagating ACK signal. Every source has a local window counter of size *WC*. This local window counter indicates how many data packets the source is allowed to send to the destination. The destination will send an acknowledgment signal when it has read $X$ data packets, where $X \leq WC$. When the source receives an acknowledge signal it increases its local window counter *WC* by $X$. By configuring the use of the acknowledgment signal and size of $X$ and *WC*, both blocking and non-blocking communications can be supported.

## 6.8 PACKET AND CIRCUIT SWITCHING

Circuit switching makes the propagation of control-free pipelined data streams possible across the network, resulting in increased energy efficiency per transported bit and throughput. Control tasks are performed at circuit set-up and tear-down, and are managed by the NI. Let us review

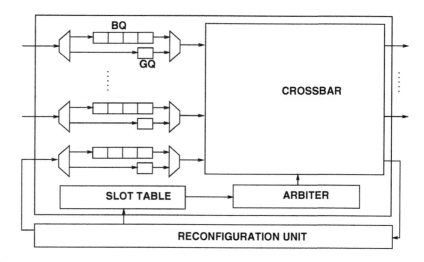

Aethereal router architecture and reconfiguration unit.

some implementation examples of how circuits can be managed by NIs. The impact on operation and architecture of the other network building blocks will be considered as well.

The typical way of setting up a circuit is to send BE programming packets through the network. Such packets can be used to reserve time slots for the access to time-division multiplexed router outputs or to permanently reserve switch resources for certain data streams.

The first approach is used in the Aethereal NoC [35], where time guarantees are provided by means of pipelined time-division multiplexing. Programming (or system) packets are used to avoid introducing an additional communication infrastructure only to program the network.

The programming functionality of the router is provided by reconfiguration units, as showed in the router architecture of Fig. 6.17. The slot table is used to switch data to the correct output while avoiding contention on a link. Every slot table $T$ has $S$ time slots (rows) and $N$ router outputs (columns). There is a logical notion of synchronicity, since all routers in the network are assumed to be in the same fixed-duration slot. In a slot $s$ at most one block of data can be read/written per input/output port. In the next slot $(s + 1 \bmod S)$ the read blocks are written to their appropriate output ports. Blocks thus propagate in a store-and-forward fashion. The latency a block incurs per router is equal to the duration of a slot, and bandwidth is guaranteed in multiples of block size per $S$ slots.

Because multiple system packets may arrive simultaneously for the reconfiguration unit (i.e., contention), they must be scheduled. This is achieved by viewing the reconfiguration unit as just another router, complete with flow control, which is placed in between the last output and input port of the router. In this way, contention on the reconfiguration unit is moved to contention on the output port.

Initially the slot table of every router is empty. There are three system packets: *SetUp*, *TearDown*, and *AckSetUp*. They are used to program the slot table in every router on their path. The SetUp packet creates a connection from a source to a destination, and travels in the direction of the data. When a SetUp packet arrives at the destination, it is successful and is acknowledged by returning an AckSetUp. TearDown packets destroy (partial) connections and can travel in either direction. SetUp packets contain the source of the data, the destination or a path to it and a slot number. Every router along the path of the SetUp packet checks if the output to the next router in the path is free in the slot indicated by the packet. If it is free, the output is reserved in that slot and the SetUp packet is forwarded with an incremented (modulo S) slot. Otherwise, the SetUp packet is discarded and a TearDown packet returns along the same path. Thus, every path must be reversible. The upstream TearDown packet frees the slot and continues with a decremented slot. Downstream TearDown packets work similarly and remove existing connections. A connection is successfully opened when an AckSetUp is received, else a TearDown is received. With minor additions, system packets can also be used to program multicast connections.

The programming model is pipelined, concurrent (multiple system packets can be active in the network simultaneously, also from the same source) and distributed (active in multiple routers). Given the distributed nature of the programming model, ensuring consistency and determinism is crucial. The outcome of programming may depend on the execution order of system packets, but is always consistent.

The SoCBus NoC architecture [22] uses a similar resource reservation scheme, but within the context of a pure circuit-switched architecture. The mechanism is called *packet connected circuit (PCC)*, to indicate the hybrid circuit-switching scheme with packet-based set-up. The network transactions consist of four to six phases dependent on whether the first try routing is successful or not. A successful transaction, see Fig. 6.18(a), has four phases:

- (I) First a request is sent from the source NI to the network. As this request finds its way through the network the route is temporarily locked and cannot be used by any other transactions.

- (II) The second phase starts when the request reaches its destination: an acknowledge is sent back along the route and the locks are changed to permanent locks.

- (III) When the acknowledge has returned to the source the third phase starts. This phase holds the actual transfer of data payload.

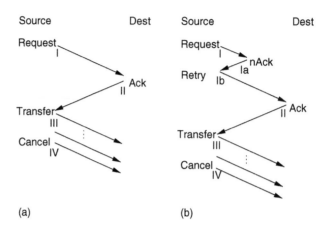

■ **FIGURE 6.18**

Communication protocol in the SoCBus NoC architecture. (a) No retry necessary. (b) One retry used.

- (IV) Finally, after the data has been transferred, a cancel request is sent that releases all resources as it follows the route.

As depicted in Fig. 6.18(b), if a route is blocked in a node the routing request is canceled in phase (Ia), and the blocking switch returns a negative acknowledge to the source. In (Ib), the source NI retries to acquire route resources at a later stage, which means that the additional two phases (NACK and retry) might need to iterate.

Till now, we have seen architectures where programming packets are injected by NIs of communicating cores. Alternatively, one processing element can be devoted to task allocation and to the configuration of communication resources, but also in this case the NI architecture of communicating cores is impacted. For instance, in Ref. [23] a tile-based architecture is proposed with centralized control. A *central coordination node* (CCN) performs system coordination functions. It performs run-time mapping of the newly arrived applications to suitable processing tiles and inter-processing communications to a concatenation of network links. The CCN does not perform run-time scheduling of individual processes and communications during execution. That is performed by the individual tiles and network routers. The CCN performs the feasibility analysis, spatial mapping, process allocation, and configuration of the tiles and the NoC before the start of an application. In particular, a circuit-switched network is used for time-guaranteed traffic and is configured through a dedicated configuration interface at each switch, accessible through a separate (BE) control network driven by the CCN (see Fig. 6.19(a)).

Tile interfaces are connected to the switches by means of data converters. The reason lies in the way circuit switching is implemented, denoted as lane-division multiplexing. This solution leverages small channels (e.g.,

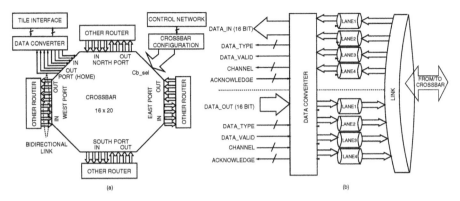

**■ FIGURE 6.19**

(a) Block diagram of the router. (b) Data converter between tile interface and router.

four bits) called *lanes*. The bidirectional link between two routers consists of a concatenation of unidirectional lanes (e.g., 2 times four lanes) as depicted in the right side of Fig. 6.19(b). This increases the flexibility as is available in time division multiplexed systems. The small lanes are connected to a tile interface via the data converter. Figure 6.19(b) depicts a data converter that converts the 16-bit data to the width of the lanes and vice versa. In lane-division multiplexing, each lane can be used for a different physical connection between processors. Note the acknowledge signal for each lane, used for end-to-end flow control as explained in Section 6.7. The width and number of lanes are adjustable parameters in the design. In the router, the four lanes of one port have to be connected with all the four lanes of all the other four ports.

## 6.9 NI ARCHITECTURE: THE AETHEREAL CASE STUDY

Aethereal offers a shared-memory abstraction to the connected modules [21]. Communication is performed using a transaction-based protocol, where master modules issue *request messages* (e.g., read and write commands at an address, possibly carrying data) that are executed by the addressed slave modules, which may respond with a *response message* (i.e., status of the command execution, and possibly data). This protocol ensures backward compatibility to existing on-chip communication protocols (e.g., AXI, OCP, and DTL), and also allows efficient implementation of future protocols, which are better suited to NoCs. Messages are transported across the network by means of packets.

The Aethereal NoC offers its services by means of *connections*. Connections allow differentiated communication services and guarantees offered

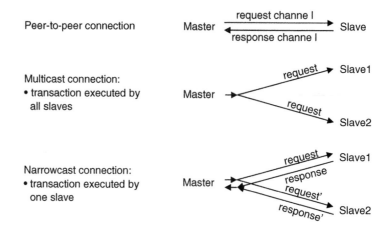

Connection support in Aethereal NI.

to the attached modules. Connections can be peer to peer (one master, one slave), multicast (one master, multiple slaves, all slaves executing each transaction, and, currently, no responses allowed to avoid merging of messages), and narrowcast (one master, multiple slaves, a transaction is executed by only one slave), as shown in Fig. 6.20 [21].

Connections are composed of unidirectional peer-to-peer channels (between a single master and a single slave). To each channel, properties are attached, such as guaranteed message delivery or not, in-order or unordered message delivery, and with or without timing guarantees. As a result, different properties can be attached to the request and response parts of a connection, or for different slaves within the same connection.

Connections can be opened and closed at any time. Opening and closing of connections takes time, and is intended to be performed at a granularity larger than individual transactions. In Aethereal, message delivery is guaranteed by not allowing network buffers to overflow. This is ensured using credit-based flow control at the link level to avoid router buffer overflow, and at the channel level (i.e., peer-to-peer between two NIs) to avoid NI buffer overflow.

Message ordering is offered natively by the channel implementation. Within a channel, messages are packetized and sent by the NI in order of their receipt from the attached core. The packets in a channel are forced on the same path in the NoC, where they are kept in order by the routers and NIs. This choice limits the flexibility in routing the packets (e.g., no dynamic routing is possible), however, simplifies significantly NI design and reduces its cost, because there is no need to reorder within the NI.

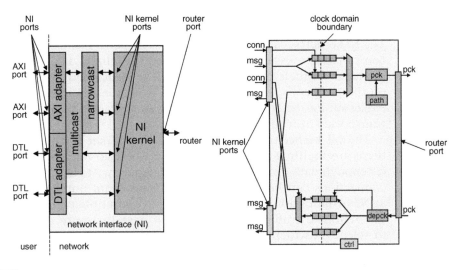

**■ FIGURE 6.21**

Architecture of the Aethereal NI. (a) NI kernel and shells. (b) NI kernel ports.

Ordering guarantees are provided only within a channel. Different channels are treated as separate entities in the NI scheduler, and they may have different routes. As a result, message reordering is possible across channels.

Support for real-time communication is achieved by providing throughput and latency guarantees. These are essential for complex real-time streaming application (e.g., high-end TV chips), because they considerably reduce the integration time, and allow design reuse. In Aethereal, throughput and latency guarantees are implemented by configuring connections as pipelined time-division multiplexed circuits over the network. Throughput guarantees are given by the number of slots reserved for a connection. A slot corresponds to a given bandwidth $B_i$ and, therefore, reserving $N$ slots for a connection results in a total bandwidth of $N \times B_i$. The latency bound is given by the waiting time until the reserved slot arrives and the number of routers data passes to reach its destination.

Reflecting the architectural template illustrated in Section 6.2, the Aethereal NI consists of two parts (see Fig. 6.21(a)):

- *The NI kernel*, which implements the channels, packetizes messages, and schedules them to the routers, implements the end-to-end flow control and clock-domain crossing.

- *The NI shells*, which implement the connections (e.g., narrowcast, multicast), transaction ordering for connections, and other higher-level issues specific to the protocol offered to the attached system cores.

### 6.9.1   NI Kernel Architecture

The NI kernel (see Fig. 6.21(b)) receives and delivers messages from system cores. These messages contain the sequentialized data provided by the modules via their protocol. The message structure may vary depending on the protocol used by the module. However, the message structure is irrelevant for the NI kernel, as it just sees messages as pieces of data to be transported over the NoC.

The NI kernel communicates with the NI shells via *ports*. Typically, there are multiple ports per NI, as an attached module does not transfer enough data to fill up the NoC link capacity. At each port, peer-to-peer connections can be configured, their maximum number being selected at NI instantiation time. A port can have multiple connections to allow differentiated traffic classes, in which case there are also *conn* signals to select on which connection a message is supplied or consumed. In Fig. 6.21(b), an example NI with two ports is shown, in which the first port can have up to two connections, and the second port can only have one connection.

A peer-to-peer connection consists of two channels: one request channel and one response channel (see Fig. 6.20). The reason for this choice is to come close to existing on-chip communication protocols (AXI, OCP, or DTL) and, consequently, have low-cost shell implementations. Alternatively, unidirectional connections (i.e., consisting of a single channel) are also possible, but they would expose a message-passing communication model which is not considered in this implementation.

Each channel uses two queues for storing messages, one in each of the two NI kernels. As there are two channels per connection, in each NI kernel, there are two message queues for each connection (one *source queue* for messages going to the NoC, and one *destination queue* for messages coming from the NoC), as previously showed in Fig. 6.15. Their size is also selected at the NI instantiation time. Queues are also used to provide the clock-domain crossing between the network and the attached modules. Each NI kernel port can, therefore, have a different clock frequency. Credit-based end-to-end flow control ensures that no data is sent unless there is enough space in the destination buffer to accommodate it.

From the source queues, data is packetized, and sent to the NoC via the router port. A packet header consists of the routing information, remote queue identifier (i.e., the queue of the remote NI in which the data will be stored), and piggybacked credits. The packetization is controlled by the *flit ctrl* module, which indicates if the produced word is a header (Fig. 6.22). The path and the remote queue IDs are taken from the *conn table*, and the credits to be reported are taken from the *credit table*.

A NI scheduler arbitrates between the channels that have data to be transmitted, as showed in Fig. 6.22.

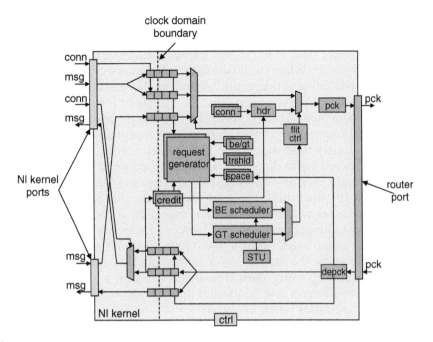

■ **FIGURE 6.22**

NI kernel architecture.

The scheduler is split in two:

1. The GT scheduler checks if the current slot is reserved for a GT channel. If the current slot is reserved, and there is sendable data in the queue for which the reservation has been made, that queue is scheduled.

2. The BE scheduler uses round-robin arbitration to select between the BE queues which have sendable data, but only when there is no queue scheduled in the GT scheduler.

In this way, the NoC provides time guarantees (i.e., bandwidth and latency), while at the same time allowing the remaining NoC capacity to be used by BE traffic.

To increase the NoC efficiency, it is preferable to send longer packets. To achieve this, some optimizations can be used. First, the decision on when the packet is finished is taken as late as possible. This allows newly arrived data to be attached to an ongoing packet, increasing its length, and optimizing NoC utilization. Second, a configurable threshold mechanism, which skips a channel as long as the sendable data is below the threshold (thresholds are stored in the *trshld table*, see Fig. 6.22). This is applicable to both BE and GT channels. To prevent starvation at user/application level (e.g., due to write data being buffered indefinitely on which the processor core waits for an acknowledge), a flush signal is also provided for each channel (and a bit in the message header) to temporarily override the threshold. When the flush signal is high for a cycle, a snapshot of its source queue filling is taken, and as long as the words in the queue at the time of flushing have not been sent, the threshold for that queue is bypassed.

The flush signal is controlled by the processor core. This is a standard technique in modern communication protocols, such as DTL (which has a similar flush signal) or AXI (which forces transmission of potentially buffered write data with an unbuffered write command). For read commands, no such flush-equivalent exists and, therefore, the NI shells always set the flush high for read commands and read data. Using thresholds increases network utilization, especially for small burst sizes, because it forces longer packets, and, hence, a lower number of packet headers. On the other hand, waiting for data to accumulate to create longer packets increases the latency, and longer packets require larger buffers. Consequently, using data thresholds involves a trade-off between network utilization on one hand, and latency/cost on the other hand, and should be used only when necessary.

A similar threshold is set for credit transmission (also in the *trshld* in Fig. 6.22). The reason is that, when there is no data on which the credits can be piggybacked, the credits are sent as empty packets, thus consuming extra bandwidth. To minimize the bandwidth consumed by credits, a credit threshold is set, which allows credits to be transmitted only when their sum is above the threshold. Similar to the data case, credits can also be flushed to prevent possible starvation.

As for the data thresholds, credit thresholds increase NoC utilization, especially for small burst sizes, because by forcing credit accumulation, less empty packets carrying only credits are generated. However, as opposed to the data thresholds, setting credit thresholds has little impact on the latency and buffer requirements. The reason is that the application has a periodic bursty behavior, and the time needed to report credits back (in one or multiple packets) is lower than the time between bursts. Consequently, credits are always reported in time, thus preventing data being buffered, and not affecting the latency.

As credits are piggybacked on packets, a queue becomes eligible for scheduling (i.e., *request generator* issues a signal for that queue to *GT scheduler* and *BE scheduler*) when either the amount of sendable data is above its data threshold or when the amount of credits is above its credit threshold. However, once a queue is selected, a packet containing the largest possible amount of credits and data will be produced.

For the incoming packets, the NI inspects the header, adds the credits to the counter space, and stores the data (without the header) in the queue specified by the queue ID field in the packet header. The data is then ready for consumption by the shells at the NI kernel ports.

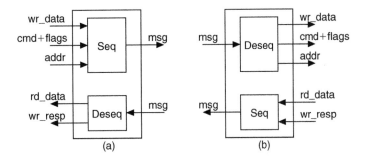

**FIGURE 6.23**

Master (a) and slave (b) NI shells.

### 6.9.2  NI Shell Architecture

Shells are typically added around the NI kernel. As an example, the NI of Fig. 6.21(a) has two DTL ports and two AXI port. All ports provide peer-to-peer connections. In addition to this, the two AXI ports provide narrowcast connections, and one DTL and one AXI port provide multicast connections. Note that these shells add specific functionality, and can be plugged in or left out at instantiation time, according to the requirements. Support for different protocols is possible due to the fact that the NI kernel is protocol agnostic.

The basic functionality of a master (slave) shell is to (de)sequentialize commands and their flags, addresses, and write data in request messages, and to (de)sequentialize messages into read data and write responses. Figure 6.23 illustrates this concept.

In Fig. 6.24(a), we show an example of 32-bit message structures for the AXI protocol resulting after sequentialization. These messages are passed from NI shells to the NI kernel. For the request message, the command (AWRITE) and all its flags are included in the first word of the message, the address (ADDR) is set in the second word, and the write data (WDATA), in the case of a write command, are appended at the end. There is one limitation in this encoding compared to the original AXI protocol: the strobe is identical for all transferred write words in a burst, while in AXI each write word has an individual strobe. If individual strobes for each word are required, this can be implemented either by extending the link width from for example, 32 to 36 bits to also accommodate the strobes or by adding extra strobe words in the message formats.

For the response message (Fig. 6.24(b)), there is a bit (R/B) specifying if the message corresponds to a read data or to a write response. R/BRESP[1:0] indicates if the transaction is successful or not. In the case of a read data response, _ALEN[3:0] is also copied in the response message by the slave AXI NI shell to reduce the NI buffering cost. Additionally, if

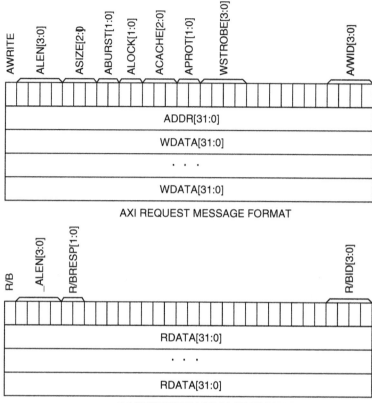

AXI REQUEST MESSAGE FORMAT

AXI RESPONSE MESSAGE FORMAT

■ **FIGURE 6.24**

Example of AXI message formats: (a) request and (b) response.

multiple connections are implemented at the NI port, a connection ID can be generated based on the transaction's address or thread ID.

Finally, the interested reader is referred to Ref. [21] for an in-depth description of narrowcast and multichannel shells.

### 6.9.3 Network Configuration

Before the Aethereal NoC can be used by an application, it must be configured. NoC (re)configuration means opening and closing connections in the system, and this could be done also while the system is running.

Opening/closing connections implies allocating/deallocating resources for communication. For example, a connection requires buffering resources, it is associated an identifier, it is configured a memory map, and it may possibly be allocated a priority and/or a portion of the bandwidth. In practice, these resources are allocated in the network components by writing registers via a control port using a standard protocol, such as AXI or DTL-MMIO.

Resource allocation requires arbitration for the case in which two connections request the same resources. This arbitration can be performed either centralized or distributed. In the centralized case, there is only one resource manager, which opens and closes all connections. In this case, the cost of managing resources is lower, and can produce better results. However, a centralized resource manager does not scale and cannot be used for large NoCs. In the distributed case, there are multiple resource managers distributed in the NoC, which arbitrate locally for their resources. This is a scalable resource management solution. However, the associated cost is higher. In the current prototype of the Aethereal NoC, a centralized configuration mechanism is used, because it is able to satisfy the needs of a small NoC (around 10 routers), and has a simpler design and lower cost. More specifically, when a centralized configuration scheme is used, routers do not require any configuration (they are stateless and identical). The reasons are as follows:

- By using source routing, the routing information is present in the packet headers.

- In the slot allocation scheme, it is enough to reserve slots in the source NI for a given path to guarantee the bandwidth for that path.

In this way, router design is simplified, leading to an approximately 30% lower-cost router, at the price of introducing headers for the GT traffic too.

In general, opening a connection between two NIs requires setting up two channels (one request channel and one response channel). For each channel, only the source NI needs to be configured. Consequently, opening a connection between two NIs results in configuring these two NIs. Setting up one channel consists of two parts:

1. Finding the free slots and selecting those to be allocated to the channel.

2. Writing the registers of the channel in the source NI.

Finding and selecting slots are performed traversing the slot tables of each link along the path from the source NI to the destination NI, and selecting the slots required to accommodate the required bandwidth. In the case not enough free slots are available, an error is returned to the programmer. Slot finding and selection is proportional to the length of the path and the slot table size.

Writing the registers of a channel implies executing write transactions over the NoC. This is again proportional to the NoC speed, NoC size

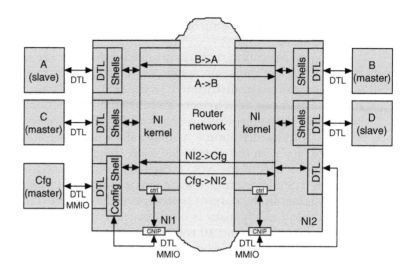

■ **FIGURE 6.25**

NI configuration mechanism.

(i.e., the distance between the NI where the configuration is performed and the NI to be configured), and the number of registers to be written (three for normal channel configuration, two in case the channel is attached to a narrowcast shell, and for slot reservation).

### 6.9.4    NI Configuration

NIs are configured via a *configuration port* (CNIP), which offers a memory-mapped view on all control registers in the NIs. This means that the registers in the NI are readable and writable by any master using standard read and write transactions. Configuration is performed using the NoC itself (i.e., there is no separate control interconnect needed for NoC configuration). Consequently, the CNIPs are connected to the NoC like any other slave (see CNIP at NI2 in Fig. 6.25). At the configuration module *Cfg*'s NI, a configuration shell (*Config Shell*) is introduced, which, based on the address, configures the local NI (NI1), or sends configuration messages via the NoC to other NIs.

Connections are set up by writing the proper values in NI registers. To configure a connection between two modules (e.g., from master B and slave A in Fig. 6.25), the NIs to which the modules are attached must be configured. These NIs are configured either directly (e.g., NI1 via CNIP), or by using a configuration connection (e.g., NI2 via the connection Cfg → NI2).

As NoC configuration is quite elaborate, and is susceptible to introducing programming errors, a high-level *application programming interface*

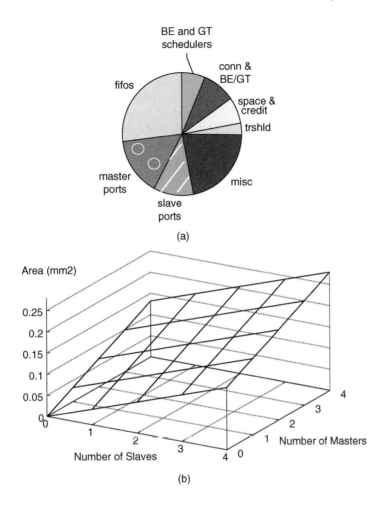

■ **FIGURE 6.26**

Area results. (a) Six ports NI area breakdown. (b) Trend with increasing number of ports.

(API) has been introduced for NoC configuration. This API provides primitives for transparently opening and closing NoC connections. Besides simplifying the NoC programming, this API allows the configuration code to be independent of the NoC implementation.

## 6.9.5  Implementation

Figure 6.26(a) shows a detailed view of the area consumption of different parts of the NI. The figure shows an NI with three single-connection master ports, four single-connection slave ports of which one is used for NI configuration, where all queues are eight words deep. Note that all queues are area-efficient *custom-made hardware FIFOs*.

One can note that for this NI instance, a large part of the area is consumed by the FIFOs (27%). Master and slave shells take 26% of the NI area. The tables for channel configuration (i.e., *conn, be/gt, trshld, space* and *credit*) occupy 19% of the NI area. The BE and GT schedulers occupy 6.5% of the NI area. The rest of 21.5% is consumed by other logic, such as packetization/depacketization, multiplexing of data, or credit management. If the NI has only single-connection ports, its area is approximately proportional to the number of ports. The non-proportional part consists of the master and slave shells, which have different costs, and the area in *misc*. Figure 6.26(b) shows the area of the NI, when we vary the number of master and slave single-connection ports.

If we consider an NI instance with four DTL ports, one for configuration (one channel to which the configuration shell is attached), two masters (one offering narrowcast with two channels), and one slave (multichannel with four channels), the total area is 0.169 mm$^2$ in 0.13-$\mu$m technology. It grows to 0.25 mm$^2$ after layout, due to the utilization (70%), and to the power ring (not normally needed when the NI is part of a larger design). The operating frequency is 500 MHz.

The latency introduced by the NI is two cycles in the DTL master shell (due to sequentialization, as part of packetization), 0–2 in the narrowcast and multicast shells (depending on the NI instance), and between one and three cycles in the NI kernels (as data needs to be aligned to a three-word flit boundary), and two clock cycles for clock-domain crossing.

Additional delay is caused by the arbitration, but it is not included in the NI latency, as it needs to be performed anyway (also in the case of a bus, arbitration is performed). The resulting latency overhead introduced by the NI is between four and ten cycles, which is pipelined to maximize throughput.

## 6.9.6  NI and the Rest of the System

The Aethereal NoC has been implemented in a companion chip for a TV system architecture [24]. TV systems tend to be partitioned into multiple chips or dies (system in a package). Ultimately, however, the multiple chips implement a single system. Hence it is important to be able to design a system as a whole, and then to be able to seamlessly distribute it over multiple chips. Alternatively, a given combination of (companion) chips should be easily combined to a working system.

For this purpose, the scope of NoCs could be extended to transparently span multiple chips. In fact, several companion chips are usually combined in an *ad hoc* manner. NoCs should make this process more structured, simpler, and faster. Networks on different chips are connected through *high-speed external links* (HSEL). A dedicated interconnect of an

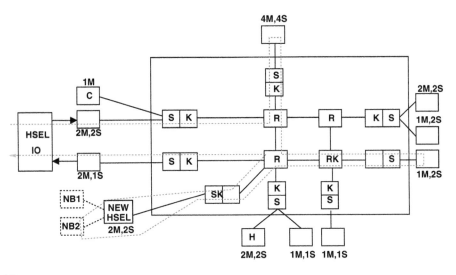

■ **FIGURE 6.27**

Architecture of the companion chip with the NoC.

existing TV companion chip has been replaced by an instance of the Aethereal NoC, with a port dedicated to an HSEL. Let us now try to capture the impact of NIs on system interconnect area and power.

The original TV companion chip contains nine system blocks for enhancing video (some are subsystems). Their processing order is fixed, except for one block (horizontal scaler H), which can appear at one of two places in the processing pipeline. Video data enters and leaves the companion chip through a single HSEL. Depending on the mode and location in the processing chain, pixel data is 8, 9, or 10 bits in 4:4:4, 4:2:2, or 4:2:0 Y:U:V format. Pixel data is augmented with 4 bits of side-band information (e.g., end of field). All blocks use a streaming data protocol (DTL peer-to-peer streaming data).

The dedicated interconnect design of this companion chip is addressed in Ref. [53]. It consists of wires for direct block-to-block connection, with multiplexers for a few bypasses. The NoC-centric redesign of system architecture is reported in Fig. 6.27. The NoC is showed at the core of the system, with its components (routers R, NI kernels K, and NI shells S). The number of master (M) and slave (S) ports are shown at each NI. The NoC enables a programmable order of hardware blocks. The SOC's input and output are shown by the HSELIO block, which directly connects to two processing blocks, for legacy reasons. A new HSEL to attach another companion chip, such as an FPGA, is directly connected to two master and two slave NI ports. The computation units include a horizontal scaler (H), two new blocks (NB1, NB2), and a control processor (C). The dashed

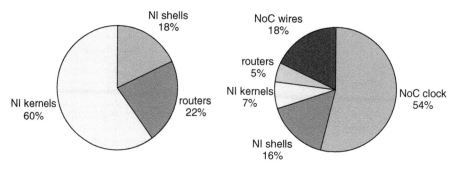

■ **FIGURE 6.28**

(a) Relative contributions to increase in area. (b) Relative contributions to increase in power dissipation.

line in Fig. 6.27 illustrates a task graph that includes a new block (NB2) on another companion chip. Further system details are reported in Ref. [24].

As a result of the (verified) NoC dimensioning and configuration, a $2 \times 2$ mesh with routers of arity 3, 4, and 5 is used. The TDMA slot table contains 20 slots. A total of 10 computation units are connected to the 8 NIs, using 38 NI ports with one connection each. One additional port per NI (8 total) is used for programming the NoC. The number of master (M) and slave (S) ports of each NI is shown in Fig. 6.27. In the extended version, the computation units run at 100 MHz, and the NoC at 300 MHz, in a CMOS 0.13-$\mu$m technology. The NoC can run up to 500 MHz but the raw link bandwidth of 1.2 Gbit/s (32 bits times 300 MHz) is sufficient for this application.

The Aethereal NoC is (re)configured at run time with the required task graph using memory-mapped IO. A control processor is directly connected to the NoC (block C in Fig. 6.27). In a definitive implementation the MIPS control processor on the main chip [53] would run the embedded reconfiguration software.

The area of the new companion chip is 4% larger than the original. Figure 6.28(a) shows that the routers cause 22% of the increase in area, the NI kernels 60%, and the NI shells 18%. The area numbers of the NI kernels and routers are produced by the Aethereal design flow for an NoC running at 500 MHz. These area numbers will therefore be an over-estimate. The routers support both GT and BE traffic. The BE traffic class is only used for configuration, and using GT also for this traffic would reduce the router area by a factor 4. The NI kernels use most area (60%) because they implement the per-connection buffers at the edge of the network, containing a total of 2608 32-bit words. There are a total of 46 NI ports each supporting one connection (38 functional ports, and one port per NI, 8 total, for programming the NoC). With two queues per port (one for each of the request and response traffic), queues are on average 28 words deep (2608 words/((38 + 8) × 2) queues).

For the most demanding task graph, the power dissipation increases by 12%. Figure 6.28(b) shows that the largest percentage (54%) is due to the

NoC clock, followed by the NoC wires (18%). Wire lengths of 0.8 mm for computation units to NI links, and intra-NoC wires of 1.1 mm were used. The high cost of clocking is caused by Aethereal's synchronous implementation, to offer performance guarantees. Note that the frequency of the NoC is constant for all task graphs, and dimensioned for the worst case (most demanding task graph). Promising recent results show that scaling the frequency and operating voltage of the NoC based on the required load per task graph can lead to average savings of 50% [54]. A 6% increase in power dissipation is acceptable for stationary systems.

Latency is not a major concern in streaming systems, such as the video-processing application. But in any case, for the most demanding task graph, the additional latency introduced by the NoC for the whole processing pipeline of the companion chip (HSEL input to HSEL output) is 126 $\mu$s, or just 10%. Part of the increase in latency is due to the NoC latency, but most is caused by the fact that the connections in the task graphs are not dimensioned for the peak bandwidth of the communicating cores, but for the average bandwidth. As a result, non-uniform processing speeds (in particular, above-average bursts) increase the total latency. Allocating more than average bandwidth will reduce the total latency. In computing the performance of the system as a whole, both the guaranteed communication performance and the computation performance must be taken into account therefore [55]. The latency discussed above does not include the time to configure the NoC, which is 118 $\mu$s.

## 6.10 NI ARCHITECTURE: THE xpipes CASE STUDY

xpipes is a SystemC library of parameterizable, synthesizable NoC components (NI switch and link modules), optimized for low-latency and high-frequency operation [44]. Communication is packet switched, with source routing (based on street-sign encoding) and wormhole switching.

The xpipes NI is designed as a bridge between an OCP interface and the NoC-switching fabric. Its purposes are the synchronization between OCP and xpipes timings, the packeting of OCP transactions into xpipes flits and vice versa, the computation of routing information, and the buffering of flits to improve performance. The xpipes NI is designed to comply with version 2.0 of the OCP specifications. In addition to the core OCP signals, support includes for example the ability to perform both non-posted or posted writes (i.e., writes with or without response) and various types of burst transactions, including reads with single request and multiple responses.

To provide complete deployment flexibility, the NI is parameterizable in both the width of OCP fields and of xpipes flits. Depending on the ratio between these parameters, a variable amount of flits is needed to encode

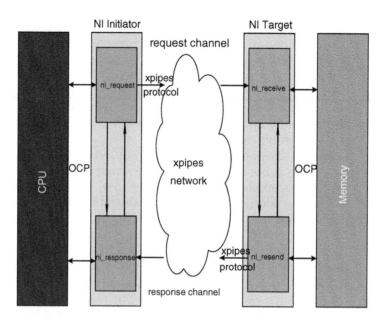

■ **FIGURE 6.29**

xpipes NIs.

an OCP transaction. For any given transaction, some fields (such as the OCP MAddr wires, specific control signals, routing information) can be transmitted just once; in contrast, other fields (such as the OCP MData or SData wires) need to be transmitted repeatedly, for example during a burst transaction. Thus, the NI is built around two registers: one holds the transaction header, while the second one holds the transaction payload. The first register samples OCP signals once per transaction, while the second is refreshed on every burst beat. A set of flits encodes the header register; subsequently, multiple sets of flits are sent toward the fabric, each encoding a snapshot of the payload register subsequent to a new burst beat. Sets of payload flits are pushed out until transaction completion. Header and payload content is never allowed to mix in the same flit, thus simplifying the required logic. Routing information is attached to the header flit of a packet by checking the transaction address against an LUT.

As shown in Fig. 6.29, two NIs are implemented in xpipes, named *initiator* (attached to system masters) and *target* (attached to system slaves). A master–slave device will need two NIs, an initiator and a target, for operation. Each NI is additionally split in two sub-modules: one for the request and one for the response channel. These sub-modules are loosely coupled: whenever a transaction requiring a response is processed by the request channel, the response channel is notified; whenever the response is received, the request channel is unblocked. The mechanism is currently supporting only one outstanding non-posted transaction. The xpipes

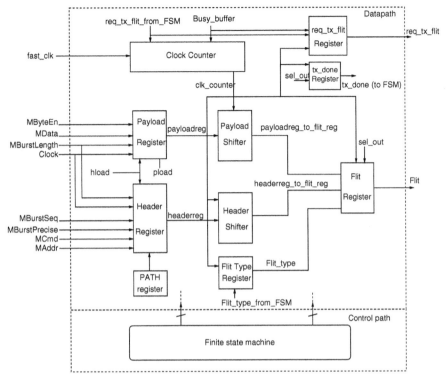

■ **FIGURE 6.30**

Architecture of the request path front-end.

interface of the NI is bidirectional; for example, the initiator NI has an output port for the request channel and one input port for the response channel (the target NI is dual). The output stage of the NI is identical to that of the xpipes switches, for increased modularity. The input stage is implemented as a simple dual-flit buffer with minimal area occupation, but still makes use of the same flow control used by the switches.

Let us now focus on the front-end block of the request path, whose detailed architecture is reported in Fig. 6.30. The back-end is represented by the output buffer of the NI, on the right-hand side of the plot. The interface signals with the output buffer are represented by:

- The *flit* signals, that is all the flits resulting from the packetization mechanism, which are input to the output buffer.

- The *req_tx_flit* signal, which is risen every time a new output flit is ready.

- The *busy_buffer* signal, which is asserted by the back-end to stop the flow of flits from the front-end.

The sub-modules of the request path front-end are described in Fig. 6.30 in a pure RTL stile, thus they are composed by a data-path and a control path. The data-path is built up of very simple elements as registers, multiplexers and some concatenation modules. The control is mainly formed by a *finite state machine* (FSM), and will not be described in detail for lack of space. The interface signals between request and response path have been also omitted for the sake of coinciseness.

The input OCP signals are sampled either by the payload register or by the header register, depending on the kind of information they carry. The header register is able to contain the header of each packet, and its width is parameterizable. The signal *hload*, coming from the FSM, indicates that the inputs can be sampled, since a new request phase is taking place. During the whole transmission of a packet, *hload* can be risen just once, for the header itself is transmitted just once whatever the OCP command to be carried out. In order to allow an easier un-packeting process at the destination NI, all the inputs forming the header fields are stored in the correct position right from the beginning. Notice that also routing information are stored in the header register, as soon as they are looked up in a table based on the destination address of the OCP transaction.

The payload register is used to store payload information, and its width depends on the width of OCP fields. Moreover, it is chosen according to the worst-case transaction, namely burst writes. Depending on the kind of transaction, not all information in the payload register are meaningful. The fields that are not necessary are zeroed out (e.g., *MData* for a read transaction). The fields of the payload register are sampled when *hload* and *pload* are driven high at the same time. Notice that the payload register can be sampled many times for the same OCP transaction, for instance for burst writes.

Both the *headerreg* and the *payloadreg* signals cross their respective shifters. This is the exact point where also clock-domain crossing takes place. In fact, the header and payload registers, together with the FSM, are clocked at the OCP clock, since they have to stay synchronized with the attached system cores. In contrast, the other blocks operate at the highest network frequency *fast_clock* or are triggered by a *clk_counter* signal which is synchronous with *fast_clock*. The *header* and *payload shifters* manage flit transmission, and output two signals (*headerreg_to_flit_reg* and *payloadreg_to_flit_reg*) that represent the intermediate flit format before the flit type is concatenated. The flit type (head, payload, or tail) is produced by the appropriate synchronous block based on the control signal *flit_type_from_FSM*.

The *flit register* block contains a multiplexer which selects which type of flit has to be transmitted. It is controlled by a signal (*sel_out*) coming out of the FSM and denoting whether header or payload flits have to be transmitted. At this stage, flits assume their final shape, by concatenating the flit type with the header or payload flits in intermediate format. This block is obviously synchronized with the xpipes clock domain.

Two further control blocks complete the architecture. The *req_tx_flit register* drives an output wire that alerts the back-end module of the request channel that a new flit is ready for transmission. This signal is activated provided the output buffer is not full (see the *busy_buffer* signal) and based on the FSM command. Finally, the *tx_done* signal is fed to the FSM to indicate that all the flits covering a packet header or payload have been successfully transmitted, and new OCP transactions can be accepted.

Let us briefly focus on the frequency decoupling support. A global parameter is defined in the architecture which determines the number of *fast_clk* cycles (clock period of the network) building up one *clock* cycle (clock period of the system cores). xpipes clock period is always greater or, at least, equal to the OCP one. Frequency decoupling is actually accomplished through the *clock counter* block. Its output *clk_counter* is updated on the positive edge of *fast_clk*, if a new flit can be sent since the output buffer can accept it (*busy_buffer*) and the flit itself is ready (*req_tx_flit_from_FSM*). All the sub-blocks sensitive to *clk_counter* can update themselves when it changes. *clk_counter*, during the transmission phase, rules which register section has to be packed within the current flit. This counter is reset when there are no more flits for the header or for the payload to be sent. Packet transmission therefore takes place as follows, once the header and payload registers have been loaded with OCP fields. If *clk_counter* is zero, inside the *header shifter* the least significant *FLITWD* bits (corresponding to the flit width minus the flit-type bits) are selected to form the partial flit. If *clk_counter* equals one, then the bits in the range $2(FLITWD - 1)$ to *FLITWD* are selected, and so on. At the end, the most significant bits will be selected. At this point, it is necessary to synchronize with the next rising edge of the OCP clock, activate the *tx_done* signal, reset the *clk_counter* and start sweeping the *payload shifter* in the same way.

With a UMC 0.13-$\mu$m technology library, a Master NI with a 4-flit output buffer (in the request path, since the response path has a fixed 2-flit input FIFO) consumes $0.029\,\text{mm}^2$, while a dual Slave NI with the same amount of buffering consumes $0.0315\,\text{mm}^2$. The operating frequency is about 1 GHz (on the network side), and the latency for traversing the NI is two clock cycles (including the output inferring stage). The OCP side can be operated at 1 GHz, or at any divider of this frequency.

A breakdown of area occupation of a Master NI is reported in Fig. 6.31(a). Most of the area is drawn by the ni_request block, since it contains the big *header* and *payload registers*, which store the entire OCP transaction fields. The output buffer (*out_buffer* in the plot) of the request path consumes a relevant fraction of the area as well. In particular, since in this instance it has a 6-flit depth, it is 3 times larger than the buffer in the *ni_buffer* block (which has a 2-flit depth). However, due to the control logic complexity, the area is almost 2.5 times as much. This confirms that most of the area of the NI is drawn by buffering resources. Similarly, the NI response block, interfacing the response path of the NI with the attached core, is dominated by the area required by the registers storing

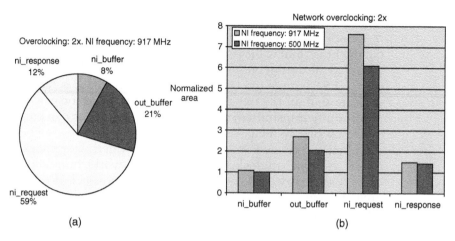

**■ FIGURE 6.31**

Area breakdown of a master NI.

OCP response fields. The needed buffering is much lower with respect to the *ni_request* block, since this time we just have to store response data and state, namely *SData* and *SResp* signals. Most of the transaction control signals have been previously discarded. The control logic complexity decreases accordingly.

In general, if the target synthesis frequency of the NI is reduced (see Fig. 6.31(b)), an area saving can be observed, due to the increased margin for the synthesis tool to perform area optimizations. Of course there will be a saturation at some point. Finally, if we keep the xpipes clock frequency constant and decrease the OCP frequency (because we are for instance attaching slow cores to the network), the area savings are marginal since the NI part operating at the reduced frequency is only a small fraction of whole area.

## 6.10.1   The NI and the Rest of the System

In order to assess the impact of the NI on a real NoC-centric system, let us analyze the system which was already illustrated in Fig. 6.12. We recall that this system implements a mesh-like topology with the xpipes NoC architecture, and exhibits 15 master NIs, 15 slave NIs, and 15 switches. Output stages of switches and NIs all have 3-flit buffers.

The area breakdown for the entire NoC is reported in Fig. 6.32(a). It shows that master NIs have almost the same area as slave NIs and a slightly lower area than switches. In fact, xpipes NIs have almost the same area as $5 \times 5$ switches, however, in the topology under test we have three $6 \times 6$ switches, eight $5 \times 5$ switches, and four $4 \times 4$ switches. Switches with arity

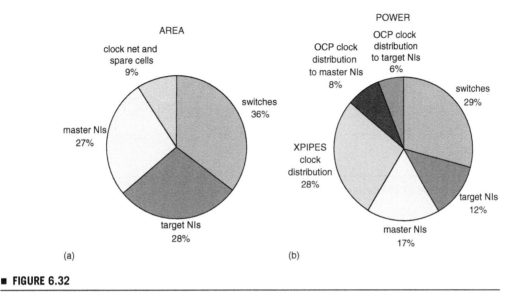

■ **FIGURE 6.32**

Area (a) and power (b) breakdown of an entire xpipes-based system.

6 explain the overall larger area of switches with respect to master or slave NIs. The remaining area is consumed by the clock distribution network and the spare cells. Notice that the area drawn by the power distribution network is not taken into account here.

As regards power consumption (see estimation flow in Ref. [25]), Fig. 6.32(b) interestingly shows that switches consume the same power as the sum of master and slave NI contributions. This is a result of the lower switching activity in NIs and of the fact that part of them works at a lower frequency than the rest of the network. As already observed for the Aethereal case study, the power consumption of the clock tree dominates overall power dissipation (42%). This stresses the importance of power minimization techniques targeting the clock distribution network and the sequential logic in the system.

## 6.11  NIs FOR ASYNCHRONOUS NoCs: THE MANGO CASE STUDY

Since Aethereal and xpipes are purely clocked designs and do not address global synchronization issues, we finally introduce an NI architecture which bridges a clocked OCP socket and an asynchronous network. MANGO (*Message-passing Asynchronous Network-on-chip providing Guaranteed services over OCP interfaces*) [56] is a clockless NoC architecture enabling GALS-type systems, which leverages the advantages of

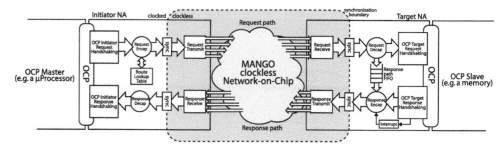

■ **FIGURE 6.33**

Master and slave NI architectures in the clockless MANGO NoC.

asynchronous implementations, namely zero dynamic idle power and low forward latency. It provides connectionless BE routing as well as connection-oriented guaranteed services (latency and bandwidth bounds).

The MANGO NI performs OCP transaction handshaking at the front-end, encapsulation of the transactions for the underlying packet-switched network and synchronization between a communicating core and the network. The architectural template of the master and slave NIs is reported in Fig. 6.33 [15]. It includes a request and response path, and the clocked and clockless parts can be clearly identified. The NI provides a number of input and output network ports, each corresponding to a time-guaranteed (GS) connection or to a BE service. Output ports are pointed to using the OCP signal *MConnID*, and several threads may use each port without restrictions.

In the initiator NI, the *OCP initiator request handshaking* module implements part of the OCP slave socket. The *Request encap* module maps the OCP signals to packet signals. For BE transmissions the routing path is read from an LUT (programmable through the OCP interface) and is appended to the payload. This procedure takes one clock cycle. There would be little performance gains here by using clockless circuits in the handshaking and encapsulation modules, since the OCP interface is clocked itself. Serialization of packet flits is done by the *Request transmit* module, in the clockless domain. This makes flit serialization independent of the OCP clock speed. Also, the synchronization overhead is minimized by synchronizing an entire packet to the clockless domain in one go, instead of one flit at a time.

In the target NI packets are disassembled in the clockless *Request receive* and decapsulated in the clocked *Request Decap* modules. The *OCP target request* and *Response handshaking* modules then conduct an OCP transaction accordingly. A packet is forwarded to the clocked part of the NI only when the entire packet has arrived. If the transaction requires a response, this latter is encapsulated by the slave NI, serialized in the clockless Response transmit module and transmitted across the network.

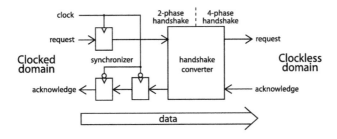

■ **FIGURE 6.34**

Synchronization module.

While a request is being processed by the slave core, the *Response path FIFO* stores the return path. Response packets are reassembled in the *Response Receive* module, decapsulated in the *Response Decap* module, and the transaction is completed by the *OCP initiator response handshaking* module.

The OCP signal *MThreadID* is stored in request packets and returned in the corresponding response. The NIs thus handle any number of outstanding transactions. If ordering is guaranteed by the network, for instance on GS connections, multiple transactions can even be initiated on each thread.

Interrupts are implemented as *virtual wires*: an interrupt triggers the slave NI to transmit an interrupt packet. Its destination is defined by configuring the slave NI with a routing path or connection ID. Upon receiving this packet, the initiator NI asserts the local interrupt signal pin.

In order to minimize the synchronization overhead, a two-phase handshake channel is implemented, employing a two-flop synchronizer [57]. As illustrated in Fig. 6.34, in the clockless domain there is a conversion to a four-phase protocol compliant with the handshaking of MANGO. A complete clock cycle is available for resolving metastability in the flip-flop. The estimated *mean-time between failure* (MTBF) at a 400 MHz OCP clock is longer than 8000 years [15].

## 6.11.1  Implementation Results

NIs with 32-bit OCP data and address fields and 32-bit network ports have been implemented using 0.13-$\mu$m standard cells. Area, port transmit speed and transaction latency values are reported in Table 6.6. The area values are purely cell area (pre-layout). Due to the GS input buffers, the area of the slave NI supporting GS ports is twice that supporting only a BE port. The area of the master NI however is not very different in the two configurations, its main part being the BE routing path LUT. A pure GS master would be smaller.

**TABLE 6.6** ▪ Standard cell implementation in $0.13\,\mu m$ technology. Worst-case process corner.

| NI type | OCP speed | Area | | Port transmit speed | | OCP read | | OCP write | | OCP interrupt | |
|---|---|---|---|---|---|---|---|---|---|---|---|
| | | Initiator | Target | Initiator | Target | BE | GS | BE | GS | BE | GS |
| BE only | 400 MHz | 0.058 mm² | 0.020 mm² | 725 MHz | 1.08 GHz | 6 clks | – | 4 clks | – | 3 clks | – |
| BE + 3GS | 400 MHz | 0.064 mm² | 0.039 mm² | 483 MHz | 633 MHz | 8 clks | 6 clks | 5 clks | 4 clks | 4 clks | 3 clks |
| BE + 3GS | 200 MHz | 0.060 mm² | 0.037 mm² | 483 MHz | 633 MHz | 6 clks | 6 clks | 3 clks | 3 clks | 3 clks | 3 clks |

The port transmit speed indicates the flit speed on the clockless output ports. Since these are independent of the OCP clock, flits can be transmitted fast, even from a slow core. The NIs with only a BE port are simpler than those with multiple GS ports, hence their port speed is higher. Finally, the table shows the latency overhead of conducting transactions through the NIs, relative to transactions of a master and a slave core connected directly. The latency is displayed in OCP cycles. The overhead in the clockless circuit is independent of the OCP clock, hence the overhead in terms of cycles is less for slower cores.

BE transactions necessitate transmitting the routing path in a header flit, thus the total latency is increased. For a slow core, however, an entire packet can be transmitted in less than one OCP clock cycle, even when needing a routing header, hence the overhead of BE transactions is not greater than that for GS transactions.

## 6.12   SUMMARY

This chapter has addressed NI architecture and design issues, leveraging a number of case studies. It has presented interface protocols at the front-end of NI architectures as the result of a long evolutionary path of SoC communication protocols. Moreover, it has showed some implementation approaches to packetization, from traditional wrapper-based solutions up to the potential solutions opened by the latest developments in processor interfaces. This chapter has then analyzed the impact that design choices taken at the NI might have on system-level performance, area and power: from packet size and flit width all the way to flow control schemes and switching techniques. Finally, three representative case studies have been presented, addressing design issues of NIs for QoS-oriented, BE, and asynchronous NoCs. Through the area and power breakdown of NI statistics, the reader has been able to understand where complexity arises in the architecture, and has been led to assess the fraction of area and power NIs consume in real platform implementations using NoCs.

## REFERENCES

[1] G. Cyr, G. Bois and M. Aboulhamid, "Generation of Processor Interface for SoC Using Standard Communication Protocol," *IEE Proceedings of the Computers and Digital Techniques*, Vol. 151, No. 5, 2004, pp. 367–376.

[2] Virtual Component Interface Standard, http://www.vsi.org

[3] M.S. Allen, W.K. Lewchuk and J.D. Coddington, "A High Performance Bus and Cache Controller for PowerPC Multiprocessing Systems," *IEEE International Conference on Computer Design: VLSI in Computers and Processors (ICCD)*, 1995, pp. 204–211.

[4] R. Hofmann and B. Drerup, "Next Generation CoreConnect Processor Local Bus Architecture," *IEEE International ASIC/SOC Conference*, 2002, pp. 221–225.

[5] A. Allison, (editor), "Hardware and Software Platforms: Embedded to Supercomputing," *Inside the New Computer Industry*, Issue 121, pp. 1–28, June 1999.

[6] *ARM7DMI Data Sheet*, Advanced RISC Machines Ltd, 1994.

[7] J. Gaisler, "A Portable and Fault-Tolerant Microprocessor Based on the SPARC v8 Architecture", *International Conference on Dependable Systems and Networks*, June 2002, pp. 409–415.

[8] B. Ackland et al., "A Single-Chip, 1.6-Billion, 16-b MAC/s Multiprocessor DSP", *IEEE Journal off Solid-State Circuits*, Vol. 35, No. 3, March 2000, pp. 412–424.

[9] P. Wodey, G. Camarroque, F. Baray, R. Hersemeule and J.P. Cousin, "LOTOS Code Generation for Model Checking of STBus Based SoC: The STBus Interconnect," *First ACM and IEEE International Conference on Formal Methods and Models for Co-Design*, June 2003, pp. 204–213.

[10] K. Keutzer, A.R. Newton, J.M. Rabaey and A. Sangiovanni-Vincentelli, "System-Level Design: Orthogonalization of Concerns and Platform-Based Design," *IEEE Transactions on Computer-Aided Design of Integrated Circuits and Systems*, Vol. 19, No. 12, December 2000, pp. 1523–1543.

[11] *The importance of sockets in SoC design*, White paper from http://www.ocpip.org/socket/whitepapers/

[12] *OCP-IP, emphOpen Core Protocol Specification*, Version 2.0, September 2003, www.ocpip.org

[13] P. Bhojwani and R. Mahapatra, "Interfacing Cores with On-Chip Packet-Switched Networks," *International Conference on VLSI Design*, January 2003, pp. 382–387.

[14] R. Mullins, J.G. Lee and S. Moore, "Selecting a Timing Regime for On-Chip Networks," *Proceedings of the 17th UK Asynchronous Forum*, September 2005.

[15] T. Bjerregard, S. Mahadevan, R.G. Olsen and J. Sparso, "An OCP Compliant Network Adapter for GALS-Based SoC Design Using the MANGO Network-on-Chip," *International Symposium on System-on-Chip*, November 2005, pp. 171–174.

[16]  E. Bolotin, I. Cidon, R. Ginosar and A. Kolodny, "QNoC: QoS Architecture and Design Process for Network on Chip," *The Journal of Systems Architecture*, Special Issue on Networks on Chip, Vol. 50, No. 2–3, February 2004, pp. 105–128.

[17]  Philips Semiconductors, *Device Transaction Level (DTL) Protocol Specification*, Version 2.2, July 2002.

[18]  S.J. Lee, K. Kim, H. Kim, N. Cho and H.J. Yoo, "Adaptive Network-on-Chip with Wave-Front Train Serialization Scheme," *Symposium on VLSI circuits*, 2005, pp. 104–107.

[19]  U. Ogras, J. Hu and R. Marculescu, "Key Research Problems in NoC Design: A Holistic Perspective," *Proceedings of the CODES ISSS*, September 2005, pp. 69–74.

[20]  E. Nilsson, M. Millberg, J. Oberg and A. Jantsch, "Load Distribution with the Proximity Congestion Awareness in a Network on Chip," *Proceedings of the Design Automation and Test in Europe Conference 2003*, 2003, pp. 1126–1127.

[21]  A. Radulescu, J. Dielissen, S.G. Pestana, O.P. Gangwal, E. Rijpkema, P. Wielage and K. Goossens, "An Efficient On-Chip NI Offering Guaranteed Services, Shared-Memory Abstraction, and Flexible Network Configuration," *IEEE Transactions on Computer-Aided Design of Integrated Circuits and Systems*, Vol. 24, No. 1, January 2005, pp. 4–17.

[22]  D. Wiklund and D. Liu, "Socbus: Switched Network on Chip for Hard Real Time Systems," *International Parallel, and Distributed Processing Symposium*, April 2003.

[23]  P.T. Wolkotte, G.J.M. Smit, G.K. Rauwerda and L.T. Smit, "An Energy-Efficient Reconfigurable Circuit-Switched Network-on-Chip," *IEEE International Parallel and Distributed Processing Symposium*, April 2005, p. 155a.

[24]  F. Steenhof, H. Duque, B. Nilsson, K. Goossens and P. Llopis "Networks on Chips for High-End Consumer-Electronics TV System Architectures," *Design Automation and Test in Europe Conference*, March 2006.

[25]  F. Angiolini, P. Meloni, S. Carta, L. Benini and L. Raffo, "Constrasting a NoC and a Traditional Interconnect Fabric with Layout Awareness," *Design Automation and Test in Europe Conference*, March 2006, pp. 124–129.

[26]  J. Nurmi, H. Tenhunen, J. Isoaho and A. Jantsch (editors), *Interconnect-Centric Design for Advanced SoC and NoC*, Kluwer Academic Publisher, Norwell, MA, April 2004.

[27]  A. Scherrer, T. Risset and A. Fraboulet, "Hardware Wrapper Classification and Requirements for On-Chip Interconnects," *Signaux, Circuits et Systmes 2004*, Monastir, Tunisie, March 2004, pp. 31–34.

[28]  P. Martin, "Design of a Virtual Component Neutral Network-on-Chip Transaction Layer," *Design Automation and Test in Europe Conference*, 2005, pp. 336–337.

[29]  E. Nilsson and J. Oeberg, "Reducing power and Latency in 2-D Mesh NoCs Using Globally Pseudochronous Locally Synchronous Clocking," *CODES ISSS*, September 2004, pp. 176–181.

[30]  S.J. Lee, K. Lee and H.J. Yoo, "Analysis and Implementation of Practical, Cost-Effective Networks on Chips," *IEEE Design and Test of Computers*, September–October 2005, pp. 422–433.

[31]  E. Svensson, O. Seger, D. Liu, D. Wiklund and S. Sathe, "Socbus: The Solution of High Communication Bandwidth On Chip and Short TTM," *Real-Time and Embedded Computing Conference*, September 2002.

[32] M. Dall'Osso, G. Biccari, L. Giovannini, D. Bertozzi and L. Benini, "xpipes: A Latency Insensitive Parameterized Network-on-Chip Architecture for Multiprocessor SoCs," *International Conference on Computer Design*, 2003, pp. 536–539.

[33] L. Scheffer, "Methodologies and Tools for Pipelined On-Chip Interconnect," *International Conference on Computer Design*, 2002, pp. 152–157.

[34] L.P. Carloni, K.L. McMillan, A.L. Sangiovanni-Vincentelli, "Theory of Latency-Insensitive Design," *IEEE Transactions on CAD of ICs and Systems*, Vol. 20, no. 9, September 2001, pp. 1059–1076.

[35] E. Rijpkema, K. Goossens, A. Radulescu, J. Dielissen, J. van Meerbergen, P. Wielage and E. Waterlander, "Trade-Offs in the Design of a Router with Both Guaranteed and Best-Effort Services for Networks on Chip," *IEE Proceedings of Computers and Digital Techniques*, Vol. 150, No. 5, September 2003, pp. 294–302.

[36] F. Angiolini, P. Meloni, D. Bertozzi, L. Benini, S. Carta and L. Raffo, "Networks on Chips: A Synthesis Perspective," *Proceedings of the Parallel Computing (ParCo) Conference*, September 2005.

[37] L. Benini and G. De Michli, "Powering Networks on Chips," *ISSS*, 2001, pp. 33–38.

[38] ARM, *AMBA 3.0 AXI Specification*, March 2004.

[39] ARM, *AMBA 2.0 AHB Specification*, May 1999.

[40] A.S. Tanenbaum, *Computer Networks*, Prentice Hall, Upper Saddle River, NJ, 1996.

[41] O.P. Gangwal, A. Radulescu, K. Goossens, S.G. Pestana and E. Rijpkema, "Building Predictable Systems on Chip: An Analysis of Guaranteed Communication in the Aethereal network on Chip," In P. van der Stok (editor), *Dynamic and Robust Streaming In and Between Connected Consumer-Electronics Devices*, Springer, Berlin, Germany, 2005.

[42] M. Millberg, E. Nilsson, R. Thid, S. Kumar and A. Jantsch, "The Nostrum Backbone – A Communication Protocol Stack for Networks on Chip," *International Conference on VLSI Design*, 2004, pp. 693–696.

[43] J.L. Ayala, M. Lopez-Vallejo, D. Bertozzi and L. Benini, "State-of-the-art SoC communication architectures," In R. Zurawski (editor), *Embedded Systems Handbook*, CRC Press, Florida, 2005.

[44] S. Stergiou, F. Angiolini, S. Carta, L. Raffo, D. Bertozzi and G. De Micheli, "xpipes Lite: A Synthesis Oriented Design Library for Networks on Chips," *Design Automation and Test in Europe Conference*, 2005, pp. 1188–1193.

[45] A. Radulescu and K. Goossens, "Communication Services for Networks on Chip," *SAMOS*, 2002, pp. 275–299.

[46] T.R. Halfhill, Tensilica's automation arrives: new design tool creates CPU extensions from C/C++ programs, Microprocessor Report, www.MPRonline.com, July 12, 2004.

[47] S. Leibson and J. Kim, "Configurable Processors: A New Era in Chip Design," *IEEE Computer*, Vol. 38, No. 7, July 2005, pp. 51–59.

[48] T. Tohara, "A New Kind of Processor Interface for a System-on-a-Chip Processor with TIE Ports and TIE queues of Xtensa LX," *Proceedings of the Innovative Architecture for Future Generation High Performance Processors and Systems (IWIA'05)*, January 2005, pp. 72–79.

[49] T.T. Ye, L. Benini and G. De Micheli, "Packetized On-Chip Interconnect Communication Analysis for MPSoC," *Design Automation and Test in Europe Conference*, 2003, pp. 10344–10349.

[50] Instruction-Level Parallelism in Shared Memory Multiprocessors, http://rsim.cs.uiuc.edu/rsim/

[51] D.A. Patterson and J. Hennessy, *Computer Organization and Design, The Hardware/Software Interface*, Morgan Kaufmann Publishers, San Francisco, CA, 1998.

[52] W.J. Dally and B. Towles, *Principles and Practices of Interconnection Networks*, Morgan Kaufmann, San Fransisco, CA, 2004.

[53] K. Goossens, O.P. Gangwal, J. Roever and A.P. Niranjan, "Interconnect and Memory Organization in SOCs for Advanced Set-Top Boxes and TV – Evolution, Analysis, and Trends," in J. Nurmi, H. Tenhunen, J. Isoaho and A. Jantsch (editors), *Interconnect-Centric Design for Advanced SoC and NoC*, Kluwer, Norwell, MA, 2004, Chapter 15, pp. 399–423.

[54] S. Murali, M. Coenen, K. Goossens, A. Radulescu and G. De Micheli, "Mapping and Configuration Methods for Multi-Use-Case Networks on Chips," *Proceedings of the Asia and South Pacific Design Automation Conference (ASP-DAC)*, 2006.

[55] M. Bekooij, O. Moreira, P. Poplavko, B. Mesman, M. Pastrnak and J. van Meerbergen, "Predictable Embedded multiprocessor System Design," *Proceedings of the International Workshop on Software and Compilers for Embedded Systems (SCOPES)*, Springer, September 2004.

[56] T. Bjerregaard and J. Sparso, "A Router Architecture for Connection-Oriented Service Guarantees in the MANGO Clockless Network-on-Chip", *Proceedings of Design, Automation and Testing in Europe Conference 2005 (DATE05)*, 2005.

[57] R. Ginosar, "Fourteen Ways to Fool Your Synchronizer," *Proceedings of the Ninth International Symposium on Asynchronous Circuits and Systems (ASYNC03)*, 2003.

[58] M. Posner and D. Mossor, "Designing Using the AMBA 3 AXI Protocol," *Easing the Design Challenges and Putting the Verification Task on a Fast Track to Success*, Synopsys, April 2005.

[59] F. Martini, *Design of a Low Latency Circuit Switched NoC Architecture for High Performance MPSoCs*, Master thesis, University of Ferrara, Italy, March 2006.

# CHAPTER 7

# NoC Programming

This chapter focuses on parallel programming for network-on-chip (NoC) platforms, with a significant number of cores operating in parallel. Raising the abstraction level for computation and communication specification seems the only way to master the complexity of mapping a large software application onto an multi-processor systems-on-chip (MPSoC). Even though the bulk of this book is on architectural and lower-level issues, high-level programming models are needed to support abstraction of hardware and software architectures.

Parallel computer architectures and parallel programming have deep roots in high-performance computing. Early programming abstractions for parallel machines go back almost 60 years. In the last half-century, the traditional dichotomy between shared memory and message passing as programming models for multi-processor systems has consolidated. For small-to-medium scale multi-processor systems consensus was reached on cache-coherent architectures based on shared memory programming model. In contrast, large-scale high-performance multi-processor systems have converged toward *non-uniform memory access* (NUMA) architectures based on *message passing* (MP) [5, 6]. As already discussed in previous chapters, several characteristics differentiate NoCs and MPSoCs from classical multiprocessing platforms, and this view must be carefully revisited.

First, the "on-chip" nature of interconnects reduces the cost of inter-processor communication. The cost of delivering a message on an on-chip network is in fact at least one order of magnitude lower (power- and performance-wise) than that of an off-chip interconnect. NoC platforms feature a growing amount of on-chip memory and the cost of on-chip memory accesses is also smaller with respect to off-chip memories. Second, NoCs are often deployed in resource-constrained, safety-critical systems. This implies that while performance is obviously important, other cost metrics such as power consumption and predictability must be considered and reflected in the programming model.

Unfortunately, it is not usually possible to optimize all these metrics concurrently, and one quantity must typically be traded off against the other. Third, unlike traditional MP systems, most NoC architectures integrate highly heterogeneous end-nodes. For instance, some platforms are a mix of standard processor cores and application-specific processors such as digital signal processors or micro-controllers [7, 8]. Conversely, other platforms are highly modular and reminiscent of traditional, homogeneous multi-processor architectures [9, 10], but they have a highly application-specific memory hierarchy and input output interface.

These issues indicate that parallel programming based on uniform, coherent shared memory abstraction is not a viable "one-size-fits-all" solution for highly parallel NoCs. This conclusion is strengthened by observing that uniform, coherent shared memory is not supported even by lower-dimension MPSoC platforms on the market today (refer to the following sections), as well as by many forward-looking research prototypes. On the contrary, evidence accumulates in favor of "communication exposed" programming models, where communication costs are made visible (at various levels of abstraction) to the programmer.

In other words, we are witnessing a paradigm shift from computation-centric programming toward communication-centric programming, where most of the developer's ingenuity and the software development system support must be focused on reducing or hiding the cost of communication. This poses significant challenges, especially when considering the inefficiency incurred when mapping high-level programming models (such as message passing) onto generic architectures, in terms of software and communication overhead.

Another important trend in parallel programming for NoC architectures is the distinction between two different types of parallelism: *instruction-level* (ILP) and *task-level* (TLP). These two types of parallelism are addressed in different ways. Fine-grained ILP discovery and exploitation is extremely difficult and labor-intensive for application programmers. Hence, much effort has been devoted to create tools for ILP extraction and hardware architectures that can dynamically extract ILP at execution time. From the programmer viewpoint, emphasis should be on adopting a programming style that does not obfuscate ILP with convoluted computation and memory access patterns. Programming languages abstractions can facilitate ILP extraction by forcing programmers to use "ILP-friendly" constructs. Almost all the commercial and research MPSoC/NoC platforms provide some degree of support for ILP-friendly programming. The key challenges are in shortening the learning curve for the programmer, in ensuring software portability under programming models which are expressive enough to allow synthetic description of complex functionality. These are conflicting requirements, and there a comprehensive solution has not emerged so far.

TLP discovery and exploitation is generally left to the programmer. Automatic TLP is at a very preliminary stage, even though some interesting solutions do exist (they are discussed later in the chapter). Most of existing software development environments provide several facilities to help programmers specify task-level parallel computation in an efficient way. To be more precise, parallel computation is only one (and possibly not the most critical) characteristic of parallel applications. In fact, parallel memory access and synchronization are extremely critical for efficiency. This is especially true in NoCs, where communication latencies are relatively short and on-chip memory is often not sufficient to store the full working set of the application. Thus, the key to an efficient parallelization is often not so much in discovering a large number of parallel threads, but in ensuring that their execution is not starved by insufficient I/O bandwidth to main memory, or slowed down by the latency of synchronization and off-chip memory accesses.

This chapter focuses primarily on software support for TLP. This choice has several motivations. First, exploitation of ILP is highly dependent on the target processor architecture, and it is very difficult to derive general approaches and guidelines. Second, ILP extraction approaches is often performed at compile time (e.g., in VLIW architectures) or at run time (e.g., in dynamically scheduled superscalar architectures) without direct involvement by the programmer. Nevertheless we shall discuss a few concrete examples of programming environments that support ILP extraction. In these environments, the programmer is offered constructs and libraries that greatly facilitate the automatic extraction, by optimizing compilers, of a high degree of ILP from sequential code.

TLP is supported in various ways. In this chapter, we distinguish three main levels of support, namely: augmentation to traditional sequential programming, communication-exposed programming and automatic parallelization. The first level is commonplace in current software development environments for MPSoCs, the second is at an advanced research stage, with several ongoing technology transfer initiatives, the third is still at an early research stage, not fully mature for industrial exploitation. The remainder of this chapter is divided in three sections, focused on the three support levels mentioned above.

## 7.1 ARCHITECTURAL TEMPLATE

Before delving into the details of TLP exploitation, we set the stage for a quantitative comparison by defining a general architectural template of a multi-core NoC platform that will be referred throughout the chapter in a few case studies. The template is shown in Fig. 7.1: it features a number of (possibly heterogeneous) processing cores, with a level-1 private fast-access memory (usually, single-cycle). Level-1 memory can be instantiated as a software-controlled scratchpad, a hardware-controlled cache or a mix of both. These cores and their level-1 memories constitute the

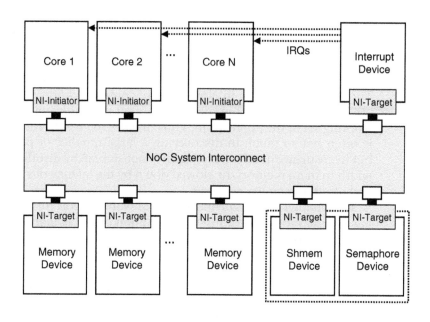

■ **FIGURE 7.1**

Multi-processor platform template.

basic computing tiles, which are tied to an NoC communication fabric through one or more network interface ports.

The simplest configuration has one initiator port, but more complex configurations, with multiple initiator and target ports are possible. The NoC is also connected to a number of targets, which represent on-chip or off-chip memories, and I/Os. Memories can be associated to a special device, which is used for synchronization. The synchronization device supports atomic read-modify operations, the basic hardware primitive for supporting mutual exclusion. The memory-synchronization device pair is needed for ensuring protected memory access when two computing tiles share one memory slave.

An important feature of the architectural template is interrupt support. Interrupts can be issued by writing to an interrupt device, which raises an interrupt to any specified target processors. Interrupt requests are issued to the interrupt device by dedicated point-to-point channels. Multiple interrupt devices can be instantiated, for better scalability. It is important to notice that interrupts can be raised by processing tiles and also by synchronization devices. Thus, it is possible to send an interrupt to a processor not only through explicit write to the interrupt device from another tile, but also when a lock on a specific memory location is freed.

This generic platform, called MPSIM [12] has been described in cycle-accurate SystemC, and it can be instantiated to model and simulate a wide variety of NoC architectures. We describe in more details two instantiations which have been used in the following sections to quantitatively analyze and compare different parallel programming models on an NoC target.

### 7.1.1 Shared Memory MPSIM

This architecture consists of a variable number of processor cores (ARM7 simulation models will be deployed for our analysis framework) and of a shared memory device to which the shared addressing space is mapped, connected via a system interconnect.

As an extension, each processor also has a private memory connected to the NoC where it can store its own local variables and data structures. Hardware semaphores and slaves for interrupt generation are also connected to system interconnect. In order to guarantee data coherence from concurrent multi-processor accesses, shared memory can be configured to be non-cacheable, but in this case it can only be accessed through NoC transactions, as no caching of shared memory data is allowed.

Alternatively, the shared memory can be declared cacheable, but in this case cache coherence has to be ensured. Hardware coherence support is based on a write-through policy, which comes into two variants: one based on an invalidate policy (*Write-Through Invalidate, WTI*), the other based on an update policy (*Write-Through Update, WTU*). In contrast with the non-cacheable shared memory platform, the cache-coherent platform imposes a very significant restriction to the interconnect architecture. Namely, a shared-bus interconnect is required because cache coherency is ensure through a snoopy protocol, which requires continuous monitoring of bus transaction.

The hardware snoop devices, for both invalidate and update case, are depicted in Fig. 7.2. The snoop devices sample the bus signals to detect the transaction which is being performed on the bus, the involved data and the originating core. The input pinout of the snoop device depends on the particular bus protocol supported in the system, and Fig. 7.2 reports the specific example of the interface with the STBus shared-bus node from STMicroelectronics [11].

When a write operation is flagged, the corresponding action is performed, for example, invalidation for the WTI policy, rewriting of the data for the WTU one. Write operations are performed in two steps. The first one is performed by the core, which drives the proper signals on the bus, while the second one is performed by the target memory, which sends its acknowledge back to the master core to notify operation completion (there can be an explicit and independent response phase in the communication protocol or a ready signal assertion in a unified bus communication phase). The write ends only when the second step is completed and when the snoop device is allowed to consistently interact with the local cache.

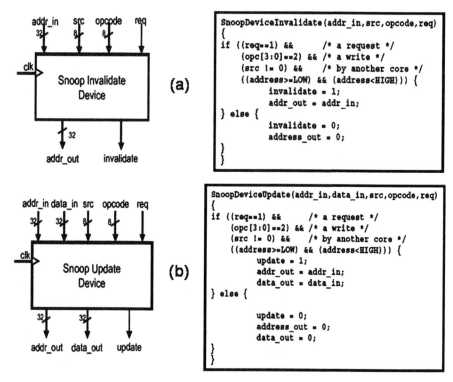

■ **FIGURE 7.2**

Interface and operations of the *snoop device* for the (a) invalidate and (b) update policies.

Of course, the snoop device must ignore write operations performed by its associated processor core. In our simulation model, synchronization between the core and the snoop device in a computation tile is handled by means of a local hardware semaphore for mutually exclusive access to the cache memory.

The template followed by this shared memory architecture reflects the design approach of many semiconductor companies to the implementation of shared memory multi-processor architectures. As an example, the MPCore processor implements the ARM11 micro-architecture and can be configured to contain between 1 and 4 processor cores, while supporting fully coherent data caches [9].

### 7.1.2 Message-Oriented Distributed Memory MPSIM

This instantiation represents a distributed memory MPSoC with light-weight hardware extensions for message passing, as depicted in Fig. 7.3.

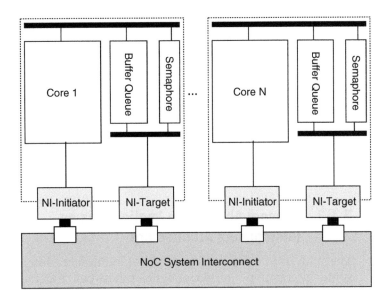

■ **FIGURE 7.3**

Message-oriented distributed memory architecture.

In the proposed architecture, messages can be directly transmitted between scratch-pad memories attached to the processor cores within each computation tile. In order to send a message, a producer writes in the message queue stored in its scratch-pad memory, without generating any traffic on the interconnect. Then, the consumer is allowed to transfer the message to its own scratch-pad, directly or via a direct memory access(DMA) controller. Scratch-pad memories are therefore connected as slave ports to the communication architecture and their memory space is visible to the other processors. As far as synchronization is concerned, when a producer intends to generate a message, it locally checks an integer semaphore which contains the number of free messages in the queue.

If enough space is available, it decrements the semaphore and stores the message in its scratch-pad. Completion of the write transaction and availability of the message is signaled to the consumer by incrementing a semaphore located in its scratch-pad memory. This single write operation goes through the NoC interconnect. Semaphores are therefore distributed among the processing elements, resulting in two advantages: the read/write traffic to the semaphores is distributed and the producer (consumer) can locally poll whether space (a message) is available, thereby reducing interconnect traffic.

A DMA engine has been attached to each core, as presented in [13], allowing efficient data transfers between the local scratch-pad and non-local memories reachable through the NoC interconnect. The DMA control logic supports multichannel programming, while the DMA transfer engine has a dedicated connection to the scratch-pad memory allowing fast data transfers from or to it.

The architectural template of the Cell Processor [14] developed by Sony, IBM and Toshiba shares many similarities with the distributed memory architecture. The Cell Processor exhibits eight vector computers equipped with local storage and connected through a data-ring-based system interconnect. The individual processing elements can use this ring-NoC to communicate with each other, and this includes the transfer of data in between the units acting as peers of the network.

### 7.1.3   Memory Abstraction Implications on Interconnect Architecture

The analysis of the various platform embodiments described above reveals some interesting inter-dependencies between the memory abstraction support and the interconnect architecture. More specifically, non cache-coherent and distributed memory architectures impose weak coupling constraints on interconnect architecture: these architectures are well-matched to both a shared-bus interconnect and a complex multi-hop network on chip. In contrast, supporting cache coherency on an NoC-based interconnect appears to be a very challenging task.

More in details, non-cacheable shared memory communication is relatively easy to support, from the functional viewpoint, on a multi-hop NoC interconnect. The shared memory bank can be connected to the NoC as a target, and the NoC will route all the shared-memory reads and writes to the corresponding end-node. Synchronization and atomicity are much more challenging. In MPSIM, for instance, synchronization is supported by a special-purpose slave featuring atomic read-modify operations. Every master willing to get atomic access to a shared memory region must first acquire a lock via a read to the special-purpose slave. In terms on NoC transactions, shared memory locking requires one or more read transaction (multiple memory reads are required in case of access contention) to the synchronization slave. Clearly, this paradigm is not very efficient nor scalable in an NoC context, because it implies a lot of inefficient reads[1] and destination congestion, as all locked accesses must go through lock acquisition on the synchronization slave. Efficiency can be improved if shared memory targets and the corresponding synchronization targets are not unique. In this way, destination congestion can be alleviated, but even in the best case, locked writing to shared memory has a latency cost which is at least three times the NoC latency (a read-modify to the synchronization slave followed by a posted write to the shared memory target).

---

[1] Remember that a read transactions incurs NoC latency twice.

Supporting the cache-coherent memory abstraction across an NoC interconnect is even more challenging. In principle, this problem has been solved by directory-based cache-coherency schemes [5], which have been proposed for large-scale multi-computers with multi-hop interconnects. These schemes have a significant hardware overhead (the directory, and directory-management logic) and they trigger a number of network transactions that are completely invisible to the programmer. Directory-based cache-coherent memory hierarchies have never been implemented in an NoC platform, hence their cost and efficiency has not been assessed. On the other hand, several MPSoCs with snoop-based cache coherency have been developed [15, 16]. This scheme requires a shared-bus interconnect, and it is the one supported in our MPSIM platform template. Clearly, it inherits the same scalability problems of any bus-based architecture, and it is not easily generalized to a scalable NoC fabric. The development of efficient cache-coherency schemes for NoC targets is still an open research topic.

Finally, the message-passing architecture is clearly well-suited to an NoC interconnect. Distributed communication FIFOs and synchronization eliminate destination-contention bottlenecks (unless they are present at the application level), and most of the communication can be done through posted writes which can be farmed off to DMA engines that run in parallel with the processors in every computational tile. It is then quite clear, from the architectural viewpoint, that the message-passing architecture is more scalable and best-matched to an NoC interconnect. However, adopting a pure message-passing architecture has severe implications on the flexibility of the programming model. For instance, applications where parallelism comes from having many workers operating in parallel on a very large common data structure (e.g., a high-definition TV frame, where each processor works on a frame window) are not easily coded using a strict message-passing semantics. These issues will be dealt in more detail in Section 7.3.2.

In the following sections, we will use the MPSIM template to quantitatively compare various programming models and the corresponding architectural support. To allow a fair comparison, we will assume that the interconnect is a shared bus. This choice reduces to some degree the competitive advantage of message-oriented architectures in terms of interconnect scalability, but it allows a more precise assessment of cost and benefits of programming models, without distortions caused by different interconnect fabrics.

## 7.2   TASK-LEVEL PARALLEL PROGRAMMING

MPSoC platforms are now commonplace in embedded computing, ranging from wireless and wireline communication [17], to multimedia [7, 18], signal processing [19] and graphics [14]. From the software view-point, there is little consensus on the programmer view offered in support of these highly parallel platforms. In many cases, very little support is offered and the programmer is in charge of explicitly managing data transfers and synchronization. Clearly, this approach is extremely labor-intensive, error-prone and leads to poorly portable software. For this reason MPSoC platform vendors are devoting an increasing amount of effort to offering more abstract programmer views through parallel programming libraries and their application programming interfaces(APIs).

The most common form of support is in essence an incremental evolution of standard sequential programming. The application developer usually start from a single-threaded application developed for a uni-processor. Porting the application to a multi-processor target is a work-intensive activity, which can be summarized as follows. First, the single-threaded implementation is analyzed in terms of its main functional bocks and their data dependencies. This analysis has three main objectives, which can be summarized as follows:

1.  Identifying the critical computational kernels of the applications. This activity is often supported by profiling tools.

2.  Identifying the data-flow between computational kernels. This is done by hand, exploiting the knowledge of the data structures and the application flow.

3.  Applying a workload partitioning strategy that splits the application in a sub-blocks that can be executed in parallel. This is the most difficult task, as a successful partitioning fully depends on the developer's ingenuity and experience.

The last two steps are far from trivial. In many cases, sequential applications do not show significant potential for parallelization, simply because they have been written in a non-modular fashion. A typical example of this problem is the common case of a single loop that creates most of the application workload. In this case, the loop must be rewritten, for instance as a set of loops operating on disjoint data sets, in order to extract some coarse-grained parallelism. More in general, many global transformations can be applied at this stage, such as merging or splitting loops, choosing suitable data-structures, replicating data, etc.

Fortunately, there are several application domains where simple coarse-grained parallelization strategies are available. For example many multimedia applications operate on a set of independent frames (e.g., a sequence of audio samples, independent video frame overlays, etc.), and frame-by-frame parallelization is possible. For instance, if encryption is performed on a data stream on fixed-sized data blocks, it is possible to instantiate multiple identical encryption tasks, each one operating in parallel on a different block.

Generally, application tasks are not fully independent, but they share and exchange data, and need to synchronize. Parallel programming API must therefore provide support for data sharing and synchronization. Programming abstraction is a key requirement: even though it is in principle possible to fully expose the hardware to the programmer and let her/him use low-level software interfaces to synchronization and communication primitives, programmer productivity would be extremely low, and the resulting software would be completely architecture-specific and nonportable. Thus, parallel programming libraries provide an abstract view of the underlying architecture, through architecture-neutral high-level primitives. The two most common abstractions offered by current libraries are *message passing* (MP) and *shared memory* (SM).

## 7.2.1 Message Passing

Message passing and shared memory correspond to "abstract machines," which are generally a much simplified model of the underlying hardware. The message-passing abstract machine has a number of independent processors with disjoint memory spaces. Communication between processors happens only via explicit messages (send–receive pairs). Special messages can be used for synchronization purposes. Message passing has first been studied in the high-performance multi-processor community, where many techniques have been developed for reducing message delivery latency. Several authors developed light-weight software wrappers (bypassing the operating system) for message-passing interface to reduce the setup cost (e.g., [20]). Others have moved message-passing primitives in hardware, ranging from dedicated instructions (e.g., [21]) to complex co-processors for accelerating message passing (e.g., [22]). Dedicated instructions coupled to a low-latency communication network support fine granularity messages, enabling massive amounts of parallelism.

### Message-passing implementation in MPSoCs

Message passing has also entered the world of embedded MPSoC platforms. In this context it is most commonly implemented on top of a shared memory architecture (e.g., **TI OMAP** [19], Philips Eclipse [8], Toshiba Kawasaki [23], Philips Nexperia [7]). This approach is a consolidated one also in the traditional high-performance parallel computing field, and numerous implementations of message-passing standards (e.g., MPI) onto shared memory machines have been reported [5].

The main issue in MPSoC architectures is that shared memory is likely to become a performance/energy bottleneck, even when DMAs are used to increase the transfer efficiency. This statement is easily understood if we think that transferring a message from a processor to another using a shared memory requires at least one read and one write to that memory. Let us consider an architecture with one single shared memory slave on the interconnect. Clearly, since every message delivery (i.e., a `send–receive` pair) require writing then reading the message (payload and some amount of wrapping information), all process-to-processor messaging traffic will have to go through a single memory, which inevitably becomes the system's bottleneck. Note that an NoC interconnect would not help in this case, because the architecture will suffer from destination contention, which cannot be alleviated by any scalable interconnect.

Synchronization poses more subtle and even more difficult challenges to the efficiency of message passing based on shared memory. This can be better understood through an example.

**Example 7.1.** Consider a send–receive pair where the sender is late, and the receiver processor is waiting for a message. Let us assume that the receiver waits on a spin lock, that is it continuously polls a synchronization variable (e.g., a semaphore mapped device), waiting for the message to arrive. Polling will create a significant amount of read traffic which will contribute to interconnect saturation. Moving from active waiting to interrupt-based wakeup will alleviate the problem, but it will increase the latency of message delivery, because of interrupt handling overhead. Moreover, synchronization operations requires locked (atomic) transactions over the communication fabric. This not only further degrades the performance, but also limits the scalability of the above implementations for more advanced communication architectures.

Several authors have recently proposed support for message passing on a distributed memory architecture. An interesting case-study is presented in Ref. [24] where a turbo-coder is mapped on a message-passing architecture. On each processing tile an I/O-device is responsible for transmitting/receiving messages from the communication architecture. Buffer underflow/overflow of the I/O-device has to be avoided in software. In Ref. [25], more generic support for message passing is provided. A light-weight co-processor, called a memory server access point, is added to each processor. The access point links the processor to its own local memory, but also to the remote memories. The synchronization is left to the message-passing protocol. The above approaches have limited support for synchronization and limited flexibility in matching the application to the communication architecture. For example, in Ref. [25] remote memories are always accessed with a DMA-like engine even though this is not the most efficient strategy for small message sizes.

Hardware acceleration in support of message passing is also proposed in the StepNP architecture [26]. In StepNP a high-level software abstraction is supported, namely message passing based on parallel communicating objects. From the programmer's viewpoint, a message is sent by calling a method of a remote object, similarly to what is done in high-level distributed programming environments like JAVA, CORBA and DCOM. Most of the operation related to translating a remote method invocation in a physical data-transfer (i.e., marshaling and un-marshaling, packetization, etc.) are implemented by hardware accelerators in the processor's network interfaces, and therefore they have a very low cost in terms of execution cycles. However, all remote method calls and returns in a system are managed by a single *object request broker* (ORB), implemented with dedicated hardware engine. This design decision, needed to support the high-level programming abstraction of the STEP-NP platform, leads to an intuitive system bottleneck, namely the ORB, which has to manage all the traffic related to all the messages sent and received by all the tasks in the system.

### Programming a message-passing NoC platform

To give the reader a more concrete understanding of the issues and challenges in supporting message passing for MPSoC, we now discuss in detail a specific case study. An implementation based on the distributed memory architecture described in Section 7.1.2 was developed, leveraging scratch-pad memories and local hardware semaphores. Moreover, full porting of standard message-passing libraries traditionally used in the parallel computing domain might cause a significant overhead in resource-constrained MPSoCs. For this reason, we developed a simple but highly optimized message-passing library, custom-tailored for the scratch-pad-based distributed memory architecture we are considering. The most important functions are listed in Table. 7.1.

Messages are managed via *message queues*. To instantiate a queue, both the producer and consumer must run an initialization routine. To initialize the producer, *sq_init_producer* is called. It takes as arguments the ID of the consumer's, the message size, the number of messages in the queue and a binary value. The last argument specifies whether the producer should poll the producer's semaphore or suspend itself until an interrupt is generated

**TABLE 7.1** ■ Simple message-passing library APIs.

| Return type | Function | Arguments |
|---|---|---|
| SQ_PRODUCER* | sq_init_producer | int consumer_id<br>int message_size<br>int total_messages<br>bool use_suspension |
| SQ_CONSUMER* | sq_init_consumer | int consumer_id<br>bool buffer_space_location<br>bool use_suspension |
| void | sq_write(_dma) | SQ_PRODUCER *queue_p<br> char *source |
| char* | sq_getToken_write | SQ_PRODUCER *queue_p |
| void | sq_putToken_write | SQ_PRODUCER *queue_p |
| char* | sq_read(_dma) | SQ_CONSUMER *queue_c |

by the semaphore. The consumer is initialized with *sq_init_consumer*. It requires the ID of the consumer's itself, the location of the read buffer and the poll/suspend flag. In detail, the second parameter indicate the address where the function sq_read will store the message transferred from the producer's message queue. This address can be located either on the private memory or on the scratchpad memory.

The producer sends a message with the *sq_write(_dma)* function. This function copies the data from *\*source* to a free message block inside the queue buffer. This transfer can either be carried out by the core or via a DMA transfer (*x_dma*). Instead of copying the data from *\*source* into a message block, the producer can decide to directly generate data in a free message block. The *sq_getToken_write* returns a free block in the queue's buffer on which the producer can operate. When data is ready, the producer should mark its availability to the consumer with *sq_putToken_write*. The consumer transfers a message from the producer's queue to a private message buffer with *void sq_read(_dma)*. Again, the transfer can be performed either by a local DMA or by the core itself.

This simple messaging library thus supports:

1. either processor or DMA-initiated data transfers to remote memories,

2. either polling-based or interrupt-based synchronization,

3. flexible allocation of the consumer's message buffer, that is on scratchpad or on a private memory at a higher level of the hierarchy.

Thanks to the high-level APIs, this flexibility can be effectively used to optimize message passing based applications.

### Message-passing performance analysis

The message-passing library is carefully optimized for the distributed memory architecture showed in Fig. 7.3. The library implementation is very lightweight, since it is based on C macros that do not introduce significant overhead. A producer–consumer exchange of data programmed via the library showed just a 1% overhead with respect to the programmer's direct transfer control.

More interestingly, the library can be used to compare alternative implementations of message-passing architectures and for fine architectural tuning. First, we should question whether this customized solution (message-oriented architecture and MP library) is competitive with an alternative solution where message passing is implemented on top of shared memory.

We are interested in the average time required for a consumer to obtain a message available on the producer's queue (Fig. 7.4), under the assumption

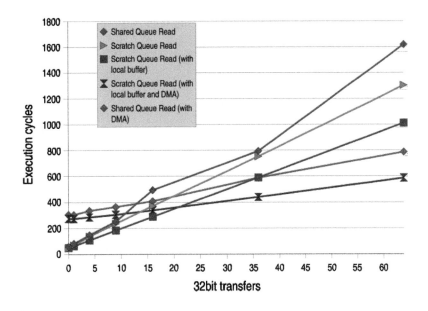

■ **FIGURE 7.4**

Analysis of read execution cycles in a producer consumer benchmark.

that the queue dimension is always sized big enough to contain all transferred messages. In other words, the destination queue never becomes full. We measure the transfer time per message (averaged over 20 messages) and explore different message sizes using either the processors or DMAs for transfers.

Performance of the distributed memory solution can be deduced from Fig. 7.4 by looking at "Scratch Queue Read" family curves. The comparison is with "Shared Queue Read" (a solution wherein the communication queue between two processors is stored in shared memory, and thus reflects message-passing implementation on top of a shared-memory architecture). We can observe that, the longer the message (see X-axis) the more the application execution time is dominated by memory transfers. Hence, performance gains of "Scratch Queue Read" become larger, due to the use of a fast, local scratchpad memory instead of a slow shared memory. Employing a DMA engine for "Shared Queue Read" does not help a lot. Please note that there are three architectural variants for "Scratch Queue Read":

(i) basic version, where data are moved from producer's scratchpad to the consumer's private memory by the consumer processor itself via single transfers;

(ii) "local buffer" version, where data are moved to the consumer's scratchpad instead of its private memory, thus cutting down on memory write costs;

**TABLE 7.2** ▪ Different message-passing implementations.

| Solution | Queue position | Transfer mode | Arrival notification |
|---|---|---|---|
| 1 | scratchpad | processor | polling |
| 2 | scratchpad | processor | interrupt |
| 3 | scratchpad | dma | polling |

(iii) "local buffer and DMA" version, where data are moved using dedicated DMA engines, thus making data transfers between scratchpad memories more efficient.

The fixed setup cost for programming DMA can clearly be observed for zero-sized messages. As soon as the message size exceeds 25 words, the increased transfer efficiency compared to explicit copying outweighs the setup cost. From these results, we clearly observe the inefficiency of message-passing implementation on top of a shared-memory architecture.

We continue our analysis to show that flexibility in the message-passing library can improve performance results. In fact, the library can exploit several features of the underlying hardware such as processor-versus DMA-driven data transfers or interrupt versus active polling. The proper architecture configuration can be selected based on the application.

Let us consider an application consisting of a functional pipeline of eight matrix multiplication tasks. Each stage of this pipeline takes a matrix as input, multiplies it with a local matrix and passes the result to the next stage. We iterate the pipeline 20 times. We run the benchmark respectively on an architecture with eight and four processors. In the first case, only one task is executed on each processor, while in the second, we added concurrency by scheduling two tasks on each core. First, we compare three different configurations of the message-oriented architecture (Table 7.2). We execute the pipeline for two matrix sizes: $8 \times 8$ and $32 \times 32$ elements. In the latter case, longer messages are transmitted.

Analyzing the results in Fig. 7.5, referred to the case where one task runs on each processor, we can observe that a DMA is not always beneficial in terms of throughput. For small messages, the overhead for setting up the DMA transfer is not justified. In case of larger messages, the DMA-based solution outperforms the processor-driven transfers. Instead, employing a DMA always leads to an energy reduction, even if the duration of the benchmark is longer, due to a more power efficient data transfer. Please note that energy of all system components (DMA included) is accounted for in the energy plot.

Furthermore, the way a consumer is notified of the arrival of a message plays an important role, performance- and energy-wise. The consumer has to wait until the producer releases the consumer's semaphore. With a single task per processor (Fig. 7.5), the overhead related to the interrupt

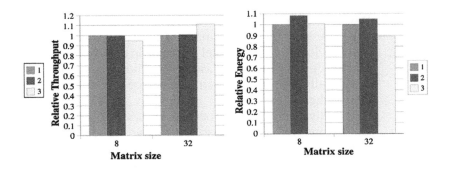

Comparison of message-passing implementations in a pipelined benchmark with eight cores from Table 7.2.

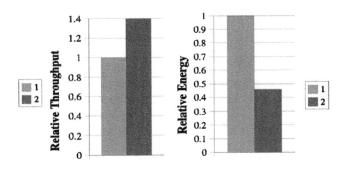

■ **FIGURE 7.6**

Task scheduling impact on synchronization in a pipelined benchmark with four cores from Table 7.2.

routine can slows down the system, depend on the communication versus computation ratio and polling is, in general, more efficient. On the contrary, with two tasks per processor (Fig. 7.6, referred to matrices of $8 \times 8$ elements) the interrupt-based approach performs better. In this case, it is more convenient to suspend the task because the concurrent task scheduled on the same processor is in "ready" state. Instead, with active polling, the processor is stalled and the other task cannot be scheduled. This simple application case study shows the implementation of message passing should be matched with application's workload. This is only possible with a flexible message-passing library.

## 7.2.2   Shared memory

The key property of the shared-memory approach is that communication occurs implicitly as a result of conventional memory access instructions.

Processes have address spaces that can be configured so that portions of them are shared, that is, mapped to a common physical location. Processes belonging to a parallel application communicate with each other by reading and writing shared variables. The presence of a memory hierarchy with locally cached data is a major source of complexity in shared-memory approaches.

Broadly speaking, approaches for solving the cache-coherence problem fall into two major classes: hardware-based and software-based. The former imposes cache coherence by adding suitable hardware which guarantees coherence of cached data, whereas the latter imposes coherence by limiting the caching of shared data. This can be done by the programmer, the compiler, or the operating system. For a survey of hardware-based cache-coherence solutions in general-purpose computing, the reader is referred to [5,27,28] software-based cache-coherence solutions are reviewed in [29].

### Shared-memory in MPSoCs

Even though message passing has received considerable attention recently, shared memory is the most common programmer abstraction in today's MPSoCs. In embedded MPSoC platforms, shared-memory coherence is often supported only through software libraries which rely on the definition of non-cacheable memory regions for shared data or on cache flushing at selected points of the execution flow. There are a few exceptions that rely on hardware cache coherence, especially for platforms which have a high degree of homogeneity in computational node architecture [10]. On the contrary, hardware cache coherency is by far the most common architectural option of homogeneous chip multi-processors which are becoming commonplace in general-purpose computing [30].

This bifurcation has taken place for two main reasons. First, most general-purpose engines are homogeneous, and all processors have the same level-1 cache architecture and memory management unit. The cores used in these chip multi-processors are often derived from older designs with cache-coherency support, thus, no extra design effort is needed to support it. Second, and most important, a general-purpose chip multi-processor is designed to replace a traditional high-performance microprocessor, hence it should run efficiently single-processor workloads. From the programmer viewpoint, this implies that the hardware architecture should provide a programming model with the smoothest transition from single thread of execution to multi-threading. Hardware-based cache coherency fully support a uniform, coherent memory space, where threads can communicate through shared variables and parallelization is therefore relatively easy.

In contrast, MPSoCs for low-power embedded applications are heterogeneous collections of programmable cores which have been carefully chosen and customized for a specific class of applications. For instance, the widespread SoC platforms for wireless handsets feature a dual-processor architecture, where a RISC microcontroller is coupled with a DSP. The microcontroller performs system control functions and manages the user interface, while the DSP runs the base-band section of the modem. These are very different workloads, and each core is highly efficient (performance and energy-wise) in executing a specific workload. Achieving the same level of performance on a general-purpose dual-processor engine would be highly energy-inefficient as it would require two very powerful general-purpose cores.

Supporting both hardware-based cache coherency and heterogeneous processing elements is a very challenging task from the architectural viewpoint. First, heterogeneous processors work almost always on disjoint data structures. Hence, most of the coherency-related operations performed by the hardware at runtime are pure overhead. Second, different types of processors have very different access patterns to global memory. Conflicting access patterns can put under severe stress a cache-coherent machine. For instance, the streaming access pattern of signal processors is highly predictable (no data-dependent control flow), but has very low locality and poor cache behavior. Last, hardware-based cache coherence heavily relies on the presence of a shared-bus interconnect. Many MPSoCs already feature a number of cores and bandwidth requirements that exceed the bandwidth limits of a shared bus. Extensions of hardware-based cache coherency to high-bandwidth parallel or multi-hop NoCs is difficult and has not been attempted yet.

Even though shared memory is considered an inherently easier architectural abstraction than message passing from the programmer's viewpoint, writing complex and correct parallel programs on a shared-memory machine is very challenging. For these reasons, a number of parallel programming libraries for shared-memory machines have been developed such as *Pthreads*. The main purpose of these libraries is to provide a high-level interface to parallel programming primitives based on shared memory, in order to enhancer the productivity of the software developer. These APIs offer three main types of functions:

1. creation, management and destruction of parallel threads of execution;
2. allocation, deallocation access of shared data-structures;
3. various forms of synchronization.

Besides facilitating the development of complex parallel applications, these libraries provide a number of side benefits: code portability across multiple hardware platforms, easier debugging and documentation, faster parallelization of existing sequential code, better upgrading of existing code base. The critical distinctive issue in embedded MPSoCs is that software abstraction is not without a cost, and library-induced overhead of hundreds of instructions are not affordable in a number of tightly performance-constrained applications. For these reasons, the libraries made available in MPSoC development toolkits offer only basic abstractions and functions. In the next subsection, we will use the MPSIM virtual platform as a case study for analyzing in detail the implementation of a shared-memory library in a resource-limited MPSoC context.

### Programming a shared-memory NoC platform

We focus on the shared-memory MPSIM platform (see Section 7.1.1). The system V IPC library, which was originally developed to write parallel applications in the Unix operating system was ported onto MPSIM. One important advantage of using a standard library available in uniprocessor Linux and Unix environments is that it becomes possible for software designers to develop their applications on host PCs and to easily port their code onto the MPSoC virtual platform for validation and fine-grained software tuning on the target architecture.

System V IPC is a communication library for heavy-weight processes based on permanent kernel resident objects. Each object is identified by a unique kernel ID. These objects can be created, accessed and manipulated only by the kernel itself, granting mutual exclusion between processes. Three different types of objects, also named *facilities*, are defined: messages queues, semaphores and shared memory. Processes can communicate through System V IPC objects using ad hoc defined APIs, that are specific for each facility. In order to use an object, users have first to invoke the "get" API of the desired facility (*msgget, semget and shmget*). This function returns the ID of an object matching with a "key" parameter, possibly creating the object if it does not exist. By using the same value of "key" ensures that communicating processes refer to the same object. The second step is to use the object invoking APIs on the object ID previously obtained. Each System V IPC object is protected from illegal accesses by a classic Unix bit protection code, like a file, which contains read/write/control rights for owner-creator/group/other users. The main features of System V facilities are reviewed next.

*Message queues* are objects similar to pipes and FIFOs, which can be used to implement message passing on a shared-memory machine. A message queue allows different processes to exchange data with each other in the form of messages in compliance with the FIFO semantic. Messages can have different sizes and different priorities. The send API (*msgsnd*) puts a message in the queue, suspending the calling process if there is no enough free space. On the other hand, the receive API (*msgrcv*) extracts from the queue the first message that satisfies the calling process requests in terms of size and priority. If there is not a valid message or if there are no messages at all the calling process is suspended until a valid message is written to the queue. A special control API (*msgctl*) allows processes to manage and delete the queue object.

*Semaphore* objects consist of a set of classic Dijkstra's semaphores. A process calling the "operation" API (semop) can wait and signal on any semaphore of the set. Moreover, System V IPC allows processes to request more than one operation on the semaphore set at the same time. That API ensures that the operations will be executed atomically. A special control API (*semctl*) allows to initialize and delete the semaphore object.

*Shared memory* is the third facility. Shared-memory objects are buffers of memory which a process can link to its own memory space through the attach API (*shmat*). All processes which have attached a shared-memory buffer see the same buffer and can share data directly reading and writing on it. As the memory spaces of the processes are different, the shared buffer could be attached by the API at different addresses for each process. Therefore, processes are not allowed to exchange pointers which refer to the shared buffer. In order to successfully share a pointer, its absolute address must be changed into an offset relative to the starting location of the shared buffer. A special control API (*shmctl*) allows processes to mark a buffer for destruction. A buffer marked for destruction is removed from the kernel when there are no more processes that are linked to it. A process can unlink a shared buffer from its memory space using the detach API (*shmdt*).

In the implementation of the System V IPC on the MPSIM shared-memory platform all objects that can be shared are stored in the shared-memory target device. A dynamic allocator was developed to efficiently support data allocation and de-allocation in shared memory. All original IPC kernel structures were optimized by removing many process/permission-related information, in order to reduce shared-memory occupancy and therefore API overhead. In the MPSIM library implementation mutual exclusion on the critical sections of an object was ensured by means of hardware mutexes that are accessible on the shared-memory space. Each IPC object is protected by a different hardware mutex, allowing parallel execution on different objects.

MPSoC platforms are typically resource-constrained. Therefore, we decided not to implement some of the features of System V IPC. The priority in the message queues facility and the atomic multi-operations on the semaphore sets have not been implemented. These features are not critical in System V IPC, so that their lack will only marginally affect code portability.

MPSoC IPC library was tested and optimized to improve performance of APIs. The length of the critical sections was reduced as much as possible in order to optimize code efficiency. Similarly, the number of shared memory accesses was significantly reduced. Moreover, in case of repeated read accesses to the same memory location, we hold the read value. Write

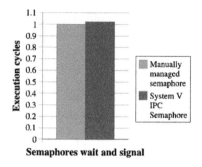

Impact of System V IPC semaphore facility on system performance.

operations were optimized avoiding to perform useless write accesses to shared memory (e.g., writing the same value).

### Shared-memory performance analysis

We characterized the overhead introduced by the library by comparing the case where the library is used to manage system resources against that where they are managed with low-level programming. Three features should be characterized: semaphores, dynamic memory allocation and message queues. We do not analyze the performance of message queues in detail: intuitively, as they implement a message-passing abstraction on a generic share memory architecture, they are much less efficient than hardware-accelerated message passing.

Focusing on the System V facilities needed to support shared-memory programming, we first consider synchronization. Real applications make significant use of the semaphore facility, and it is therefore important to assess the cost incurred by the library in managing this facility.

Consider an *ad hoc* benchmark where two tasks are running onto two different processors: the first one periodically releases a certain semaphore, while the second one is waiting on that semaphore. We measured the time to perform signal and wait over 40 iterations. As shown in Fig. 7.7, the overhead for using System V IPC with respect to the manual management of the hardware semaphores is negligible (only 2%).

The System V shared-memory facility can be characterized by analyzing allocation and deallocation costs. Two steps are required to allocate shared memory:

1.  A call to the `int shmget(key_t key, int size, int shmflg)` function. This function assigns a unique identifier to the memory area to be allocated, and it creates an association in a index table between the identifier and the address of the corresponding memory

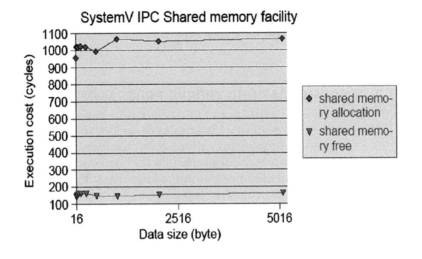

■ **FIGURE 7.8**

Cost of shared-memory allocation and deallocation in System V.

area. It also checks if there is enough shared memory available to satify the request.

2.  A call to the `void *shmat (int shmid, const void *shmaddr, int shmflg)` which returns a pointer to the shared memory area corresponding to the identifier obtained by `shmget`.

De-allocation frees the table entry and makes the corresponding memory area available for future usage. These operations, and in particular the management of the shared-memory objects table, have a non-negligible overhead, in terms of executed operations. This is evident from the plot in Fig. 7.8. Note also that there is a remarkable asymmetry between the cost of de-allocation and the cost of allocation. This is because de-allocation simply implies marking one of the shared-memory object table and the corresponding memory area as available. No checking of ID uniqueness and memory availability is required.

The cost analysis of Fig. 7.8 is done by averaging the cost of a number (20) of memory allocations and de-allocation done by a single processor. In complex applications with significant memory allocation and de-allocation, the globally shared nature of the table creates a destination bottleneck, as it must be accessed in a mutually exclusive fashion by all allocation and de-allocation request, hence the allocation and de-allocation cost is likely to increase due to destination congestion.

These results point out to the non-negligible cost of supporting high-level shared-memory abstractions. Several optimizations are possible in the implementation NoC-oriented allocators, such as lock-free distributed approaches [44], but clearly programmers should be aware of the fact that dynamic allocation and de-allocation of shared memory is an expensive operation, especially when shared memory is centralized.

## 7.3  COMMUNICATION-EXPOSED PROGRAMMING

Parallel programming libraries as described in the previous sections are a viable short-term answer to the quest for increased software productivity for NoC platforms, but ultimately they are only partial extensions to traditional sequential programming languages. If we reason more abstractly about parallel programming environments as syntactic frameworks for expressing concurrency, today we are in parallel computing at a level similar to sequential programming at the dawn of the structured programming era. Locks and threads are basic structured constructs of concurrency. Similarly to what happened to sequential programming languages with the object-oriented revolution, there is a need for higher-level abstractions for building large-scale concurrent programs.

The evolution toward high-level concurrent software development environments is going to be a difficult but unavoidable step, because parallelism exploitation will be essential for achieving higher performance. Commercial and systems programming languages will be assessed based on their support for concurrent programming. Existing languages, such as C, are likely to be extended with concurrency support beyond simple library extensions. Languages that will not evolve to support parallel programming will gradually become useless in high-performance application. On the other hand, parallel programming is demonstrably more difficult than sequential. For example, simultaneous analysis of program context and synchronization is proved to be undecidable [31]. Moreover, human programmers experience many difficulties in reasoning in terms of concurrency, especially when analyzing the side effects of the many possible interleaving of parallel tasks.

A well-tried approach to raise the abstraction level without compromising efficiency is to reduce the expressiveness of the language. Taking a more positive viewpoint, this approach implies focusing the software development framework on a model of computation which is both abstract and efficient. The price to be paid is in lack of generality and flexibility, but this is an acceptable loss if the model of computation is well-suited to describe a wide spectrum of market-relevant applications.

The most successful example of this approach in the area of parallel programming for NoCs is *stream computing*. The key idea in stream computing is to organize an application into streams and computational kernels to expose its inherent locality and concurrency. Streams represent the flow of data, while kernels are computational tasks that manipulate and transform data. Many signal- and media-processing algorithms can easily be seen as sequences of transformations applied on a data stream.

In this case, streams are a natural way of expressing the functionality of the application, while at the same time, they match very well the hardware execution engines.

More formally, a stream computation can be represented by a directed graph, where nodes represent computations and edge represent data communication. Both nodes and edges are annotated. Nodes are annotated with the function that they execute, while edges are annotated with the data they carry. The stream model fully exposes inter-task communication. Intuitively, this is a very desirable feature because it forces the programmer to focus on communication, not only on computation. By analyzing the communication edges in a stream computation, it is possible to obtain precise estimates of the communication requirements for a given application. This greatly simplifies analysis and mapping of application onto parallel architectures. In fact, one of the most challenging issues in parallel programming is the analysis of communication, which is critical for performance optimization, but it is often obscured by the computation-oriented semantic of traditional programming languages.

Stream-oriented models of computations have been studied since the early 1960s (the interested reader is referred to the good survey by Stephen [32]), under the name of *dataflow*. Numerous languages (e.g., LUCID) have been developed to support dataflow semantics. Probably, the best-known semantic for dataflow has been proposed by Kahn, and it is known as *kahn process network* (KPN) [32]. In the KPN formalism, computational nodes perform arbitrary computations and communicate through unbounded FIFOs associated with the edges. A computation consumes one or more data tokens (of arbitrary size) on the input FIFOs and produces one or more data tokens on the output. Since FIFOs are unbounded, there is no problem in blocking or overflow of full FIFOs. Computational nodes can be blocked only on empty FIFOs. The KPN have been proved to be Turing-equivalent, hence they are expressive as any sequential semantic, but they have a much more explicit model for parallel execution of tasks and inter-task communication. One important result proved by Kahn is that the result of the execution of a KPN is invariant in the time order (schedule) of the node computations, provided that the partial order rules implicitly created by the FIFO semantics on the channels are respected.

Clearly KPNs are a theoretical model, which is not implementable in practice. The most obvious unrealistic assumption is that of unbounded FIFOs. Any real-life system must use bounded FIFOs. Ensuring schedulability of a computation with bounded FIFO space is a much tougher theoretical problem, and it often solved by restricting the execution semantic of the KPN model. One of the most well-known KPN specializations featuring an efficient schedulability analysis is *Synchronous Dataflow* (SDF). In SDF computational nodes (called actors) have static, non-uniform rates of execution (firing), firing is atomic and data driven. This model is implementable on real single- and multi-processors and its properties have been studied extensively [33]. Unfortunately, SDF is not Turing equivalent, in other words there are computations that can be expressed by KPN and cannot be performed in and SDF model.

The mathematical properties of dataflow have been extensively studied by the theoretical computer science community [32], and the importance of a solid theoretical foundation for program verification is widely recognized and understood. Dataflow/streaming languages and programming environments move from these solid theoretical foundations, but they usually make compromises to provide flexibility and to ease code development.

## 7.3.1   Graphical dataflow programming environments

Many commercial environments supporting stream programming are available and widely adopted (e.g., Mathlab/Simulink, SPW, COSSAP, etc.) and even more viable prototypes have been demonstrated in research (e.g., Ptolemy). Developing streaming applications in these environments is relatively straightforward. The programmer usually works using a *graphic user interface* (GUI) to specify the dataflow graph nodes and dependencies. A large library of computational nodes, implementing common processing tasks is available (for instance, an FIR filter block, with programmable coefficients). For instance, the Mathlab/Simulink environment provides hundreds of pre-defined actors that can be readily instantiated without any code development. When the behavior of an actor is not available in the library, the programmer moves to a text-based interface where it can specify the node behavior, using pre-specified function calls to read input tokens and produce output tokens. Computation is usually specified in an imperative language, such as C.

Graphical dataflow environments are very effective in enhancing the productivity of the software developer, and this explains their widespread usage in software prototyping. However, in practice they have not been used for final product-code implementation, because they generate quite inefficient code for the target MPSoC architectures. In order to understand the reasons for this shortcoming, we need to look deeper in implementation details.

Probably the most critical issue in translating dataflow specifications into efficient NoC platform code is memory management. Let us consider as an example a video-processing pipeline that operates on a large image frame (which can reach multi-megabyte size for high-definition color images). Each actor in the pipeline performs processing on an incoming frame and passes it to the following actor. Let us assume that, we parallelize the execution by allocating each frame-processing actor on a different processor on a multi-core architecture. The token-passing semantic hides a performance pitfall: passing a frame token implicitly specifies a multi-megabyte data transfer. If the processors have private memories, passing a frame requires a very expensive memory-to-memory

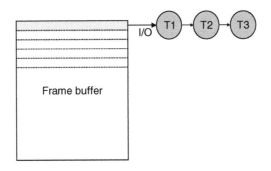

Partitioning a large memory frame to improve streaming performance.

copy. Even worse, if the frame does not fit into the on-chip private memory of each processor, an external memory overflow area will have to be defined, and each token passing will then imply a significant amount of off-chip memory traffic, which is extremely slow and energy-inefficient. Moreover, it creates a destination contention bottleneck that no NoC architecture can alleviate, as all background memory transfers must go through a single NoC target interface to off-chip memory. As a result, the convenient token-passing abstraction breaks down in terms of efficiency, as execution becomes severely memory and communication bound.

Intuitively, it is quite clear that this memory bottleneck can be eased if computation is specified in a more memory-aware fashion. For instance, the computation can be performed in sub-sections of the image by finer-granularity task, as shown in Fig. 7.9. In this case, the full frame resides in main (off-chip) memory, and the image is processed one sub-unit ad a time (e.g., stripe and sub-square). Tokens have much finer granularity and they can be passed along without clogging the on-chip interconnects and the Ios. Unfortunately, traditional prototyping-oriented dataflow environments do not offer analysis and optimization features that help programmers in this difficult balancing act. Hence, the traditional design flow is that an executable specification obtained in a dataflow programming environment will have to manually translated in efficient code in a slow and painful manual tuning process.

Another critical issue with dataflow programming is global data-dependent control flow. Control flow (conditionals, loops) can be specified in a dataflow environment as a part of actor behavior. For instance, conditional execution of an actor's computation can be specified by guarding its code with a conditional. However, in many practical applications a single control-flow decision impacts the execution of a number of actors. Consider, for example, MPEG video stream processing: different computations are performed on different frames, depending on the frame type (B, I, T). Even worse, asynchronous events (e.g., the user pressing a button) may imply significant changes in the computation to be performed by actors. The "pure" dataflow semantics have been extended to deal with global control flow and asynchronous events, but the efficient mapping of these extensions on target platform hardware is still not mature.

## 7.3.2 Stream Languages

Stream languages were developed with the purpose of closing the gap between abstraction and efficiency. In other words, they aim at providing high-level abstraction to the programmer for controlling stream manipulation and processing. At the same time they syntactically enforce a set of rules to facilitate the creation of complex programs that can be compiled very efficiently onto the hardware platform targets. On one hand, developing a complex stream-processing application using a stream language should be as fast and productive as using a graphical prototyping environment. On the other hand, the compilation of such a program should produce a very efficient executable without needing slow and error-prone manual translation.

The basic rationale of this approach is to create languages that are a better match for communication-dominated NoC-based platforms than traditional imperative languages such as plain C or C++. This matching is obtained by making parallelism and communication explicit at the language level, while at the same time providing high-level primitives to manage them. Efficient automated generation of executable code is facilitated by restricting the freedom given to the programmer in accessing global resources (e.g., global variables), and in specifying arbitrarily complex control and data flow. In order to give concreteness to these observations, we will now analyze in some details one of the most well-developed stream languages, StreamIt, proposed by the MIT Group lead by S. Amarasinghe [34].

### A case study: StreamIt

StreamIt is a language explicitly designed for stream application programming. It defines a set constructs that expose the parallelism and communication of streaming applications, but it avoids explicit references to the topology of the target architecture and the granularity of its computational engines. A program consists of a number of simple blocks which communicate data in a limited set of patterns. The atomic block in StreamIt is called *Filter*: it specifies a single-input–single-output computations. A filter body is specified using a single-threaded JAVA-like procedural language. This is the *work* function, which specifies the Filter's steady-state computation. An example of a simple filter is shown in Fig. 7.10.

A Filter communicates with connected blocks through input and output FIFO queues. These channels support three primitive access functions: (i) *pop()*, which removes an item from the output end of the channel and

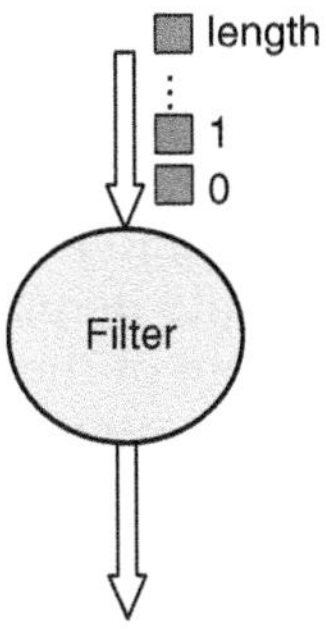

```
float->float filter LowPassFilter (int N, float freq) {
    float[N] weights;

    init {
        weights = calcWeights(N, freq);
    }

    work push 1 pop 1 peek N {
        float result = 0;
        for (int i=0; i<weights.length; i++) {
            result += weights[i] * peek(i);
        }
        push(result);
        pop();
    }
}
```

■ **FIGURE 7.10**

Filter construct in StreamIT.

returns its value, (ii) *peek(i)*, which returns the value of the item $i$ spaces from the end of the FIFO, without removing it and (iii) *push(x)* writes the value of $x$ to the front of the channel. If $x$ is an object, a copy is enqueued on the channel. Filter input–output operations have statically (compile-time) specified rates of production and consumption of FIFOs elements. Note that the work function is triggered when enough samples are in the input FIFO to allow all input queue peek operations specified by its body. Filters also contain initialization functions, which serve two purposes: they set the initial state of the Filter (such as, for instance, the taps coefficients in a FIR filter), second, they specify the Filter's I/O types and data rates to the compiler.

Filters are compositional. They can be connected via three composite blocks: pipelines, split-joins and feedback loops. Each of these structures also has a single entry and exit points, allowing modular recursive composition. The *Pipeline* construct builds a sequence of streams. It has an initialization function, whose primary purpose is to add component streams in a specified sequence. There is no work function, because the behavior of a pipeline is fully specified by its components. The SplitJoin construct specifies independent parallel streams that diverge from a common splitter will eventually merge into a common joiner. Similarly to *Pipeline*, the components of a SplitJoin are specified with successive calls to add from the init function. The splitter specifies how to distribute items from the input of the SplitJoin to the components. Three types of split are supported: (i) *Duplicate*, which replicates each data item and sends a copy to each parallel stream, (ii) *RoundRobin*, which assigns a weight to each parallel stream and sends a number of items to each stream equal to its weight before moving to the next and (iii) *Null*, which means that the parallel components do not require inputs. The joiner indicates how the outputs of the parallel streams should be interleaved on the output channel of the SplitJoin. There are two joiner types: *RoundRobin* and *Null*.

The *FeedbackLoop* construct enable the specification cycles in a stream graph. It contains: (i) a body stream, which is the block around which the "feedback path" will be closed, (ii) a loop stream, to perform computation on the feedback path, (iii) a splitter, to distribute items between the feedback path and the output channel at the bottom of the loop, (iii) a joiner, which merges items between the feedback path and the input channel. All these components are specified in the initialization function. Upon initialization, there are no items on the feedback path and they can be provided by an *initPath* function.

An important characteristic of the StreamIt language is that it cannot specify arbitrary data flow graph. Only hierarchical composition of pipelines, splitjoins and feedback loops can be specified. This greatly facilitates compiler analysis and optimization, while forcing the programmer to build well-organized code.

StreamIt also support constructs for passing irregular, low-volume control information, called *messages*, between filters and streams. It is possible to send a message from within the body of a filter work function to modify a parameter setting in other filters. The sender continues execution while the message is being delivered: message delivery is asynchronous, and it returns no value. One important characteristics of the message delivery abstraction is that it is possible to specify latency bounds for the delivery times. Recall that execution is parallel and asynchronous in dataflow languages, and there is no global notion of time. Hence, latencies can be specified only relatively to data token. It is therefore possible to specify a latency bound in terms of number of data token starting from a reference token. For instance, it is possible to specify that a message arrives to the receivers work function $K$ evaluations after the arrival of the data token that has triggered message issue by the sender. Various types of messages are supported, including broadcast messages and re-initialization messages.

From this cursory view of the StreamIt language, we can draw a few observations. First, data communication between filters and filter compositions are completely explicit can be analyzed at compile time. Second, data production and consumption is statically defined at compile time. This greatly eases allocation and scheduling of hardware resources at compilation time (e.g., FIFO sizing an channel sizing). Moreover, the language's strictly defined firing rules make it much easier to schedule computations and to guard them. Overall, the language is expressive enough to model non-trivial computations, and its strong structure makes it much easier to generate highly efficient executable on an NoC-based tiled architecture.

However, there are some limitations that clearly point to the need for significant additional work in this area. First, there is no way to keep the size of data tokens under control. Hence, the programmer can easily specify a behavior where very large data tokens are passed around. If data tokens cannot fit on local storage associated to computational nodes, a lot of traffic from–to global memories will be created, with obvious efficiency losses. Second, the language constructs are quite limited and a lot of the behavior has to be fully at compile time. Data-dependent global control flow is very difficult to express (messages help, but they are cumbersome). It is quite clear than many choices have been made to limit expressiveness in an effort to ease the job of the compiler. The suitability of StreamIt to describe truly complex application behaviors is therefore not yet proved.

## 7.4 COMPUTER-AIDED SOFTWARE DEVELOPMENT TOOLS

Parallel programming critically requires tool support because of its unfamiliarity and intrinsic difficulty. Programming tools are not limited to compilers for efficient executable generation. They should help in systematically finding defects and performance bottlenecks, as well as debugging and testing programs. Without these tools, parallelism is likely become an obstacle that reduces developer and tester productivity and makes development more expensive and of lower quality.

### 7.4.1 Compilation

Compilers for high-performance processors have traditionally focused on ILP extraction. This approach has been followed also in embedded computing for signal and media processing. These application domains are characterized by vast amounts of ILP and word-level parallelism, as well as by regular data access patterns. For this reason, processor architectures for signal and media processing usually feature a large number of parallel execution units and follow architectural templates reminiscent of general-purpose VLIW and vector processors [35]. These processors heavily rely on compiler support to keep execution units busy and achieve massive speedups over single-instruction pipelined execution.

Compilers for data-parallel processors have evolved very rapidly, and they feature aggressive allocation and scheduling algorithms that achieve utilizations which are comparable to hand-optimized code [36]. A survey of this interesting field is outside the scope of the chapter, and the reader is referred to one of the many surveys and textbooks available [37, 38]. One interesting point is worth mentioning, however. Even within a single-processor engine, architectures are becoming increasing interconnect dominated. More specifically, the communication between execution units the register file is the communication bottleneck that limits scalability in the number of functional units that can be successfully operated in parallel. For this reason processor architectures are evolving toward clustered organizations, with multiple register files and complex interconnects.

These in-processor interconnects, also called *scalar operand networks* [39] have strong similarities with chip-level NoCs, in that they need to be scalable and to provide huge bandwidth. However, since their main purpose is to deliver instruction operands to functional units and results to register files, their latency is extremely tightly bound (e.g., one or two clock cycles). Hence, computation and communication cannot be decoupled via a network interface and its connection-oriented services, which would impose unacceptable latency penalties. Consequently, scalar operand networks are tightly coupled with the instruction set architecture, and they are viewed as explicit resources during compilation. Communication links are allocated and scheduled exactly like functional units and registers [40]. From the topological viewpoint, scalar operand networks are usually full or partial crossbars, even though multi-hop topologies will most likely appear in the future. As the focus of this book is on chip-level NoCs, we refer the interested reader to the many interesting references in the literature that deal with scalar operand networks architecture and scaling trends [39].

**Coarse-grain parallelism extraction**

While compilers for ILP extraction are now mature and successful in industrial applications, the massive amount of research work on TLP discovery and extraction has not enjoyed a similar degree of success. The most common approaches in the industrial practice rely on explicit directives, inserted by the programmer to drive the compiler toward good task-level parallelizations. This pragmatic semi-automated approach can be very successful, even though its implementation is quite challenging.

One of the main issues in semi-automated TLP is the difficulty for programmer in understanding and using program annotations for parallelism when specifying computation in a sequential fashion. In other words, if the programming language has an implicit sequential semantics, the programmer is required to manage continuous shifts between a sequential and a parallel programming model, which ultimately lead to inefficient or incorrect code. As a result the learning curve for the programmer is quite steep and parallelization is done as an afterthought, by modifying sequential code. This process is inefficient and often leads to sub-optimal results.

An example of semi-automated approach is OpenMP [41]. OpenMP provides a set of pragmas that aim at giving the compilation system directives on how to automatically generate a multi-threaded execution flow. The programmer inserts pragmas, such as `#pragma Omp parallel`, immediately before loops that should be parallelized, and the compilation system decides how to split the target loop in parallel threads that are then distributed among the available processors. Thus, the programmer is relieved from the burden of managing data sharing and synchronization, while the compiler is facilitated in focusing efforts where the programmer expects most advantages from parallelization. OpenMP is gaining acceptance in high-performance chip multi-processors, such as IBM's Cell.

**Example 7.2.** Cell uses OpenMP pragmas to guide parallelization decisions. It then uses the pre-existing parallelization infrastructure of the proprietary IBM XL compiler. In the first pass of the compiler machine-independent optimizations are applied to the functions outlined by the programmer. In the second pass of optimization, the target functions are cloned, then optimized codes are generated for both the master processor (SPE) and the slave data-parallel processors (PPEs). The Master Thread runs on the PPE processor and manages work sharing and synchronization. Note that cloning of the call sub-graph of the target functions occurs in the second pass of optimization, after interprocedural analysis. Thus, the compiler can link the appropriate versions for all the library function calls for the SPE and the PPE.

An important issue is the limited-size local memory of the SPEs, which must accommodate both code and data. Hence, there is always a possibility that a single SPE-allocated block will be too large to fit. Since OpenMP generally is used to drive parallelization of tight loop nests, code spilling is not very common. Data spilling is instead a much more critical problem. Fortunately data is managed explicitly and the optimizing compiler can in principle insert all the necessary data manipulation function (e.g., DMA-accelerated data transfers) in an automated fashion. This is however most easily said than done, and in reality a sub-optimal data organization choices cannot be fully remedied by background DMA-based memory transfers.

The critical challenge for the success of OpenMP and similar approaches is in the efficient implementation of the parallelization step. In an NoC architecture, communication latencies and network topology should be accounted when deciding the degree of parallelization of a loop (declared by the programmer as "parallelizable"). In other words, it may be more efficient to parallelize the loop to a limited degree, if then the various parallel threads can be mapped on a set of topologically close computational units. Aggressive parallelization may be inefficient if it requires long-range data communication across many network switches. Clearly, a significant amount of research and development work is required to develop NoC (or more generally communication) aware software parallelization tools, that can account for NoC communication latencies, and can at the same time interact with NoC mapping to influence the way parallel threads are allocated onto the network topology. At the same time, OpenMP appears somewhat limited in its expressiveness for what concerns the important issue of data allocation and partitioning: in other words, it is based on an uniform memory access abstraction which does not fit well to NoC-based distributed architectures.

## 7.4.2  Testing and debugging

Program testing and debugging is of the utmost importance. In fact, application parallelism introduces new types of programming errors, in addition the usual ones in sequential code. Incorrect synchronization can cause data races and livelocks which are extremely difficult to find and to understand, since they appear as non-deterministic and hard to reproduce. Conventional debugging, based on breakpointing and re-execution, does not work in this context, because error manifestation depends on subtle timing relationships between parallel tasks. The simple insertion of a breakpoint or of instrumentation code changes the timing of a program and may completely hide many serious bugs. Even if the problem can be reproduced, its diagnosis can be extremely difficult to show, since the state of the system is distributed and extremely difficult to analyze.

Language support for development of correct-by-construction programs is therefore as critical as language support for efficient compilation. Low-level hardware-related features for communication and synchronization should not be exposed to the programmer, not only because of code portability issues, but above all for avoiding the creation of intricate program logic which becomes impossible to debug. Clearly, even though well-designed languages can greatly help in writing correct code, systematic defect detection tools are extremely valuable. One promising approach is to use static program analysis to systematically explore all possible executions traces of a program. In this way it becomes possible to find errors that are hidden very deep in the program logic, and are almost impossible to reproduce by standard testing.

Similar techniques are used with an increasing degree of success in hardware verification, and are known as "formal verification" techniques. Hardware is highly parallel and complex, and it is notoriously hard to verify, but unfortunately, software verification is even more difficult. One of the main challenges is the size of the state space (which relates to the number of possible execution trace) of programs. Its size is formidable, especially if we consider parallel programs (remember that the size of the state space for two programs running in parallel is the product of the two component's state spaces). Hence, its complete exploration is hopeless, and in many case we must limit ourselves to formally checking only some program properties [31].

Since formal software verification is not a panacea, developers critically require traditional debuggers to explore and understand the complex behavior of their parallel programs. Two main approaches are used in this area. First, tracing and logging tools aim at keeping trace, and tagging messages and shared data accesses by various threads, allowing the developer to directly observe a parallel program partially ordered execution. Important features that require further developments are: the ability to follow causality trails across threads (such as chains of accesses to distributed objects), to replay messages in queues and reorder them if necessary, step through asynchronous call patterns (including callbacks). The second approach is reverse execution, which permits a programmer to back up in a programs execution history and re-execute some code. This technique is expensive and complex, but it is becoming viable thanks to the increased capabilities of execution cores, which have a large amount of shadow storage and are increasingly used as virtual processors, emulating via micro-code well-known instruction sets.

Software testing will also have to undergo substantial extensions. Parallel programs have non-deterministic behaviors and are therefore much more difficult to test. Basic code coverage metrics (e.g., branch or statement coverage), will have to be extended to take into account multiple concurrently executing threads. Single-thread coverage is very optimistic, as it completely neglects the notion of global state required by multiple execution threads. Thus, stress tests will have to be augmented by more systematic semi-formal techniques (e.g., model-checking-like assertions).

In embedded systems, system simulation and emulation techniques are extremely useful and relevant for debugging. Simulation relies on the concept of virtual platform [43], as exemplified in Section 7.1 which are run on powerful host machines. When a parallel application is run on a virtual platform, its execution can be very carefully monitored, stopped and rolled back in ways that are not possible when debugging on the target hardware, and the developer has much more control on fine-grain details of the execution. The main shortcoming of virtual platforms is that they executed ad orders-of-magnitude slower rates than actual hardware, and reproducing complex incorrect behaviors with billion-instruction traces may require a huge amount of time. Emulation platforms greatly accelerate simulation speed, but they are extremely expensive and traditionally they have been used only for hardware debugging. However, new approaches to affordable MPSoC platform emulation are emerging that show promise [42].

## 7.5  SUMMARY

In this chapter, we have discussed software abstractions and tools targeting scalable NoC platforms. The focus of our treatment has been first on state-of-the art solutions, based on coarse-grained manual parallelization of applications, and on library support for NoC programming. The architectural implications of various NoC programming models have been explored. Looking forward we have surveyed stream-oriented programming, which appears to offer well-matched abstractions to the software developer targeting an NoC architecture. Finally, we surveyed programming environments and outlined the many research and development challenges that must be addressed to develop effective software development frameworks for NoCs.

Software abstractions and software development support for NoCs are still immature, as the corresponding hardware architectures are just now emerging from the laboratories. It should be clear however, that NoCs will be successful in practice only if they can be efficiently programmed. For this reason we believe that the topics surveyed in this chapter will witness significant growth in research interest and results in the next few years.

# REFERENCES

[1] F. Boekhorst, "Ambient Intelligence, the Next Paradigm for Consumer Electronics: How will it Affect Silicon?," *International Solid-State Circuits Conference*, Vol. 1, 2002, pp. 28–31.

[2] G. Declerck, "A Look into the Future of Nanoelectronics," *IEEE Symposium on VLSI Technology*, 2005, pp. 6–10.

[3] W. Weber, J. Rabaey and E. Aarts (Eds.), *Ambient Intelligence*. Springer, Berlin, Germany, 2005.

[4] S. Borkar, et al., "Platform 2015: Intel Processor and Platform Evolution for the Next Decade," *INTEL White Paper* 2005.

[5] D. Culler and J. Singh, *Parallel Computer Architecture: A Hardware/Software Approach*, Morgan Kaufmann Publishers, 1999.

[6] L. Hennessy and D. Patterson, *Computer Architecture – A Quantitative Approach*, 3rd edition, Morgan Kaufmann Publishers, 2003.

[7] Philips Semiconductor, *Philips Nexperia Platform*, www.semiconductors.philips.com

[8] M. Rutten, et al., "Eclipse: Heterogeneous Multiprocessor Architecture for Flexible Media Processing," *International Conference on Parallel and Distributed Processing*, 2002, pp. 39–50.

[9] ARM Ltd, *MPCore Multiprocessors Family*, www.arm.com

[10] B. Ackland, et al., "A Single Chip, 1.6 Billion, 16-b MAC/s Multiprocessor DSP," *IEEE Journal of Solid State Circuits*, Vol. 35, No. 3, 2000, pp. 412–424.

[11] G. Strano, S. Tiralongo and C. Pistritto, "OCP/STBUS Plug-in Methodology," *GSPX Conference* 2004.

[12] M. Loghi, F. Angiolini, D. Bertozzi, L. Benini and R. Zafalon, "Analyzing On-Chip Communication in a MPSoC Environment," *Design and Test in Europe Conference (DATE)* 2004, pp. 752–757.

[13] F. Poletti, P. Marchal, D. Atienza, L. Benini, F. Catthoor and J. M. Mendias, "An Integrated Hardware/Software Approach For Run-Time Scratchpad Management," in *Design Automation Conference*, Vol. 2, 2004, pp. 238–243.

[14] D. Pham, et al., "The Design and Implementation of a First-generation CELL Processor," in *IEEE International Solid-State Circuits Conference*, Vol. 1, 2005, pp. 184–592.

[15] L. Hammond, et al., "The Stanford Hydra CMP," *IEEE Micro*, Vol. 20, No. 2, 2000, pp. 71–84.

[16] L. Barroso, et al., "Piranha: A Scalable Architecture Based on Single-chip Multiprocessing," in *International Symposium on Computer Architecture*, 2000, pp. 282–293.

[17] Intel Semiconductor, *IXP2850 Network Processor*, www.intel.com

[18] STMicroelectronics, *Nomadik Platform*, www.st.com

[19] Texas Instruments, "OMAP5910 Platform", www.ti.com

[20] M. Banikazemi, R. Govindaraju, R. Blackmore and D. Panda. "MP-LAPI: An Efficient Implementation of MPI for IBM RS/6000 SP systems", *IEEE transactions Parallel and Distributed Systems*, Vol. 12, No. 10, 2001, pp. 1081–1093.

[21] W. Lee, W. Dally, S. Keckler, N. Carter and A. Chang, "An Efficient Protected Message Interface," *IEEE Computer*, Vol. 31, No. 11, 1998, pp. 68–75.

[22] U. Ramachandran, M. Solomon and M. Vernon, "Hardware Support for Interprocess Communication," *IEEE transactions Parallel and Distributed Systems*, Vol. 1, No. 3, 1990, pp. 318–329.

[23] H. Arakida et al., "A 160 mW, 80 nA Standby, MPEG-4 Audiovisual LSI 16Mb Embedded DRAM and a 5 GOPS Adaptive Post Filter," *IEEE International Solid-State Circuits Conference* 2003, pp. 62–63.

[24] F. Gilbert, M. Thul and N. When, "Communication Centric Architectures for Turbo-decoding on Embedded Multiprocessors," *Design and Test in Europe Conference* 2003, pp. 356–351.

[25] S. Hand, A. Baghdadi, M. Bonacio, S. Chae and A. Jerraya, "An Efficient Scalable and Flexible Data Transfer Architectures for Multiprocessor SoC with Massive Distributed Memory," *Design Automation Conference* 2004, pp. 250–255.

[26] P. Paulin, C. Pilkington, E. Bensoudane, "StepNP: A system-level Exploration Platform for Network Processors," *IEEE Design and Test of Computers*, Vol. 19, No. 6, 2002, pp. 17–26.

[27] P. Stenström, "A Survey of Cache Coherence Schemes for Multiprocessors," *IEEE Computer*, Vol. 23, No. 6, 1990, pp. 12–24.

[28] M. Tomasevic, V.M. Milutinovic, "Hardware Approaches to Cache Coherence in Shared-Memory Multiprocessors," *IEEE Micro*, Vol. 14, No. 5–6, 1994, pp. 52–59.

[29] I. Tartalja, V.M. Milutinovic, "Classifying Software-Based Cache Coherence Solutions," *IEEE Software*, Vol. 14, No. 3, 1997, pp. 90–101.

[30] P. Kongetira, K. Aingaran and K. Olukotun, "Niagara: a 32-way Multithreaded Sparc Processor," *IEEE Micro*, Vol. 25, No. 2, 2005, pp. 21–29.

[31] H. Sutter and J. Larus, "Software and the Concurrency Revolution," *ACM Queue*, Vol. 3, No. 7, 2005, pp. 54–62.

[32] R. Stephens, "A Survey of Stream Processing," *Acta Informatica*, Vol. 34, No. 7, 1997, pp. 491–541.

[33] E. Lee and D. Messerschmitt, "Pipeline Interleaved Programmable DSP's: Synchronous Data Flow Programming," *IEEE Transactions on Signal Processing*, Vol. 35, No. 9, 1987, pp. 1334–1345.

[34] W. Thies, et al., "Language and Compiler Design for Streaming Applications," *IEEE International Parallel and Distributed Processing Symposium*, 2004, pp. 201.

[35] Silicon Hive, *AVISPA-CH*, www.silicon-hive.com

[36] M. Bekoij, *Constraint Driven Operation Assignment for Retargetable VLIW Compilers*, Ph.D. Dissertation, 2004.

[37] R. Allen, K. Kennedy, *Optimizing Compilers for Modern Architectures: A Dependence-based Approach*, Morgan-Kaufman, 2001.

[38] P. Faraboschi, J. Fisher and C. Young, "Instruction Scheduling for Instruction Level Parallel Processors," *Proceedings of the IEEE*, vol. 89, No. 11, 2001, pp. 1638–1659.

[39] M. Taylor, W. Lee, S. Amarasinghe and A. Agarwal, "Scalar Operand Networks," *IEEE Transactions on Parallel and Distributed Systems*, Vol. 16, No. 2, 2005, pp. 145–162.

[40] P. Mattson, et al., "Communication Scheduling," *ACM Conference on Architectural Support for Programming Languages and Operating Systems*, 2000, pp. 82–92.

[41] M. Sato, "OpenMP: Parallel Programming API for Shared Memory Multiprocessors and On-chip Multiprocessors," *IEEE International Symposium on System Synthesis*, 2002, pp. 109–111.

[42] N. Genko, D. Atienza, G. De Micheli, J. Mendias, R. Hermida, F. Catthoor, "A Complete Network-on-chip Emulation Framework," *Design, Automation and Test in Europe*, Vol. 1, 2005, pp. 246–251.

[43] C. Shin, et al., "Fast Exploration of Parameterized Bus Architecture for Communication-centric SoC Design," *Design, Automation and Test in Europe*, Vol. 1, 2004, pp. 352–357.

[44] M. Michael, "Scalable, Lock-free Dynamic memory Allocation," *ACM Conference on Programming Languages Design and Implementation*, 2004, pp.110–122.

# CHAPTER 8

# DESIGN METHODOLOGIES AND CAD TOOL FLOWS FOR NoCs*

Designing *networks on chips* (NoCs) is a complex process and spans several abstraction levels, ranging from the transaction to the physical levels. Design choices are difficult to make, because most figures of merit of the network depend highly on high-level decisions on architectures and protocols. Yet these decisions can only be validated while considering physical layer measures, such as delays on interconnection links. Thus, potential design closure issues may require designers to explore various configurations with different parameters in the search for those that satisfy the network and overall system specifications. *Computer-aided design* (CAD) tools are therefore very useful to shorten the design time and provide design closure.

The major steps involved in designing NoCs include the following:

- Analyzing and characterizing application traffic.

- Synthesizing the NoC topology for the application.

- Mapping and binding of the cores with the NoC components.

- Finding paths for the traffic flows and reserving resources across the NoC.

- Determining NoC architectural parameters, such as the data width of the links, buffer sizes, and frequency of operation.

- Verifying the designed NoC for correctness and performance.

In order to achieve efficient NoC designs, many of these steps are performed together. As an example, mapping components and selecting paths for the different traffic flows can be done jointly during topology synthesis. To achieve design closure, it is important to automate and integrate the different design steps. Thus, the different design phases are coupled

* We acknowledge the contributions of Srinivasan Murali, Stanford University, Palo Alto, CA, USA; Kees Goossens, Philips Research, The Netherlands and Andreas Hansson, Technical University Eindhoven, The Netherlands.

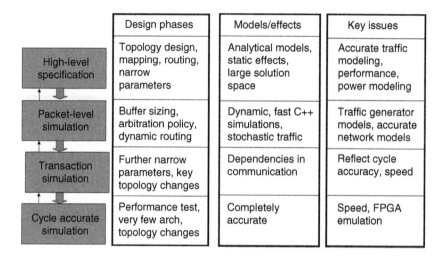

| | Design phases | Models/effects | Key issues |
|---|---|---|---|
| High-level specification | Topology design, mapping, routing, narrow parameters | Analytical models, static effects, large solution space | Accurate traffic modeling, performance, power modeling |
| Packet-level simulation | Buffer sizing, arbitration policy, dynamic routing | Dynamic, fast C++ simulations, stochastic traffic | Traffic generator models, accurate network models |
| Transaction simulation | Further narrow parameters, key topology changes | Dependencies in communication | Reflect cycle accuracy, speed |
| Cycle accurate simulation | Performance test, very few arch, topology changes | Completely accurate | Speed, FPGA emulation |

**■ FIGURE 8.1**

A layered flow for NoC design.

by providing information feedback. For example, if simulation shows that performance targets are not met, the information can be fed back to the topology synthesis which can then produce a NoC with network topology supporting higher throughput.

As the problem of designing the NoC involves several steps and each step requires different models of the system, a layered design flow is adopted (Fig. 8.1). At the top most layer of the flow, the communication characteristics of the application are abstracted by high-level models. For example, the average rate of traffic flow across the various cores in the design can be used to represent the application traffic characteristics. At this layer, based on the size and switching activity of the various NoC components, analytical models are used to characterize the power consumption of the NoC. The performance of the NoC in this layer is typically modeled by two parameters: bandwidth and hop-delay for communication. The bandwidth available across an NoC link reflects the average traffic rate that can be sustained by the link. The hop-delay metric reflects the average latency for communication in the NoC under zero-load conditions (where different traffic streams do not contend for the resources).

Clearly, models used at this layer are not as accurate as the other layers. Moreover, at this layer we cannot observe any dynamic and temporal effects such as contention of packets for different resources, as the models are static in nature. However, most of the key design steps such as topology synthesis have a very large solution space that can only be tackled at this abstraction level. The use of analytical models helps in pruning infeasible and inefficient solutions quickly. Thus, all the major steps such as topology synthesis, mapping, routing, resource reservation, and narrowing architectural parameters to a range are performed at this layer. The major challenge to be addressed here is that the models used for abstracting the traffic patterns, power consumption, and performance of the NoC should closely match the actual values obtained at the lower layers.

Once the NoC is designed using static information, in the next layer, a packet-level simulation of the NoC is carried out. The objective here is to observe dynamic effects that can be used to set several architectural parameters such as buffer sizes, arbitration policies, and routing policies (in case of dynamic routing). To achieve fast simulation, the application traffic is modeled by means of stochastic traffic generators or trace-driven traffic injectors. Normally, the simulation environment is modeled at flit-accurate level (i.e., a packet is segmented into multiple flits) in software language (such as C++) and simulation times are in the order of minutes. The key challenges at this layer are that the stochastic traffic should match the actual application traffic and the simulation models of the network components should match the characteristics of the hardware architecture of the components.

In the next layer, transaction-level simulation of the NoC by using HDL languages, such as SystemC, is performed. A key feature that can be modeled at this level is dependencies between the different messages. While designing at this layer, some changes to the topology and some tuning of architectural parameters can be performed. It is worthwhile to mention here that in many systems (e.g., [8]) transaction-based simulation is integrated with packet-level simulation.

In the final layer, cycle-accurate RTL simulations of the NoC are carried out. For complex *system-on-chips* (SoC) designs, the simulations may take several hours to complete. The simulations are usually used to verify performance metrics and to carry out system validation. To reduce the simulation time, hardware emulation can also be carried out.

Tools for addressing the various steps in the design of NoCs have been researched and developed recently. Thus, this area has not yet reached maturity and some aspects of the design process have received more attention than others. CAD support for NoCs can be roughly divided into analysis and simulation tools, synthesis and optimization tools, and toolkits for bus design. It is the purpose of this chapter to provide an overview of the tools that are specific to NoCs, and to explain how they can complement, and be interfaced with generic SoC design tools.

## 8.1 NETWORK ANALYSIS AND SIMULATION

Analysis and simulation of NoCs require both general-purpose and specialized tools. Indeed network simulation can be performed using tools such as ns2 [29] for packet-level simulations or using a SystemC simulator for transaction and cycle-accurate levels. As usual in system design, there is a trade-off between abstraction level and accuracy, and accurate models require long simulation times.

Specialized tools for NoC go hand in hand with consistent analysis methodologies that recognize the inherent nature of the network. Today the modeling and simulation of on-chip networks and their integration into a single simulation environment, combining processing elements and communication primitives are still an open research area. In the rest of this section, we will see the modeling environments and then several models that have been used to capture the application traffic behavior.

## 8.1.1 NoC Modeling Environments

For the sake of simplicity and efficiency, the design of the computation and communication architectures are usually treated as orthogonal issues [3]. During the computational architecture design, the application tasks are mapped onto processor and hardware cores. This step is also known as hardware-software partitioning or allocation, as the decision on whether a task is to be executed by a general-purpose processor or a dedicated hardware core is taken here. The goal of the communication designer is to take up the mapped system and to design the best NoC architecture for the system.

The communication characteristics can be obtained by simulation when the application code is available (before starting the NoC design). In these initial simulations, the communication architecture can be abstracted by using high-level simulation models. When simulation is performed at RTL cycle-accurate level, a perfect non-blocking communication architecture (such as a crossbar) can be utilized. Monitors, designed in hardware or software, can be added to such a design to collect the traffic information. From such a set-up, we can obtain the traffic characteristics of the application under *ideal* conditions, where contention for the communication architecture resources is not present. In case of long simulation time, hardware emulation of the design can be performed to speed up the traffic collection process.

Some dedicated modeling environments and simulators have been developed for several projects, such as the Æthereal [8], and Nostrum [13]. These environments aim at high-level modeling of NoCs and at supporting NoC design by iterative refinement. Conversely, the MPARM simulator [2] is a general-purpose multi-processor simulator that is particularly effective in analyzing the communication infrastructure interspersed among computing nodes.

### The MPARM environment

The name MPARM relates to multi-processing with the ARM processor cores. Nevertheless the simulator, originally developed for ARM, now supports other cores, such as ST LX 220, CoWare Lisatek cores, PowerPC750, and

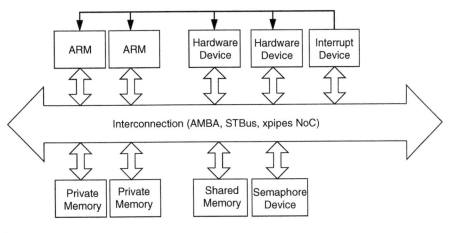

The MPARM architecture.

MIPS R3000. Without loss of generality, we will refer to processors as ARM cores from now on.

The architecture of the MPARM platform is shown in Fig. 8.2. It is representative of a large class of multi-processor SoC platforms. The platform consists of:

- A configurable number of 32-bit ARM processors.

- The processors' private memories.

- A configurable number of hardware devices or traffic generators.

- A shared memory.

- A hardware interrupt unit.

- A hardware semaphore module.

- An interconnection fabric that connects all the units.

The interconnection fabric can be an AMBA (*advanced microcontroller bus architecture*) AHB (*advanced high-performance bus*) bus [27], STBus with arbitrary topology [28], or a xpipes NoC [25], resulting in different instances of the platform.

Processor cores are modeled by means of an adapted version of a GPL-licensed ARM *instruction set simulator* (ISS) called SWARM [30] and written in C++. The hardware models of all the components are described in SystemC and the ARM ISS is embedded into a SystemC wrapper.

Several memory devices can be instantiated in the platform, which can be used by the processors and hardware units for private accesses or for shared accesses. Their latencies can be configured to explore interconnection performance under several conditions. The interrupt device allows processors to send interrupt signals to each other. This hardware primitive is needed for inter-processor communication and is mapped in the global addressing space. For an interrupt to be generated, a write must be issued to a proper address of the device. The semaphore device is also needed for synchronization among processors: it implements *test and set* operations, the basic requirements to have semaphores. Using the MPARM environment, traffic patterns that are representative of typical MPSoC loads, such as in multimedia chips and I/O controllers can be generated.

The RTEMS operating system has been ported to MPARM to provide system software support for multi-processing, as well as native calls for communication and synchronization in multi-processor environments. Software tasks can run on top of RTEMS to exercise the transfer functions on the NoC links.

Hardware and software traffic monitors are available in MPARM, which allows the user to obtain the traffic characteristics of the applications. Overall, MPARM allows the designer to test NoC configurations on functional application traffic by running software tasks on the cores. Thus, it is very useful to compare various configurations and various network parameters for a given application and network architecture. It can also be used to validate the choice of specific cores within an on-chip networked environment.

### NoC emulation

A major pitfall of cycle-accurate NoC simulation is that the simulation times are large. Even though the NoC can be simulated at the transaction level, the effect of particular choices on the NoC architecture and protocol parameters can be validated only by running software tasks of significant length and interaction. Thus, simulation time becomes a bottleneck for NoC analysis.

NoC emulation on programmable hardware (e.g., *field-programmable gate arrays* (FPGAs)) offers a way to speed up the NoC analysis time. Indeed the NoC hardware components can be mapped into logic gates and the network protocols can be either emulated in programmable hardware or in software, thus leaving a fair degree of flexibility on the NoC to be analyzed. There have been several approaches to NoC emulation [7, 16], and emulation can lead to several orders of magnitude speed-up when compared to cycle-accurate simulations [7]. The key issues in performing the emulation are as follows:

- ▪ Size and speed of the emulated NoC. Some large NoCs cannot be mapped onto a single FPGA for gate capacity reasons. In this case, they need to be split over onto multiple FPGAs, which affect the operational speed. Nevertheless, even NoCs mapped to a single FPGA operate one order of magnitude slower when compared to an (*application-specific integrated circuits*) (ASIC) implementation.

- Flexibility of testing various implementations. Since NoC emulation is effective to test various NoC configurations, it is important to consider the time and effort spent in reconfiguring the target FPGA platform. NoC emulation can be kept flexible by keeping some aspects or parameters of the NoC in software, thus requiring fewer hardware reconfigurations [7, 16].

It is also very interesting to notice that FPGA themselves can be seen as on-chip networks, even though they are usually designed with the purpose of servicing various applications. Thus, NoC emulation can be seen as the process of mapping a virtual NoC architecture on another physical NoC architecture. When this mapping can be efficiently done, then FPGAs can provide a good underlying computational substrate. We expect future FPGAs to evolve significantly, by incorporating more programmable cores and more elaborate interconnection means. Thus, FPGAs are likely to be NoC platforms suitable for many applications.

## 8.1.2  Application Traffic Models

From the initial simulation or emulation of the application, several different traffic models can be constructed. The simplest traffic model, used by several early works [10, 11], abstracts the communication behavior by means of the average rate of traffic flow across the different cores. As an example, let us consider an SoC design: the *Video Object Plane Decoder* (VOPD) (Fig. 8.3). Each block in the figure represents a processor, hardware, or a memory core. The average rate of traffic flow across the different cores can be abstracted by a graph referred to as the *Communication*

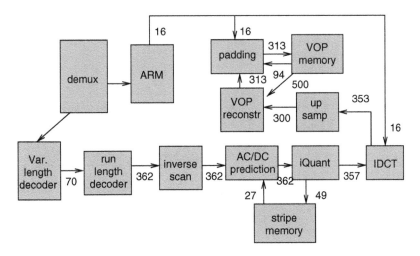

**■ FIGURE 8.3**

An example SoC application (VOPD) with the bandwidth (in MB/S) of communication between the cores annotated across the edges.

*Architecture Graph* (CAG) or the *core* graph. The core graph for the VOPD application is presented in Fig. 8.4. Other variations of the basic model are also used, such as using peak rate of transfer instead of the average rate [9].

In reality, application traffic exhibits three major characteristics: much of the application traffic in SoCs are bursty in nature, the different traffic streams have different delay/jitter constraints, and there are multiple priority levels for the different streams. Traffic models that consider all these effects have been developed in Ref. [19].

A major issue about traffic modeling is that the traffic characteristics can vary widely from one simulation run to another, depending on the inputs to the system. Thus, there should be a mechanism that ensures that each core sends traffic so that the network elements can support the traffic and the delay constraints are met. For this purpose, traffic shapers or traffic regulators that are widely used in *asynchronous transfer mode* (ATM) networks to guarantee *quality of service* (QoS) to applications can be used for NoC designs [19].

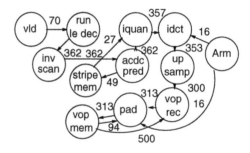

■ **FIGURE 8.4**

The core graph representing the communication between the cores of the VOPD application.

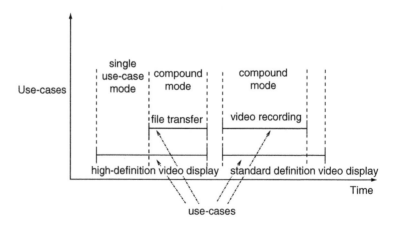

■ **FIGURE 8.5**

A single SoC supports multiple applications or use-cases.

Another interesting aspect is that a single SoC can support multiple *modes* or *use-cases*. As an example, a subset of the use-cases supported by set-top-box SoCs is presented in Fig. 8.5. The communication characteristics of the different use-cases on an SoC vary considerably and the traffic models should capture the individual use-case characteristics. For designing such systems, several core graphs can be constructed, which need to be considered during the NoC design process. A detailed approach to model and analyze such multiple use-case designs is presented in Ref. [20].

## 8.2  NETWORK SYNTHESIS AND OPTIMIZATION

SoCs that use an NoC infrastructure are economically feasible if they can be used in several product variants, and if the design can be reused in different application areas. On the other hand, successful products must provide good performance characteristics, thus requiring dedicated solutions that are tailored to specific needs. As a consequence, the NoC design challenge lies in the capability to design hardware optimized, customizable platforms for each application domain.

Computer-aided synthesis of NoCs is particularly important in the case of application-specific SoCs, which usually comprise computing and storage arrays of various dimensions as well as links with various capacity requirements. Moreover, designers may use NoC synthesis as a means for constructing solutions with various characteristics that can be compared effectively only when a detailed model is available. Thus, synthesis of NoCs can be used for comparing prototypes. Needless to say, synthesis may also be very efficient for designing NoCs with regular topologies as, for example, multi-processing systems with homogeneous cores.

Design and optimization can be achieved by facilitating the integration of domain-specific computation resources in a plug-and-play design style. Standard interface sockets such as *Virtual Component Interface* (VCI) [33] and *Open Core Protocol* (OCP) [31] have been developed for this purpose and support the use of a common NoC as the basis for system integration. A relevant task of these interfaces is to make the NoC adaptive to the different features of the integrated cores (e.g., data and address bus width).

Synthesis of NoCs can be divided into architectural exploration and instantiation. The former step deals with the synthesis of appropriate topology, mapping, and setting up of network parameters for a given application. The latter step consists of generating a network instance, possibly by using a specific library of NoC components.

### 8.2.1  NoC Architecture Exploration

Several of the major NoC design steps such as topology synthesis, mapping, routing, resource reservation, and setting of parameters are performed as part of the architecture exploration phase.

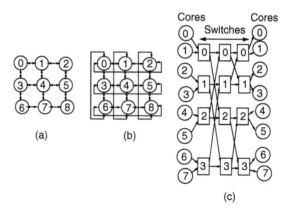

Standard NoC topologies: (a) mesh, (b) torus, and (c) butterfly.

During NoC topology synthesis and mapping, the number and size of the NoC components and their connectivity with each other and with the cores are determined. There are several different standard topologies that can be used for the NoC design, such as mesh, torus, butterfly, to name a few (Fig. 8.6). On the other hand, a custom irregular topology can also be synthesized for the particular application. A major design decision to be made by the designer is whether he or she wants to opt for a standard topology or for an application-specific custom topology.

The advantages of using standard topologies are that for systems with homogeneous cores, the interconnect structure can be well controlled. Moreover, they have nice topological properties such as connectivity between all pair of cores, which can be useful when a current design is extended to future platforms. But, if the application consists of hetero-geneous cores, an irregular application-specific topology would almost always be more power and performance efficient.

When a standard topology is used for the NoC, the major design issue is to efficiently map or bind the cores onto the different NoC components. The communication between the various cores can be represented by the core graph and the topology can be abstracted by a topology graph. Then the NoC mapping problem can be viewed as a graph mapping problem, where the core graph is to be mapped onto the topology graph, mini-mizing some design objective (such as the hop-delay for communication) and satisfying design constraints (such as the bandwidth constraints). The topology mapping problem is NP-Hard and several heuristic solutions have been proposed to address the problem [9, 11, 17, 19].

When a custom application-specific topology is to be designed, the number of switches and *network interfaces* (NIs), their size and the con-nectivity between them and with the cores need to be determined. When designing custom topologies, it is important to remove both message- and routing-dependent deadlocks in the network. During the design process, it is also important to take the wiring complexity of the topology into account. For this, a fast floorplanner can be integrated into the topology synthesis process. The application-specific topology design process has been addressed by several research works [1, 10, 21].

The path selection and resource reservation steps are usually performed within the topology mapping and synthesis step, as the information from these steps are needed to evaluate whether a mapping satisfies the design constraints and optimizes the design objectives. The topology synthesis step is repeated by varying the different architectural parameters, such as the NoC operating frequency and link data width. Typically, the designer needs to prune the set of architectural parameters to a small set of discrete values, so that the entire design process completes in reasonable time. As an example, the set of possible link-width values can be limited to multiples of 16 bits, or to match the data width of the cores.

### 8.2.2 NoC Instantiation

Once the NoC architecture is defined, the next step is to generate the RTL code of the network components using a library of soft macros for the components. A few libraries of components have been developed for this purpose. Notable examples are xpipes [4], xpipeslite [25], Proteo [24], Bone [14, 15], and Æthereal [8].

Proteo is a set of soft macros, that is, a set of hardware models that describe the basic components of an NoC that are parameterized. The models are compatible with the VSIA standard. By interfacing these models, NoCs with different desired topologies can be readily assembled. The Bone project aims at providing design components for NoCs. These components have been implemented at the circuit level, and some chip implementations using the Bone methodology have been achieved, as reported in Chapter 9. The details of the xpipes and Æthereal libraries, along with their tool flows are explained in the next section.

## 8.3 DESIGN FLOWS FOR NoCs

In this section, we focus in detail on two of the CAD tool flows for designing NoCs: the NetChip tool flow that uses the xpipes library, and the Æthereal tool flow. Both the tool flows automate most of the steps involved in designing NoCs from application specification to RTL code generation.

### 8.3.1 NetChip

NetChip is a design flow [3] for domain-specific NoCs, which automates most of the complex and time-intensive design steps in NoC synthesis.

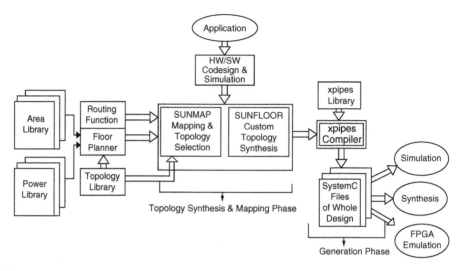

Design flow of NetChip.

It provides design support for application-specific regular and irregular network topologies, and therefore lends itself to the implementation of both homogeneous and heterogeneous system interconnects. NetChip assumes that the application has already been mapped onto cores by using pre-existing tools and the resulting cores together with their communication requirements are taken as an input. The tool-assisted design and generation of a customized NoC-based communication architecture is the ultimate goal of NetChip and is achieved by means of two major design activities: *topology synthesis and mapping* and *topology generation*. NetChip leverages three tools: SUNMAP [17], which performs the mapping and topology selection for regular topologies; SUNFLOOR [18], which performs the synthesis and mapping for custom (irregular) topologies; and xpipesCompiler [12], which performs the topology generation function.

The design flow of NetChip is presented in Fig. 8.7. NetChip has a *graphical user interface* (*GUI*) designed in TCL/TK for entering, visualizing, and modifying the input core graph. Based on the user's choice, either a custom topology is synthesized using SUNFLOOR or mapping onto a regular topology is performed using SUNMAP. The SystemC design of the synthesized topology is then generated by the xpipesCompiler tool. The network generated by the NetChip tool is highly optimized for that particular application's traffic characteristics.

## SUNMAP

SUNMAP maps the input core graph onto various standard topologies (*mesh, torus, hypercube, Clos,* and *butterfly*) defined in the topology library, which can be augmented by user-defined templates. It explores various design objectives such as minimizing average hop-delay, area, and power dissipation. The tool also supports different routing functions: *dimension ordered, minimum path, traffic splitting across minimum paths*, and *traffic splitting across all paths*. For each mapping, the bandwidth and area constraints are evaluated, so that only feasible mappings are chosen. The area–power models and a floorplanner are built into SUNMAP, so that area–power estimates can be incorporated early in the mapping process. For a chosen design objective and routing function, the best feasible mappings onto various topologies are obtained. From the different topologies in the library, the topology that best optimizes the user objectives is chosen. The design file describing the selected topology and routing information is automatically generated.

## SUNFLOOR

The SUNFLOOR tool is used to synthesize a custom irregular topology that is tailor-made for a specific application. It supports two objective functions: minimizing network power consumption and hop-delay for data transfer. The designer can optimize for one of the two objectives or a linear combination of both. The topology design process supports constraints on several parameters such as the hop-delay (when the objective is power minimization), network power consumption (when the objective is hop-delay minimization), design area, and total wire-length. SUNFLOOR also uses a floorplanner during the synthesis process to estimate the design area and wire-lengths. The wire-length estimates from the floorplan are used to evaluate whether the designed NoC satisfies the target frequency of operation and to compute the power consumption of the wires. The use of floorplan information during synthesis step helps achieving faster design closure between the high-level design and the physical design. There are two kinds of deadlocks that can occur in an NoC: routing deadlocks and message-level deadlocks. Freedom from both these deadlocks is critical for proper NoC operation. Methods to find deadlock free paths for routing packets are integrated with the topology design process in the SUNFLOOR tool. By mapping the request and response transactions onto separate resources, message-level deadlocks are also avoided by the tool. The tool uses accurate analytical models for power consumption and area of the network components. The power consumption values are obtained from layouts with back-annotated resistance, capacitance information, and from the switching activity of the components. The tool also tunes several NoC architectural parameters (such as the NoC operating frequency, link width) in the design process. The design file describing the synthesized topology and routing information is automatically generated.

In the *topology generation* phase, NetChip reads the topology and routing information file and generates SystemC description of network

```
// The IP core characteristics are defined here

// The parameters include the core identifier, switch identifier to which
// core is connected, the clock frequency division between the network and the
// cores, the number of NI buffers, type of core

core(core_0, switch_0,  1, 6, initiator);

// private memory and the addressing range

core(pm_1,   switch_0,  1, 6, target:0x00);

// shared memory, semaphore memory and interrupt device and their memory ranges

core(shm_2, switch_1,  1, 6, target:0x19-fixed);
core(smm_3, switch_1,  1, 6, target:0x20-fixed);
core(int_4, switch_1,  1, 6, target:0x21-fixed);

// switches: switch identifier, input ports, output ports, number of output buffers

switch(switch_0,  7, 7, 6);
switch(switch_1,  4, 4, 6);

// links

link(link0, switch_0,  switch_1);
link(link1, switch_1,  switch_0);

// routing information for the connections

// parameters include: source, destination, switches traversed

route(core_0, pm_1,   switches:0);
route(core_0, shm_2, switches:0,1);
route(core_0, smm_3, switches:0,1);
route(core_0, int_4, switches:0,1);
route(pm_3,   core_0, switches:0);
route(shm_2, core_0, switches:1,0);
route(smm_3, core_0, switches:1,0);
route(int_4, core_0, switches:1,0);
```

■ **FIGURE 8.8**

Example topology and routing specification generated by SUNFLOOR and read by xpipesCompiler.

components for the topology using xpipesCompiler. An example input file describing a custom NoC is presented in Fig. 8.8. The xpipesCompiler instantiates a network of building blocks from the xpipes library, which consists of composable soft macros (switches, NIs, and links) described in SystemC at the cycle-accurate level.

## xpipes

The xpipesCompiler generates the network using the xpipes library, which consists of highly parameterized network components that can be tailored to the communication needs of the selected architecture [25].

The `xpipes` components target heterogeneous packet-switched NoCs, thanks to the aggressive design of the network components for high performance and to their instantiation time flexibility. The high degree of parameterization of `xpipes` components is achieved by using both global network-specific parameters and local block-specific parameters. The former ones include flit size, address space of the cores, maximum number of hops between any two nodes, maximum number of bits allocated with in a packet for flow control, etc. On the other hand, specific parameters of the NI are: type of interface (master, slave, or both), flit buffer size at the output port, content of routing tables for source-based routing, other interface parameters to the cores such as number of address/data lines, maximum burst length, etc. Parameterization of the switches mainly regards the number of their I/O ports and the amount of buffering. Finally, the length of each individual link can be specified in terms of number of pipeline stages.

The `xpipes` NI uses OCP [31] as point-to-point communication protocol with the cores, and takes care of protocol conversion to adapt to the network protocol. Packets are split into flits and a flit-type field allows to distinguish between header and payload flits. Two different clock signals can be attached to NIs: one to drive the NI front-end (OCP interface) and the other to drive the NI back-end (pipes interface). The `xpipes` clock frequency must be an integer multiple of the OCP one. This arrangement allows the NoC to run at a fast clock even though some or all of the attached *Internet Protocol* (IP) cores are slower, which is crucial to keep transaction latency low. Since each IP core can run at a different divider of the `xpipes` frequency, mixed-clock platforms are possible.

The NoC backbone relies on wormhole switching and static routing. Routes are obtained by the NI by accessing a look-up table based on the destination address. Each route is represented by a set of direction bits. Each switch directs the flits belonging to a certain packet to the particular output port, based on the direction bits. This routing algorithm allows a lightweight switch implementation as no dynamic decisions have to be taken at the switching nodes. The architecture supports several flow control strategies to manage the buffering resources: ACK/NACK, Stall/Go, and link-level flow control.

Inter-block links are a critical component of NoCs, given the technology trends for global wires. The problem of signal propagation delay is, or will soon become, critical. For this reason, `xpipes` supports link pipelining, that is the interleaving of logical buffers along links. Proper flow control protocols are implemented in link transmitters and receivers (NIs and switches) to make the link latency transparent to the surrounding logic. Therefore, the overall platform can run at a fast clock frequency, without the longest wires being a global speed limiter. Only the links which are too long for single-cycle propagation will pay a latency penalty.

### 8.3.2 Æthereal **Design Flow**

Æthereal offers a flexible operational design flow to dimension and generate application-specific NoC instances and configurations. Figure 8.9 shows the Noc design flow with all input files underlined at the left-hand side. The tools that comprise the flow are shown by boxes. Their respective functionality is explained further below.

The application's requirements are specified in Excel, as described in Section 8.3.3. Application-specific NoC topologies, comprising custom-generated instances of routers and NIs, are automatically selected by the design tool. An IP port to NI port mapping is also computed. These steps are covered in section *NoC generation and configuration*. The result is SystemC and synthesizable RTL VHDL, compliant with the Philips back-end flow.

The NoC hardware is run-time (re)programmable to support different task graphs. The configuration to program the network is generated in XML for simulation and in C for embedded processors that program the NoC using *memory-mapped* IO (MMIO). To afford simulation of NoC hardware and configuration at all times, custom traffic generators are created to mimic the IP behavior.

The design flow is split into separate tools for several reasons. First, breaking the design flow in smaller steps simplifies steering or overriding heuristics used in each of the individual tools, enhancing user control. Second, it reduces the complexity of the optimization problem, and simpler, faster heuristics can be used. Higher-level optimization loops involving multiple tools can then be easily added, for example, the dashed arrow labeled "smallest mesh loop" in Fig. 8.9. Third, parts of the flow can be more easily customized, added, or replaced by the user to tailor the flow or improve its performance. Finally, redundancy in the sense of checking what should be generated automatically and correct by construction, such as simulation and performance verification of guaranteed connections, minimizes impact of potential programming errors, and acts as a safety net when allowing the user to manually create or modify intermediate results.

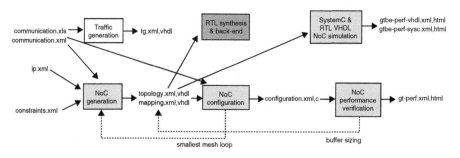

■ **FIGURE 8.9**

The Æthereal NoC design flow.

The Æthereal design flow addresses two key problems in NoC-based SoC design: the need for tools to quickly and efficiently generate application-specific NoCs, and the requirement for SoC and NoC performance validation. The latter is a unique feature of the Æthereal design flow and is further detailed in section *NoC verification*.

### 8.3.3 Input Specification

The starting point of the design flow is the description of the application's communication requirements (communication.xls,xml). An application consists of a number of task graphs, or use-cases with a number of tasks communicating using the NoC.

Figure 8.10 shows an example specification in Microsoft Excel. This is the *de facto* format for design documentation, and readily available from SoC designers. The Excel document is translated to XML, which the user can also write directly. A use-case is specified as a list of connections. A connection specifies a communication between a master port and a slave port, the required (minimum) bandwidth, the (maximum) allowed latency, and burst size for read and/or write data, and the traffic class (best-effort or guaranteed).

The second input file is the specification of the architecture around the NoC. The ip.xml file, an example of which is shown in Fig. 8.11, contains a list of all IPs connected to the NoC and the IP ports. Each port has a number of attributes, such as protocol (*Advanced eXtensible Interface* (AXI), various *Device Transaction Level* (DTL) profiles), and data word width.

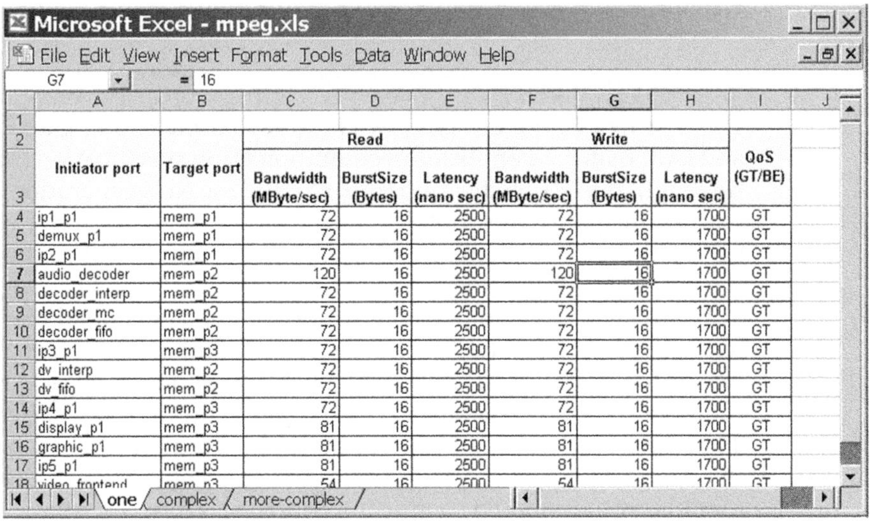
Microsoft Excel - mpeg.xls
File Edit View Insert Format Tools Data Window Help
G7 = 16

| Initiator port | Target port | Read Bandwidth (MByte/sec) | Read BurstSize (Bytes) | Read Latency (nano sec) | Write Bandwidth (MByte/sec) | Write BurstSize (Bytes) | Write Latency (nano sec) | QoS (GT/BE) |
|---|---|---|---|---|---|---|---|---|
| ip1_p1 | mem_p1 | 72 | 16 | 2500 | 72 | 16 | 1700 | GT |
| demux_p1 | mem_p1 | 72 | 16 | 2500 | 72 | 16 | 1700 | GT |
| ip2_p1 | mem_p1 | 72 | 16 | 2500 | 72 | 16 | 1700 | GT |
| audio_decoder | mem_p2 | 120 | 16 | 2500 | 120 | 16 | 1700 | GT |
| decoder_interp | mem_p2 | 72 | 16 | 2500 | 72 | 16 | 1700 | GT |
| decoder_mc | mem_p2 | 72 | 16 | 2500 | 72 | 16 | 1700 | GT |
| decoder_fifo | mem_p2 | 72 | 16 | 2500 | 72 | 16 | 1700 | GT |
| ip3_p1 | mem_p3 | 72 | 16 | 2500 | 72 | 16 | 1700 | GT |
| dv_interp | mem_p2 | 72 | 16 | 2500 | 72 | 16 | 1700 | GT |
| dv_fifo | mem_p2 | 72 | 16 | 2500 | 72 | 16 | 1700 | GT |
| ip4_p1 | mem_p3 | 72 | 16 | 2500 | 72 | 16 | 1700 | GT |
| display_p1 | mem_p3 | 81 | 16 | 2500 | 81 | 16 | 1700 | GT |
| graphic_p1 | mem_p3 | 81 | 16 | 2500 | 81 | 16 | 1700 | GT |
| ip5_p1 | mem_p3 | 81 | 16 | 2500 | 81 | 16 | 1700 | GT |
| video_frontend | mem_p3 | 54 | 16 | 2500 | 54 | 16 | 1700 | GT |

one / complex / more-complex

■ **FIGURE 8.10**

Application communication.

```
<architecture id="MPEG">
  <IP id="disp">
    <initiator id="p1"    protocol="MMBD" word="32"/>
  </IP>
  <IP id="decoder">
    <initiator id="mc"    protocol="MMBD" word="32"/>
    <initiator id="fifo" protocol="MMBD" word="32"/>
  </IP>
```

■ **FIGURE 8.11**

Part of the IP description.

```
<AENetwork id="MPEG" flitClk="6" slots="128">
  <AERouter id="R0000" iq="8">
    <AEPort id="NI"    link="L_0000" />
    <AEPort id="South" link="L_0000_0100" />
    <AEPort id="West"  link="L_0000_0001" />
  </AERouter>
  <AENI id="NI0101">
    <AEPort  id="Router"  link="L_0101" />
    <SlaveP  id="CONFIG"  conn="1" iq="4"  oq="4"/>
    <MasterP id="disp.p1" conn="1" iq="40" oq="21"/>
    ...
  </AENI>
```

■ **FIGURE 8.12**

Part of the topology description.

The ip.xml file is used to generate the right protocol-conversion shells for NIs [22].

### NoC generation and configuration

The first tool in the design flow is the NoC dimensioning and generation and tool. Using the application communication specification (communication.xml), the IP specification (ip.xml), and the NoC generation constraints (constraints.xml), this tool defines the design-time hardware components (in the form of topology.xml file): that is the number of routers, NIs, and the interconnect topology.

Figure 8.12 shows a partial topology.xml of a generated mesh. Parameters are specified for the NoC (flit duration, number of slots in TDMA table), for each router (arity, best-effort buffer size), and for each NI instance (number of NI ports, connections per port, buffer sizes per connection). To reduce NoC cost, all routers and NIs are dimensioned precisely for the application, giving many different router and NI instances per NoC. Modular router and NI architectures are therefore essential.

A synthesizable RTL VHDL description of the NoC, compatible with the standard philips back-end design flow, is also produced and an area estimate of the NoC is given [22, 23].

```
<Connection master="decoder.mc" cidm="0"
            slave="mem.p2" cids="2">
  <Request type="GT" path="3 1 0" credits="33"
           slots="22 23 24 25 26 27 28 29 30 31"/>
  <Response type="GT" slots="7 8 9 10 11 12 13"
            path="2 1 0" credits="21"/>
</Connection>
```

```
open_connection ("decoder.mc",0,"mem.p2",2,
                 "GT","22-31","3 1 0",33,
                 "GT","7-13","2 1 0",21);
```

■ **FIGURE 8.13**

Part of the configuration (XML and C).

The buffers in the NIs are dimensioned to avoid stalling of data by hiding the round-trip delay of credits, and to compensate for differences in burst sizes. Although they are part of the hardware, they depend on the configuration and are therefore computed later and back annotated in the topology.xml as indicated by the dashed arrow labeled "buffer sizing" in Fig. 8.9.

A second output is the mapping.xml file, containing the assignment of IP ports to NI ports. The mapping has a significant impact on the size of the NoC and its performance, as shown in Refs [8, 9]. The constraints.xml file allows the user to influence the mapping by specifying if sets of ports must be mapped on the same NI or must be mapped on different NIs, to reflect for example floorplanning constraints.

Using the design-time hardware (topology.xml and mapping.xml) and the communication specification, the NoC configuration tool computes the run-time software that contains all the information to program the hardware. The configuration.xml file contains the values for all programmable registers of the NIs, such as connection identifiers, and for each connection, the path from master to slave port, and flow control credits. For connections with guaranteed services, also TDMA slots are specified [23]. Figure 8.13 shows a partial configuration in XML and C. Section *NoC simulation* describes how these files are used by the SystemC and RTL VHDL simulations.

As an alternative to the separate tools for NoC generation and configuration, the Æthereal design flow also provides a unified single-objective algorithm, called *Unified MApping, Routing, and Slot allocation* (UMARS) [9]. UMARS couples path selection, mapping, and TDMA time-slot allocation, aiming to minimize the cost (area and power) of the NoC required to meet the application requirements. When this approach is applied to the example of the MPEG decoder SoC, the NoC area is reduced by 33%, power by 35% and worst-case latency by a factor of four when compared to the original multi-step flow.

Using the topology and configuration the worst-case and average energy consumption of the NoC can be estimated [5]. This, and the area estimate

| Guaranteed Throughput Verification Results - Microsoft Internet Explorer |
|---|

File  Edit  View  Favorites  Tools  Help

### - Guaranteed Throughput Verification Results for GT Connections-

| ConnId | Trans | Slot Table Size = 8 | | Throughput (Mbytes/sec) | | Latency (ns) | | BufferSize (Words) | | | | | | | | | | | | |
|---|---|---|---|---|---|---|---|---|---|---|---|---|---|---|---|---|---|---|---|---|
| | | Forward Allocated Slots | Reverse Allocated Slots | Spec | Avail | Spec | Max | Forward Master | | | Forward Slave | | | Reverse Slave | | | Reverse Master | | |
| | | | | | | | | Spec | Max | Slack | Spec | Max | Slack | Spec | Max | Slack | Spec | Max | Slack |
| 0 | read | 1 | 1 | 72.00 | 166.67 | 2500.00 | 624.00 | 16 | 16 | 0 | 4 | 3 | 1 | 8 | 8 | 0 | 4 | 3 | 1 |
| 0 | write | 1 | 1 | 72.00 | 94.67 | 1700.00 | 408.00 | 16 | 16 | 0 | 4 | 3 | 1 | 8 | 8 | 0 | 4 | 3 | 1 |
| 1 | read | 1 | 1 | 72.00 | 166.67 | 2500.00 | 624.00 | 16 | 16 | 0 | 4 | 3 | 1 | 8 | 8 | 0 | 4 | 3 | 1 |
| 1 | write | 1 | 1 | 72.00 | 94.67 | 1700.00 | 408.00 | 16 | 16 | 0 | 4 | 3 | 1 | 8 | 8 | 0 | 4 | 3 | 1 |
| 2 | read | 1 | 1 | 72.00 | 166.67 | 2500.00 | 624.00 | 16 | 16 | 0 | 4 | 3 | 1 | 8 | 8 | 0 | 4 | 3 | 1 |
| 2 | write | 1 | 1 | 72.00 | 94.67 | 1700.00 | 408.00 | 16 | 16 | 0 | 4 | 3 | 1 | 8 | 8 | 0 | 4 | 3 | 1 |
| 3 | read | 2 | 1 | 120.00 | 166.67 | 2500.00 | 432.00 | 16 | 16 | 0 | 5 | 5 | 0 | 8 | 8 | 0 | 4 | 3 | 1 |
| 3 | write | 2 | 1 | 120.00 | 296.67 | 1700.00 | 216.00 | 16 | 16 | 0 | 5 | 5 | 0 | 8 | 8 | 0 | 4 | 3 | 1 |
| 4 | read | 1 | 1 | 72.00 | 166.67 | 2500.00 | 624.00 | 16 | 16 | 0 | 4 | 3 | 1 | 8 | 8 | 0 | 4 | 3 | 1 |
| 4 | write | 1 | 1 | 72.00 | 94.67 | 1700.00 | 408.00 | 16 | 16 | 0 | 4 | 3 | 1 | 8 | 8 | 0 | 4 | 3 | 1 |

■ FIGURE 8.14

Output of the NoC verification step for the MPEG example.

of the NoC, computed from the topology, are important indicators of NoC cost. The verification and simulation steps, discussed below, compute the NoC performance metrics.

### NoC verification

A unique feature of the Æthereal design flow is the automatic performance verification: given the NoC hardware and configuration, the guaranteed minimum throughput, maximum latency, and minimum buffer sizes are analytically computed for all guaranteed connections [6]. The guaranteed communication services of Æthereal are essential to achieve this. Any NoC instance and configuration can be verified, whether automatically or manually created. Analytical performance verification eliminates lengthy simulations for the guaranteed connections, and hence reduces verification time.

Given the topology.xml, mapping.xml, and configuration.xml files the verification tool computes the worst-case throughput, latency, and buffer sizes per connection. These are compared to the requirements (communication.xml) and shown in an intuitive color-coded table (gt-perf.xml,html) as seen in Fig. 8.14.

### NoC simulation

The verification tool works for guaranteed connections only, and not for best-effort connections. Moreover, it computes worst case, not actual or average figures. To assess the average performance of both guaranteed and best-effort connections for a particular execution trace, two types of simulation are supported: RTL VHDL and SystemC. The former is a bit and cycle-accurate simulation of the RTL VHDL implementation. In the SystemC simulation, the NoC is simulated at the flit level, and the IP to NI

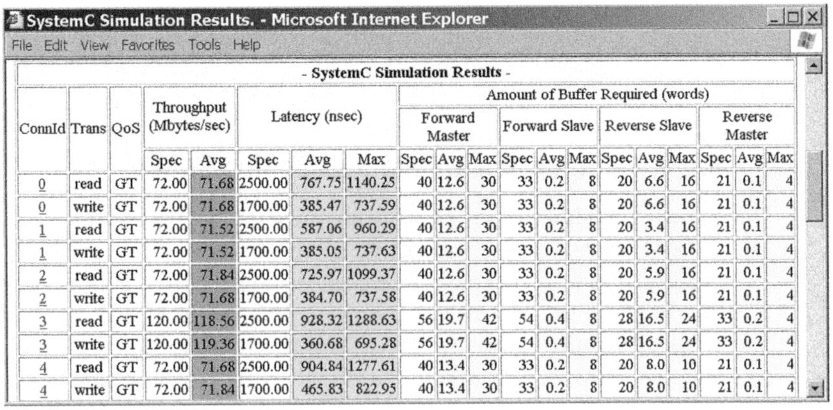

**SystemC Simulation Results. - Microsoft Internet Explorer**

File Edit View Favorites Tools Help

- SystemC Simulation Results -

| ConnId | Trans | QoS | Throughput (Mbytes/sec) | | Latency (nsec) | | | Amount of Buffer Required (words) | | | | | | | | | | | | |
|---|---|---|---|---|---|---|---|---|---|---|---|---|---|---|---|---|---|---|---|---|
| | | | | | | | | Forward Master | | | Forward Slave | | | Reverse Slave | | | Reverse Master | | |
| | | | Spec | Avg | Spec | Avg | Max | Spec | Avg | Max | Spec | Avg | Max | Spec | Avg | Max | Spec | Avg | Max |
| 0 | read | GT | 72.00 | 71.68 | 2500.00 | 767.75 | 1140.25 | 40 | 12.6 | 30 | 33 | 0.2 | 8 | 20 | 6.6 | 16 | 21 | 0.1 | 4 |
| 0 | write | GT | 72.00 | 71.68 | 1700.00 | 385.47 | 737.59 | 40 | 12.6 | 30 | 33 | 0.2 | 8 | 20 | 6.6 | 16 | 21 | 0.1 | 4 |
| 1 | read | GT | 72.00 | 71.52 | 2500.00 | 587.06 | 960.29 | 40 | 12.6 | 30 | 33 | 0.2 | 8 | 20 | 3.4 | 16 | 21 | 0.1 | 4 |
| 1 | write | GT | 72.00 | 71.52 | 1700.00 | 385.05 | 737.63 | 40 | 12.6 | 30 | 33 | 0.2 | 8 | 20 | 3.4 | 16 | 21 | 0.1 | 4 |
| 2 | read | GT | 72.00 | 71.84 | 2500.00 | 725.97 | 1099.37 | 40 | 12.6 | 30 | 33 | 0.2 | 8 | 20 | 5.9 | 16 | 21 | 0.1 | 4 |
| 2 | write | GT | 72.00 | 71.68 | 1700.00 | 384.70 | 737.58 | 40 | 12.6 | 30 | 33 | 0.2 | 8 | 20 | 5.9 | 16 | 21 | 0.1 | 4 |
| 3 | read | GT | 120.00 | 118.56 | 2500.00 | 928.32 | 1288.63 | 56 | 19.7 | 42 | 54 | 0.4 | 8 | 28 | 16.5 | 24 | 33 | 0.2 | 4 |
| 3 | write | GT | 120.00 | 119.36 | 1700.00 | 360.68 | 695.28 | 56 | 19.7 | 42 | 54 | 0.4 | 8 | 28 | 16.5 | 24 | 33 | 0.2 | 4 |
| 4 | read | GT | 72.00 | 71.68 | 2500.00 | 904.84 | 1277.61 | 40 | 13.4 | 30 | 33 | 0.2 | 8 | 20 | 8.0 | 10 | 21 | 0.1 | 4 |
| 4 | write | GT | 72.00 | 71.84 | 1700.00 | 465.83 | 822.95 | 40 | 13.4 | 30 | 33 | 0.2 | 8 | 20 | 8.0 | 10 | 21 | 0.1 | 4 |

■ **FIGURE 8.15**

SystemC simulation output for the MPEG example.

interface is at the transaction level, making it orders of magnitude faster than the VHDL simulation.

The Æthereal NoC is configured through MMIO NoC ports (the id="CONFIG" port in Fig. 8.12), using the NoC itself [22]. In the VHDL simulation, a behavioral IP configures the NoC according to configuration.c file. The SystemC simulation can use either configuration.xml directly, or model the MMIO programming by running the C program on an ARM processor model.

To enable early SoC simulation to evaluate the NoC performance, our design flow produces XML-configurable SystemC and VHDL traffic generators (see Fig. 8.9). All the IP core for which no model or implementation is available, are modeled with traffic generators. The traffic is generated according to the IP requirements as given in the communication.xls file. Hence, at all times, a complete SoC can be simulated in both SystemC and VHDL.

The traffic generators also automate the measurement of throughput and latency of all connections in the NoC, for both SystemC and VHDL, and NI buffer statistics (SystemC only). The SystemC and/or VHDL results are reported in color-coded tables (gtbe-*-perf.xml,html), as shown in Fig. 8.15. Due to simulation artifacts (use of random numbers in traffic generators, run-in effects due to insufficient simulation lengths, etc.) a guaranteed connection may not use its required bandwidth (e.g., write data of connection 0).

## 8.4 TOOL KITS FOR DESIGNING BUS-BASED INTERCONNECT

A few companies provide integrated solutions for SoC communication. These solutions include busses, multi-layer busses, and their extensions.

Specific CAD tools are needed for the design with these units. We will briefly cover the tool support for Sonics' Silicon Backplane [32] and STM's StBus.

### 8.4.1  Silicon Backplane and Sonics Studio

Sonics provides a set of communication infrastructures dubbed Smart interconnects. The objective of the Sonics products is to shorten SoC design time and increasing design predictability. Within Sonics products, the SiliconBackplaneIII is a highly configurable, scalable SoC inter-block communication system that integrally manages data, control, debug, and test flows. It enables designers to design cores independently from their NoC, thus enhancing modularity and component reuse.

In particular, SiliconBackplane III, using patented protocols, is an on-chip communications structure that guarantees end-to-end performance by managing all the data, control, and test flows between all cores in an SoC. Through hardware enhancements, SiliconBackplane III has been optimized for use within advanced Digital Set-Top-Box and Digital TV applications. An example of the application of SiliconBackplane is shown in Fig. 8.16, which describes the main interconnect of a complex multimedia SoC. A configurable agent within the Smart Interconnect cleanly decouples the functionality of each core from the communication among cores and therefore enables each core to be rapidly reused without rework in subsequent, or parallel, SoC designs. This shortens development time and costs considerably. Communication between agents and cores occurs through sockets using the industry standard OCP [31] protocol.

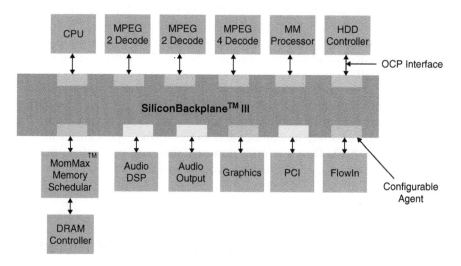

■ **FIGURE 8.16**

Sonics SiliconBackplane III.

For greater reuse of larger portions of an SoC, SiliconBackplane III enables hierarchical platform-based design in the form of *tiles*, where a tile is a collection of functions requiring minimal assistance from the rest of the die and frequently includes an embedded processing unit, local memory, and relevant I/O resources. The tile architecture facilitates its use multiple times in the same SoC, or across an SoC product family, without rework.

SiliconBackplane III can manage the high frame rate video, 3D graphics and 2D still image data streams found in multimedia applications. SiliconBackplane III concurrently handles multiple asymmetrical and variable length data flows at the maximum throughput of the memory channel with an aggregate bandwidth of up to 4 GB/s. Its levels of QoS can be modified by tailoring of data channel characteristics. They report that SiliconBackplane III delivers a sustained level of 90% interconnect utilization in the extremely demanding traffic conditions created by multiple image-based data flows between the multimedia SoC and its memory subsystem. A new product, SonicsMX, is a SMART Interconnect designed for low-latency and power-sensitive SoC applications for mobile multimedia applications. SonicsMX has the physical structures, advanced protocols, and extensive power management capabilities necessary to overcome data flow and other design challenges in the mobile multimedia applications. SonicsMX supports crossbar and shared link, or hybrid topologies within a multithreaded and non-blocking architecture. It has many features such as fine-grained clock gating, clock and voltage management, mixed latency requirements, QoS management, security protection, and error management [26]. SonicsMX also supports the OCP-IP 2.0 socket protocol to decouple cores from the interconnects.

The SonicsStudio development environment is used to support NoC design using Sonics' products. The tool flow of SonicsStudio is shown in Fig. 8.17. It includes graphical and command line based tools and utilities that provide the SoC architect and designer a single environment within which the entire SoC can be assembled, configured and the netlist can be generated. With SonicsStudio, the designer can instantiate several SoC cores, configure the interconnect, create *bus functional models* for those cores not yet available, place monitors, stimulate the SoC components and analyze the performance of the interconnect. The tools support fast changes to the interconnect, so that the entire SoC can be re-architected for subsequent analysis. In this way, the SoC architect can rapidly decide on the optimal SoC architecture that meets design and customer requirements.

SonicsStudio streamlines physical integration and validation for SoCs that incorporate Sonics products. Seamlessly integrated with tools for simulation, design synthesis, floorplanning, and timing analysis, Sonics-Studio eliminates time-consuming design work by automating synthesis script and constraint creation, timing analysis input preparation, and design-for-test management.

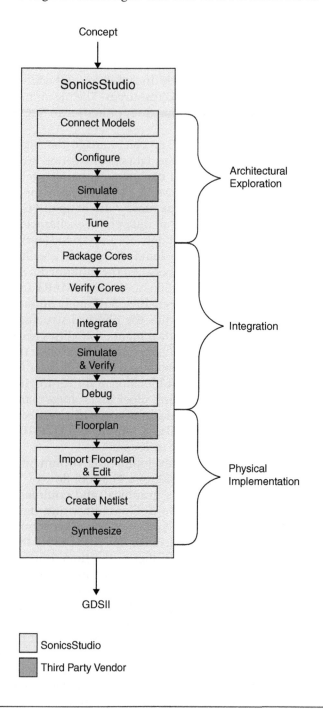

Sonics SonicStudio.

SonicsStudio also provides rapid feedback to optimize vendor library selection, interconnect configuration, and operating frequency selection. The output of the tool chain is a mapped netlist that is combined with additional elements, such as clock distribution, test, and pad structures, to complete the SoC design.

## 8.4.2  **STMicroelectronics** STBus

The STBus [28] is not only a communication system characterized by protocols, interfaces, transaction sets, and IPs, but also a technology allowing designers to implement communication networks for SoCs. It provides a development environment, including tools for *system-level design* (SLD) and architectural exploration, *silicon design* (SD), physical implementation, and verification.

**Basics on the** STBus

The STBus is a set of protocols, interfaces, and architectural specifications defined to implement the communication network of digital systems such as microcontrollers for different applications (set-top box, digital camera, MPEG decoder, and GPS). There are three different types of protocols supported by the STBus architecture (namely Type 1, Type 2, and Type 3), each one associated with a different interface and providing different performance levels:

1. *Type 1 protocol*: It is a simple synchronous handshake protocol, based on *request/grant* signals (RG protocol), with a limited set of command types available. It is suitable for accessing registers and slow peripherals. The protocol does not support pipelined transactions on the bus.

2. *Type 2 protocol*: It is more efficient than Type 1, because it supports split transactions and pipelining of transactions on the bus. The transaction set includes read/write operations with different sizes (up to 64 bytes) and also supports specialized operations like *Read–Modify–Write* and *Swap*, to name a few. Transactions may also be grouped together into chunks, which ensure that the data stream is not interrupted and the slave is allocated to that particular stream. Such a scheme is typically suitable for accessing data from and sending data to external memory controllers. A limitation of this protocol is that the transactions must be maintained in-order on the bus. This protocol is equivalent to the basic *request/grant/valid* (RGV) protocol.

3. *Type 3 protocol*: It is the most efficient protocol of the three, as it adds support for out-of-order transactions and asymmetric

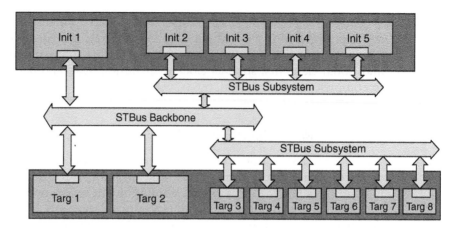

Example of an STBus interconnect.

communication (where the length of the request packet can be different from the length of response packet) on top of what is already provided by the Type 2 protocol. It can therefore be used by CPUs, multi-channel DMAs, and DDR controllers.

The STBus architecture is modular and allows masters (Init in Fig. 8.18) and slaves (Targ in Fig. 8.18) of any protocol type and data size to communicate with each other. A wide variety of arbitration policies is also available that helps the meet the different system requirements. These include bandwidth limited arbitration, latency-based arbitration, LRU, priority-based arbitration, and others. The STBus architecture builds upon a configurable switch fabric which can be instantiated multiple times to create a hierarchical interconnect structure. The architecture also includes type converters and size converters that can be used to interface heterogeneous network domains working under different protocols (i.e., Type 1, Type 2, and Type 3) and/or different data-path widths together.

The STBus now comes with an integrated development environment (the STBusGenKit) which enables users to draw, check, simulate, and generate the RTL for their interconnect. STBusGenKit supports the whole design flow, starting from the system-level parametric network specification, all the way down to the design mapping, and global interconnect floorplanning.

### STBusGenKit

The STBus development system includes a powerful and versatile tool kit that allows the users to define and implement an STBus interconnect system in an easy way. As shown in Fig. 8.19, the STBus tool kit consists of

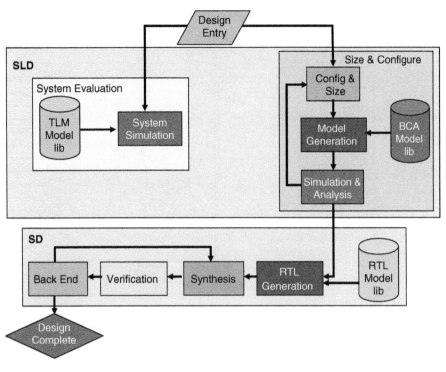

STM GenKit flow diagram.

two main tools: the first is the SLD tool that supports the definition of the STBus architecture from the user specifications. The second one is the SD tool kit that supports the generation of the RTL design (VHDL) and the gate-level netlist for the interconnect architecture. Other plug-ins are also present in the flow, such as the tools Formality [34] used for equivalence checking, and Verisity [35] used for timing analysis and verification.

### The SLD tool

The SLD tool is based on CoWare N2C [36], an environment that helps SoC designers to bring an idea or concept from its specification to its implementation on silicon. In this step, the designer specifies the system, the partitioning of the system, and the hardware implementation. In fact, the choice of the architecture (number and kind of components and configuration of the parameters for each component) is the first important step in the creation of the interconnect. The SLD tool developed by ST designers is an entire environment that helps SoC designers to simulate different architectures, analyze the performance (through a traffic analyzer), and to choose the best architecture. It offers a useful GUI for performing the

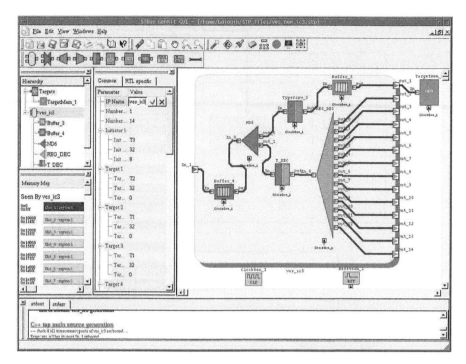

■ **FIGURE 8.20**

Screenshot of STBus Genkit GUI.

analysis (showed in Fig. 8.20). Systems can be created graphically by using *drag and drop* operations from the items on a graphical toolbar. Designers can use the toolbar to choose the type of objects (such as nodes, converters, reset, and clock generators) to be added to the model schematic and they can link the objects together. The parameters of the objects can be easily configured by the user graphically. The SLD tool also has a design toolbar to check the performance of the design and to run simulations. A message window, present at the bottom of the main window displays various messages (errors, warnings, progress reports, and so on). The output of this flow is a file describing the entire interconnect system that is read by the SD tool kit.

**The SD tool**

The SD tool kit is based on Synopsys CoreAssembler [37]. This tool takes the interconnect, specification from the SLD tool. It then assembles the IPs, the interconnect; and the other hardware blocks together. The tool allows the integrator to assemble various formats of IP: they can be imported as simple RTL; they can be SPIRIT IP (XML-based); or they can be part of an existing library of IP modules packaged as coreKits. A coreKit

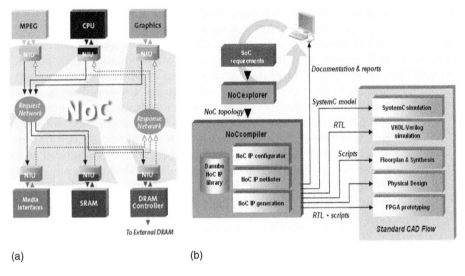

(a)                                    (b)

■ **FIGURE 8.21**

(a) An example of an SoC using Danube NoC IP library and (b) Danube NoC configuration and design flow.

contains, besides the source code, many other items that are required to make the IP as a black box.

### 8.4.3 Arteris

Arteris, an intellectual-property vendor, introduced tools to design customized NoC [38]. Its NoC solutions are comprised of an IP library of fundamental NoC units, design tools for exploring various potential topologies and a compilation software used to configure and generate the completed NoC in the form of synthesizable RTL. Arteris has divided the functionality of the NoC across three layers: transaction, transport, and physical layers. Danube, the first commercially available Arteris NoC IP library, consists of three types of units matching the three layers of NoC: Network interface units, switch, and physical links. Figure 8.21(a) shows an example of an SoC making use of Danube NoC IP library. Based on packet-based *NoC Transaction and Transport Protocol* (NTTP), it is compatible with other on-chip socket standards (AMBA AHB, AMBA AXI, OCP 2.0), while the point-to-point physical implementation uses the *globally asynchronous locally synchronous* (GALS) paradigm, allowing operating frequencies of 750 MHz or more in 90 nm silicon process with standard cell libraries and EDA tools. The NoC Interface Units (NIUs) convert IP core transactions to/from NTTP packets, decode addresses to assign packet routes, perform clock conversion, and provide error detection, reporting and logging. Packets are routed between NIUs by a user-defined topology of switches and configurable, synchronous or mesochronous 32- or 64-bit links supporting the GALS paradigm.

The `NoCexplorer` and `NoCcompiler` are used to customize a `Danube` NoC instance for integration within an application-specific SoC as shown in Fig. 8.21(b). The NoC can be configured based on the system objectives and topology requirements. To create an NoC topology, the `NoCexplorer` tool captures the data flow requirements of the IP blocks attached to the NoC and allows the designer to analyze various NoC topologies for performance and area estimation. In the `NoCcompiler` design tool, NoC units selected from Arteris NoC IP libraries are configured and connected according to the topology selection of the previous exploration phase and the specifications of the IP cores attached to the NoC. The cycle-accurate SystemC model and synthesizable RTL are then generated for FPGA prototyping or integration within the SoC using standard design flow.

## 8.5 SUMMARY

The complexity of NoC design motivates the use of CAD tools for analysis and synthesis, which can both be performed at various levels of abstraction. These tools fit into layered flows that allow designers to achieve NoCs by stepwise refinement. Analysis is typically done by simulation or by emulation on FPGAs, where the latter is particularly useful to reduce the evaluation time. NoC synthesis is particularly useful to achieve irregular topologies which can interconnect heterogeneous components and links with various requirements and parameters. Synthesis and optimization can be used to traverse the large design space of NoCs, in the search for architectures, protocols, and parameters that optimize performance and power consumption. A few synthesis tool flows are now available and they have been used to design prototype chips, which are described in the following chapter.

## REFERENCES

[1]  T. Ahonen, D.A. Siguenza-Tortosa, H. Bin and J. Nurmi, "Topology Optimization for Application Specific Networks on Chip," *SLIP, International Workshop on System Level Interconnect Prediction*, February 2004, pp. 53–60.

[2]  L. Benini, D. Bertozzi, A. Bogliolo, F. Menichelli and M. Olivieri, "MPARM: Exploring the Multi-processor SoC Design Space with SystemC," *Journal of VLSI Signal Processing*, Vol. 41, 2005, pp. 169–182.

[3]  D. Bertozzi, A. Jalabert, S. Murali, R. Tamhankar, S. Stergiou, L. Benini and G. De Micheli, "NoC Synthesis Flow for Customized Domain Specific Multiprocessor Systems-on-Chip," *IEEE Transactions on Parallel and Distributed Systems*, Vol. 16, No. 2, February 2005, pp. 113–129.

[4] M. Dall'Osso, G. Biccari, L. Giovannini, D. Bertozzi and L. Benini, "xpipes: A Latency Insensitive Parameterized Network-on-Chip Architecture for Multi-processor SoCs," *International Conference on Computer Design*, 2003, pp. 536–539.

[5] J. Dielissen, A. Rădulescu and K. Goossens, Power measurements and analysis of a network on chip. Technical Note 2005/00282, Philips Research, April 2005.

[6] O.P. Gangwal, A. Rădulescu, K. Goossens, S. González Pestana and E. Rijpkema, "Building Predictable Systems on Chip: An Analysis of Guaranteed Communication in the Æthereal Network on Chip," in P. van der Stok, (editor), *Dynamic and Robust Streaming In and Between Connected Consumer-Electronics Devices*, Volume 3 of *Philips Research Book Series*, Chapter 1, pp. 1–36. Springer, Berlin, Germany, 2005.

[7] N. Genko, D. Atienza, J. Mendias, R. Hermida, G. De Micheli and F. Catthoor, "A Complete Network-on-Chip Emulation Framework," *DATE, International Conference on Design and Test Europe*, 2005, pp. 246–251.

[8] K. Goossens, J. Dielissen, O.P. Gangwal, S. González Pestana, A. Rădulescu and E. Rijpkema, "A Design Flow for Application-Specific Networks on Chip with Guaranteed Performance to Accelerate SoC Design and Verification," *Proceedings of the Design, Automation and Test in Europe Conference and Exhibition (DATE)*, March 2005, pp. 1182–1187.

[9] A. Hansson, K. Goossens and A. Rădulescu, "A Unified Approach to Constrained Mapping and Routing on Network-on-Chip Architectures," *International Conference on Hardware/Software Codesign and System Synthesis (CODES+ISSS)*, September 2005.

[10] W.H. Ho and T.M. Pinkston, "A Methodology for Designing Efficient On-Chip Interconnects on Well-Behaved Communication Patterns," *HPCA, International Symposium on High-Performance Computer Architecture*, February 2003, pp. 377–388.

[11] J. Hu and R. Marculescu, "Exploiting the Routing Flexibility for Energy/Performance Aware Mapping of Regular NoC Architectures," *DATE, International Conference on Design and Test Europe*, March 2003, pp. 10688–10693.

[12] A. Jalabert, S. Murali, L. Benini and G. De Micheli, "xpipesCompiler: A Tool for Instantiating Application Specific Networks on Chip," *DATE, International Conference on Design and Test Europe*, February 2004, pp. 884–889.

[13] S. Kumar, A. Jantsch, J. Soininen, M. Forsell, M. Millberg, J. Oberg, K. Tiensyrj and A. Hemani, "A Network on Chip Architecture and Design Methodology," *Proceedings of the IEEE Computer Society Annual Symposium on VLSI*, April 2002, pp. 105–112.

[14] S.-Y. Lee, S.-J. Song, K. Lee, J.-H. Woo, S.-E. Kim, B.-G. Nam and H.-J. Yoo, "An 800 MHz Star-Connected On-Chip Network for Application to Systems on a Chip," *IEEE Solid-State Circuits Conference*, 2003, pp. 468–469.

[15] S.-Y. Lee, K. Lee, S.-J. Song and H.J. Yoo, "Packet-Switched On-Chip Interconnection Network for System-on-Chip Applications," *IEEE Transactions on Circuits and Systems, Part II: Express Briefs*, Vol. 52, No. 6, June 2005, pp. 308–312.

[16] T. Marescaux, J. Mignolet, A. Bartic, W. Moffat, D. Verkest, S. Vernalde and R. Lauwereins, "NoC As a Hw Component of an OS for Reconfigurable Systems," *FPL, Field Programmable Logic*, 2003, pp. 595–605.

[17] S. Murali and G. De Micheli, "SUNMAP: A Tool for Automatic Topology Selection and Generation for NoCs," *DAC, Design Automation Conference*, June 2004, pp. 914–919.

[18] S. Murali, P. Meloni, F. Angiolini, D. Atienza, S. Carta, L. Benini, G. De Micheli, L. Raffo, "Designing Message Dependent Deadlock Free Networks on Chips for Application Specific Systems on Chips", *Proceedings of the VLSI-SoC*, 2006.

[19] S. Murali and L. Benini and G. De Micheli, "Mapping and Physical Planning of Networks on Chip Architectures with Quality-of-Service Guarantees," *ASPDAC, Asia South Pacific Design Automation Conference*, January 2005, pp. 27–32.

[20] S. Murali, M. Coenen, A. Rădulescu, K. Goossens and G. De Micheli, "A Methodology for Mapping Multiple Use-Cases on to Networks on Chip," *DATE, International Conference on Design and Test Europe*, March 2006.

[21] A. Pinto, L. Carloni and A. Sangiovanni-Vincentelli, "Efficient Synthesis of Networks on Chip," *ICCD, International Conference on Computer Design*, October 2003, pp. 146–150.

[22] A. Rădulescu, J. Dielissen, S. González Pestana, O.P. Gangwal, E. Rijpkema, P. Wielage and K. Goossens, "An Efficient On-Chip Network Interface Offering Guaranteed Services, Shared-Memory Abstraction, and flexible Network Programming," *IEEE Transactions on CAD of Integrated Circuits and Systems*, Vol. 21, No. 1, January 2005, pp. 4–17.

[23] E. Rijpkema, K.G.W. Goossens, A. Rădulescu, J. Dielissen, J. van Meerbergen, P. Wielage and E. Waterlander, "Trade Offs in the Design of a Router with Both Guaranteed and Best-Effort Services for Networks on Chip," *Proceedings of the Design, Automation and Test in Europe Conference and Exhibition (DATE)*, March 2003, pp. 350–355.

[24] D. Siguenza-Tortosa and J. Nurmi, "Proteo: A New Approach to Network on Chip," *CSN, Conference on Communication Systems and Network*, 2002.

[25] S. Stergios, F. Angiolini, D. Bertozzi, S. Carta, L. Raffo and G. De Micheli, "xpipesLite: A Synthesis-Oriented Design Flow for Networks on Chip," *DATE, International Conference on Design and Test Europe*, pp. 1188–1193.

[26] W. Weber, et al., "A Quality-of-Service Mechanism for Interconnection Networks is System-on-Chips," *Proceedings of the Design, Automation and Test in Europe Conference*, 2005, pp. 1232–1237.

[27] http://www.arm.com/products/solutions/AMBAOverview.html

[28] http://www.st.com/stonline/prodpres/dedicate/soc/cores/stbus.htm

[29] http://www.isi.edu/nsnam/ns/

[30] www.swarm.org

[31] www.ocpip.org

[32] www.sonicsinc.com

[33] www.vsi.org

[34] http://www.synopsys.com/products/verification/verification.html

[35] http://www.cadence.com/verisity/

[36] http://www.coware.com/

[37] http://www.synopsys.com/products/designware/core_assembler.html

[38] http://www.arteris.com

# CHAPTER 9

# DESIGNS AND IMPLEMENTATIONS OF NOC-BASED SOCS*

*Network-on-chips* (NoC) architectures are emerging as a strong candidate for the highly scalable, reliable, and modular on-chip communication infrastructure platform to implement high-performance SoCs. There have been many architectural and theoretical studies on NoCs such as design methodology, topology exploration, *quality-of-service* (QoS) guarantee and low-power design. All of those are discussed in the previous chapters in this book. In this last chapter, we would like to introduce the silicon chip implementation trials for NoC-based *Systems-on-chip* (SoCs). We will discuss both academic and industrial design efforts, even though academic design are dominant. This is because NoC technology is still maturing and for the time being commercial SoC products have relied on mature and well-established bus-based interconnects. However, we have seen in the previous chapter that several industrial efforts are under way to develop the standardized protocol specifications, EDA tool chains and *Internet Protocol* (IP) library supports for next-generation NoC-based products.

## 9.1 KAIST BONE SERIES

For the unique purpose of realizing the new NoC technology through implementation, the BONE (Basic On-chip Network) project was launched in 2002 at KAIST (Korea Advanced Institute of Science and Technology, Daejeon, Korea). As the results of the project, new NoC techniques and the implementations have been published and demonstrated in every year as summarized in Fig. 9.1.

This project covers circuit-level design, architectural researches and system integration on an NoC platform. In this section, each generation will be overviewed and for the more detail information on the BONE, refer to website: http://ssl.kaist.ac.kr/ocn/

---

* This chapter was provided by Hoi-Jun Yoo, Kangmin Lee, Se-Joong Lee and Kwanho Kim of the Korea Advanced Institute of Science and Technology, Republic of Korea.

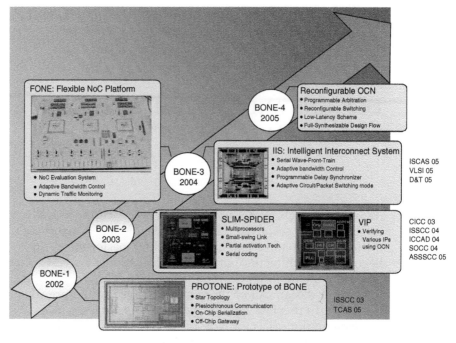

▪ **FIGURE 9.1**

BONE series road map since 2002.

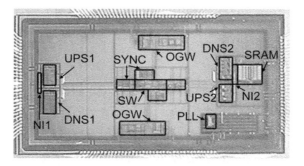

▪ **FIGURE 9.2**

Die photograph of BONE-1 [6].

### 9.1.1  BONE-1: Prototype of On-Chip Network (PROTON)

To demonstrate feasibility of the OCN architecture, a test chip, PRO-TON, is implemented using 0.38 μm CMOS technology. The BONE-1 of Fig. 9.2 is designed with two physical layer features: high-speed (800 MHz) mesochronous communication and *on-chip serialization* (OCS). Using 4:1 serialization, 80b packets are transferred through 20b links. The 4:1

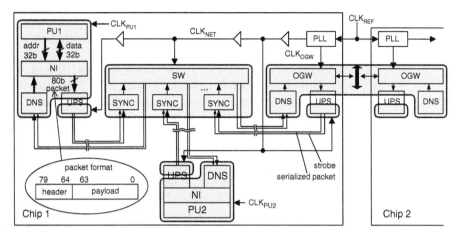

Overall architecture of the OCN for BONE-1.

serialization reduces the network area of BONE-1 by 57%, making it practical to be used in SoC design [7]. The distributed NoC building blocks are not globally synchronized, and the packet transfer is performed with mesochronous communication. Since the mesochronous communication eliminates burden of global synchronization, the high-speed clocking, 800 MHz, is possible. The implementation and its successful operation demonstrates that high-performance on-chip serialized networking with mesochronous communication is practically feasible. The chip size is 10.8 mm × 6.0 mm, the number of transistors is 81,000, and power consumption is 264 mW at 800 MHz 2.3 V operation.

### Overall architecture

Figure 9.3 shows an overall block diagram of the OCN with BONE-1. It consists of *network interface* (NI), *Up_Sampler* (UPS), link wires, FIFO synchronizer with a queuing buffer (SYNC), switch, *Down_Sampler* (DNS) and *off-chip gateway* (OGW). Using an address and/or a data output(s) of a *processing unit* (PU), a corresponding NI generates a packet. A packet is an 80b bit-stream consisting of a 16b header and a 64b payload. In the header, routing information is encoded into 16b, and the switches route a packet according to the header information. The packet also can be transferred to another chip through the OGW. An NI has an address map which defines a destination PU according to an address, and a header of a packet is generated based on the address map. The packet header size is fixed to reduce latency and hardware complexity of header parsing logic.

The UPS, link wires, SYNC, switch and DNS are called *core network*. The core network delivers a packet to a correct destination PU. The core network is serialized in order to lower wiring complexity by reducing the number of link wires. Instead, it uses its own high-frequency clock to sustain bandwidth of the network system. The UPS and the DNS interface the high-speed serialized core network to the NI.

Clock domains of the OCN are illustrated as dotted gray box in Fig. 9.3. Without global synchronization, each PU can use its dedicated clock source constituting a distinct clock domain. The clock for OGWs ($CLK_{OGW}$) is slow enough to do chip-to-chip signaling, and it is synchronized with the reference clock ($CLK_{REF}$) so that all the OGWs connected together are synchronized with each other. The $CLK_{NET}$ is distributed without any skew compensation, and packet is transmitted using mesochronous communication. When a packet traverses from one clock domain to another, a *strobe signal* (STB) is sent together with the packet as a timing reference. And, the DNS and the SYNC compensate phase difference between the STB signal and local clock. This architecture eliminates burden of global synchronization, thus enables the use of high-speed clock, that is 800 MHz in this work, which facilitates performance enhancement of an on-chip system.

**Packet routing scheme**

A packet is transferred to a destination according to the RI field. A response packet from the destination, if necessary, returns to the source along the reverse path. The RI contains a series of the port indexes which are called output port indexes (OPIs). An $OPI_n$ is an index of an output port to which a packet must be routed at the $n$th switch. For a $k \times k$ switch, $k - 1$ port indices are required because the port where a packet has entered does not use port index. Figure 9.4 shows the packet routing process using the RI field. When a packet arrives at a switch, the switch checks an OPI at the head of the packet header. Then, the header is left-shifted to locate the next OPI at the most significant bit part of the packet header, and the port index of the input port (IPI) is attached to the tail. The IPI is a port index which is viewed by the output port. Through this RI modification, each switch can find its OPI at the head of the RI field, which enables cut-through switching. And an NI of a destination PU obtains a header for the return path by bitwise flipping the header. An IPI is a flipped index of an input port, so that the bitwise flipping outputs correct OPIs.

This scheme enables switches not to have routing tables which contain topology-dependent data. Therefore, network design is decoupled from system design, and the network can be easily ported to a system. In conventional routing schemes used in many literatures, routing information header contains source and destination addresses, and each switch has its own routing table. In such a typical scheme, configuration of the routing table depends on the OCN topology, therefore, overall SoC architecture should be determined to fill out the routing table. In other words, the routing table must be re-designed at every use. This means the network design is not completely decoupled from system design. A programmable

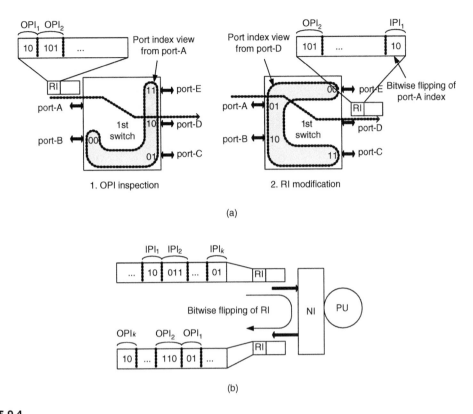

**■ FIGURE 9.4**

(a) Packet routing using RI and (b) RI generation for reverse path.

routing table may be a solution for the decoupling issue. However, it requires a kind of content addressable memory, whose area and power overheads are considerable. It also requires additional protocol burden to program the table. A disadvantage of the BONE routing method is the difficulty in supporting switch-level adaptive routing. If NIs gather traffic information, however, adaptive routing in packet transaction level is possible.

The number of PUs that can be covered by a 14b RI field depends on the network topology while the conventional scheme is fixed to $2^7$. In a *hierarchical star* (H-star) topology, the proposed routing method covers more than $2^8$ PUs regardless of the levels of hierarchy. In a mesh topology, the RI field can cover up to seven switch hops or a $4 \times 4$ mesh network, which is still reasonable for on-chip networking.

**Off-chip connectivity**

OGWs provide chip-to-chip packet transaction without any additional external component. And the inter-chip communication does not require

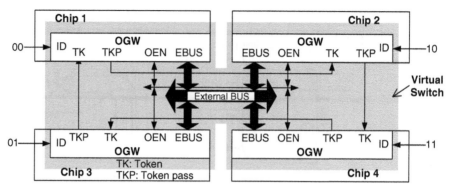

■ **FIGURE 9.5**

Chip-to-chip connection using OGWs.

a specific header type, but it uses the same format used in the core network. Figure 9.5 shows a configuration of multi-chip interconnection using OGWs. The OGWs are connected together constituting a large virtual switch, and inter-chip packet transaction is performed through the virtual switch. An OGW acts like a switch port, and the external bus functions as a switch fabric. Each OGW has its own identification index (ID) which is used like a port index of a switch. Accesses to the bus are arbitrated based on token-ring methodology

If an OGW having a packet to transmit gets a token, it outputs the packet on the bus with assertion of the OEN signal. Other OGWs monitor the OEN, and if it is asserted, check whether the OPI of the packet is matched with their IDs. If they are matched, the OGW accepts the packet and performs header modification.

Using the bus architecture with token-based arbitration, chip-to-chip interconnection can be accomplished only by wires without any external switch or arbiter.

### 9.1.2 BONE-2: Low-Power NoC and Network-in-Package for Multimedia SoC Applications

In large-scale SoCs, the power consumption on the communication infrastructure should be minimized for reliable, feasible and cost-efficient chip implementations. In the BONE-2 project, a hierarchically star-connected NoC is designed and implemented with various low-power techniques. The fabricated chip [3] contains heterogeneous IPs such as two RISC processors, multiple memory arrays, *field-programmable gate arrays* (FPGA), off-chip NIs and 1.6 GHz PLL. The integrated on-chip network provides 89.6-Gb/s aggregate bandwidth and consumes less than 51 mW at full traffic condition. On the other hand, the previous work, BONE-1 [6], consumes 264 mW with 51-Gb/s bandwidth. The ratio of power consumption

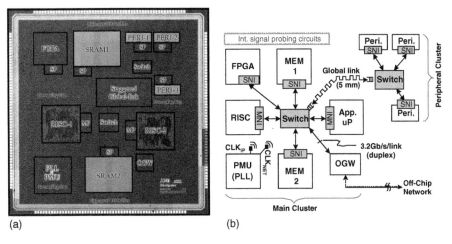

**■ FIGURE 9.6**

BONE-2: (a) chip microphotography and (b) its block organization diagram.

to provide bandwidth of this work is reduced by 10 times from the previous work. In this section, the proposed various low-power techniques for on-chip networks will be explained briefly (Fig. 9.6).

Complicated SoCs with large chip size such as embedded memory logic systems often suffer from their low-yield and high-cost problems. The NoCs have such a modular structure that large system can be divided into several parts or several chips to mitigate such yield and cost problems. In this work, four NoCs are mounted on a single package for larger system integration to form a *networks-in-package* (NiP). The chip-to-chip interconnections in a package exploit low-resistive PCB printed wires for low-latency and low-jitter off-chip communications [2].

For the design of NoCs, there are many levels of design choices to be decided such as a communication protocol, a network topology, a switching style, buffer-depth (in router), a clock synchronization method, a signaling scheme and so on. In this section, a brief summary about what is decided and why such decisions are made at each design stage will be presented based on *low-power consumption of the SoC*.

### Topology selection

The first phase for NoC architecture design is choosing the most suitable NoC topology. In this work, the power and area cost of the most popular topologies such as a multilayer bus, a 2D mesh and a newly proposed H-star topology [5] are examined. According to the analytical evaluation as shown in Fig. 9.7, H-star topology shows the lowest energy consumption under not only uniform traffic but also non-uniform localized traffic condition. The area cost of the H-star is much lower than mesh

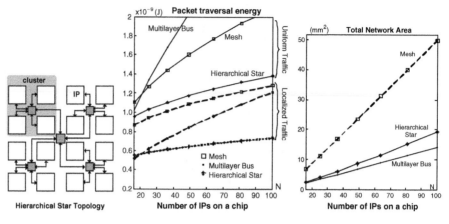

■ **FIGURE 9.7**

Energy and area cost analysis of multilayer bus, mesh and H-star topologies.

and comparable to that of the bus topology. Therefore, the H-star topology is chosen for the NoC platform. It is found cost efficient and has less switching hops.

### Signal synchronization and serialization

A state-of-the-art SoC is a heterogeneous multiprocessing system with multiple timing references, because of the difficulty of global synchronization as well as third-party PUs which are using independent clock frequency scaling. To cope with multiple clock domains, in this implementation, each PU operates with its own clock, $CLK_{Pui}$, but it communicates with a single clock, $CLK_{NET}$. NI changes the timing reference from the $CLK_{Pui}$ to the $CLK_{NET}$ and vice versa. The $CLK_{NET}$ is not synchronized over the chip so that the communications become mesochronous condition – same frequency but different skew. For the mesochronous communications, a source-synchronous scheme is adopted, where an STB goes along with the packet data. The STB is used as timing reference to latch the packet data at the receiver terminal. A packet consisting of 16-bit header, 32-bit address and 32-bit data fields is serialized into 8-bit channel to reduce the network area and power consumption.

### Low-swing signaling on global links

The global link connecting switches are usually a few millimeters long. Therefore, it suffers from longer latency and higher-power consumption than a local link does and makes cross-chip communication increasingly expensive. Low-swing signaling can alleviate the energy consumption significantly and overdriving signaling improves its delay.

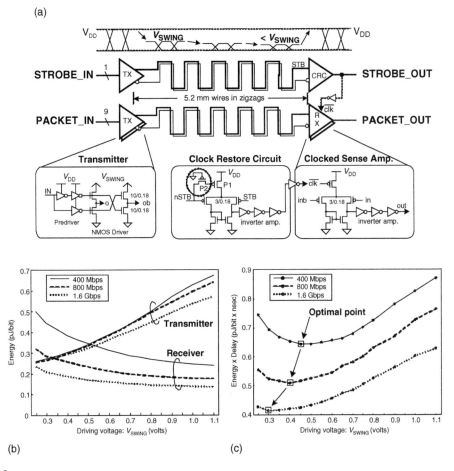

**■ FIGURE 9.8**

Low-swing signaling: circuits and $V_{SWING}$ optimization. (a) Low-swing signaling and its transceiver circuits, (b) Energy consumption versus voltage swing and (c) Energy and delay product.

Figure 9.8(a) shows the implemented differential low-swing signaling. Global wires are laid out in zigzags to emulate a long link as long as 5.2 mm without repeaters. To find out the optimum voltage swing, post-layout simulations are performed with a precise capacitance and resistance wire model [3]. Figure 9.8(b) shows the power consumption on a transmitter and a receiver according to the $V_{SWING}$. Figure 9.8(c) shows energy and delay product. The delay from a transmitter to a receiver is about 0.9 ns and its variations according to the $V_{SWING}$ or signal rates are as small as ±40 ps. As shown in the figure, the optimal voltage swing is 0.45, 0.40 and 0.30 V at 400 Mbps, 800 Mbps and 1.6 Gbps signal rates, respectively. At each operating mode, the driving voltage scales to the optimal voltage level obtained above. Due to the low-swing signaling, the power dissipation on the global link is reduced to one-third of that on a full-swing repeated link. In addition, there are no area-consuming repeaters on the wires.

### Crossbar partial activation technique

Crossbar is widely used in the router as the switching fabric. In the BONE-2, *Crossbar Partial Activation Technique* (CPAT) is used to reduce the power consumption on the crossbar [3]. The CPAT removes unnecessary activation by using tri-state buffers and multiplexers. As a result, 22% power saving is obtained on $8 \times 8$ crossbar.

### Low-energy coding on on-chip serial link

On-chip source-synchronous serial communication has many advantages over multi-bit parallel communication in the aspects of skew, crosstalk, area cost, wiring efficiency and clock synchronization. However, the serial wire tends to dissipate more energy than parallel bus due to the bit multiplexing. In BONE-2, a novel coding method, SiLENT [4], is proposed to reduce the transmission energy of the serial communication by minimizing the number of transitions on the serial wire. The transition reduction coding can save significant amount of the communication energy for multimedia applications. It reduces maximum 77% of energy for instruction memory access and 40–50% of energy for data memory access in a 3D graphics application.

### Operating frequency scaling

PLL generates internal clocks such as a 100 MHz clock for main cluster PUs, a 50 MHz clock for peripheral cluster units, and a 1.6 GHz network clock for switches and NIs. The clock frequencies are scalable for power management modes, that is 100/50/1600 MHz for FAST mode, 50/25/800 MHz for NORMAL mode and 25/12.5/400 MHz for SLOW mode.

### Design flow and methodology

The NoC was designed by semi-custom method. Integrated processors are synthesized, memories are compiled by SRAM compiler, and the on-chip networks are full-custom designed for low power and high performance. Processors and memories are obtained from vendors as IPs and reused by attaching the NI and wrappers. Figure 9.9 shows the design flow: EDA tools, design stage and the output deliverables at each stage. It took 6 months from architecture sketch to tape-out with seven engineers, which manifests the short development time for complicated SoC by using NoC.

### Chip implementation

With the proposed NoC architecture, protocol and low-power techniques, a multimedia SoC is implemented as a prototype. The block diagram is

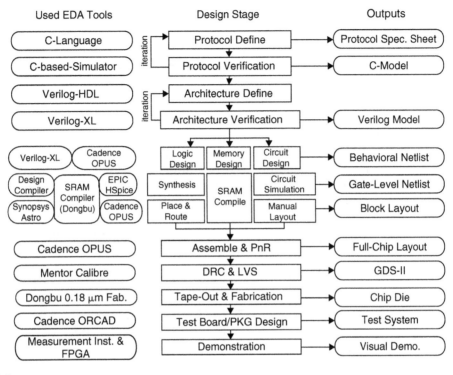

| Used EDA Tools | Design Stage | Outputs |

**■ FIGURE 9.9**

Chip and system design flow.

shown in Fig. 9.6(b). The chip integrates two clusters: a main cluster and a peripheral cluster. The main cluster contains two RISC processors, on-chip FPGA, two 64 kb SRAM and an OGW. Two RISC processors emulate multiprocessor systems. The OGW [6] enables seamless off-chip communications with other NoCs on the same package or boards in order to compose larger-scale systems. By using the OGW, a PU on a die can communicate with other PUs on the other dies without protocol conversion. The peripheral cluster contains three memories to emulate peripheral slave units. The two clusters are interconnected with each other through 5.2 mm global link which uses low-swing signaling to reduce power consumption and also differential signaling for higher signal-to-noise ratio. PLL generates scalable clocks for PUs and networks. PU clocks are not synchronized each other for the emulation of systems with multiple timing references. Therefore, no effort is needed for clock skew minimization. The on-chip network supports 3.2 Gb vtd/s communication bandwidth for each PU and 11.2 Gb/s aggregate bandwidth at FAST mode. The chip is implemented using 0.18 μm CMOS process with 6-Al metal layers and its die area takes $5 \times 5\,mm^2$. The die photograph is shown in Fig. 9.6(a).

The on-chip network power dissipates 51 mW at FAST mode with full traffic condition. By adopting the proposed techniques such as low-swing signaling, crossbar partial activation and serial-link coding, the overall power consumption is reduced by 43%.

**Networks-in-package**

Four NoCs are mounted on a single 676-BGA package as shown in Fig. 9.10(d). The NiP needs four isolated supply voltages: 1.8 V for digital logic, Quiet 1.8 V for analog circuits, 3.3 V for I/O and sub-0.6 V for low-swing links. The operating frequencies are different for each module: 50 MHz for peripheral logic, 100 MHz for processors, 800 MHz for scheduler and 1.6 GHz for on-chip networks.

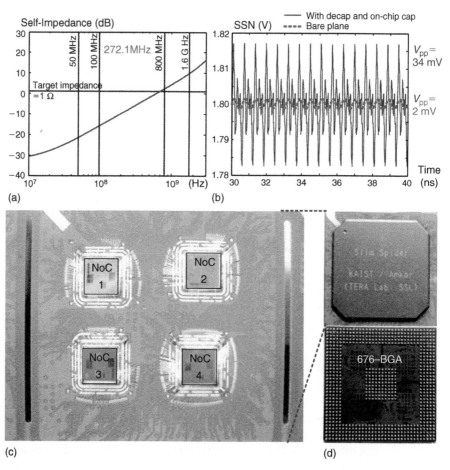

■ **FIGURE 9.10**

NiP simulation (TLM, SSN) and NiP photograph. (a) TLM result of P/G network, (b) SSN result of P/G network, (c) implemented NiP photograph and (d) after molding.

The important issue for the package design is the power integrity, that is a design of *power* and *ground* (P/G) networks. No significant resonance should occur on the P/G plane of the package at the operating frequency. Otherwise small noise from signals or external system causes so significant P/G noise that P/G network becomes unstable. In order to analyze the power integrity, *transmission line matrix* (TLM) method and *simultaneous switching noise* (SSN) analysis are performed (see Fig. 9.10(a) and (b)). Decoupling capacitors for each power voltage is integrated at the proper positions: five for logic power, two for I/O power and one for analog power. Each capacitor has 10 nF capacitance and 600 pH effective series inductance properties. Figure 9.10(a) shows the self-impedance of the power plane. Target value of the self-impedance of the power plane is 1 Ω. The solid line is for bare P/G plane and the dotted line is for the proposed decouple capacitors insertion. The impedance of the bare plane shows inductance characteristics and exceeds the target impedance at 800 MHz and 1.6 GHz. After the insertion of the decoupling capacitance, the impedance resonance occurs at 272.1 MHz that comes from the L of bare plane and C of the inserted capacitors. As a result, the self-impedance at the target frequency becomes lower than 1 Ω. Figure 9.10(b) shows the SSN analysis results when 1.6 GHz input signals are input. The switching noise is reduced from 34 to 2 mV due to the decoupling capacitor insertion on package and on-die. Ground lines are inserted between high-frequency signal lines to eliminate crosstalk.

**Measurements and demonstration of system operation**

Figure 9.11(a) shows the measured packet signals on the network at FAST mode, and Fig. 9.11(b) shows the measured packets with and without the proposed SiLENT coding, respectively, while a 3D graphics application is running on the system [9]. Transitions on a channel are reduced from 134 to 79 after the coding.

A demonstration system is developed with the implemented NiP for multimedia applications. Figure 9.12(a) shows the demonstration system which consists of NiP-board on top layer, video board on bottom layer and LCD panel module. The system demonstrates image processing and animation processing on the LCD through networks on the chip and networks in the package. The Fig. 9.12(b) shows packet transactions between two NoCs on the NiP where the two NoCs are running at different clock frequencies, for example 400 and 274 MHz.

**BONE-2 chip summary**

A low-power packet-switched NoCs and NiP with H-star topology are designed and implemented for high-performance SoC platform. The chip contains two RISC processors for multiprocessor emulation, two 64 kb

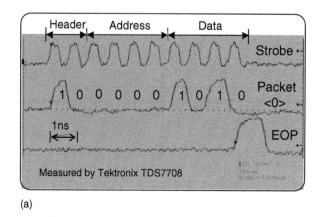

(a)

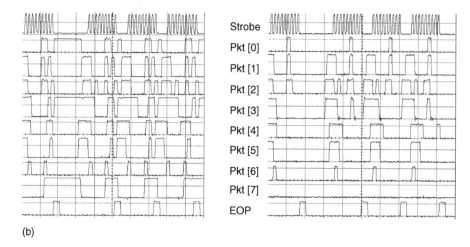

(b)

■ **FIGURE 9.11**

(a) Measured packet signals and (b) the effect of SiLENT coding (without SiLENT coding (134 transitions) and with SiLENT coding (79 transitions)).

SRAMs, on-chip FPGA, OGW for off-chip NI, three 4 kb SRAM for peripheral logic emulation, 1.6 GHz PLL for internal clock generation and on-chip networks connecting those PUs. On-chip network channel is serialized from 80 bits onto 8 bits to reduce the network area significantly. Source-synchronous signaling enables plesiochronous communications between PUs running at different clock frequencies. Low-power consumption is achieved by applying various techniques such as a lower-swing signaling link, crossbar partial activation, low-energy serial-link coding and clock frequency scaling. The chip integrates 2.5 million transistors and consumes less than 160 mW and the on-chip network consumes less than 51 mW delivering 11.2 Gb/s aggregated network bandwidth. The $5 \times 5$ mm$^2$

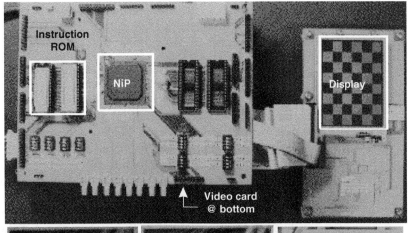

(a)

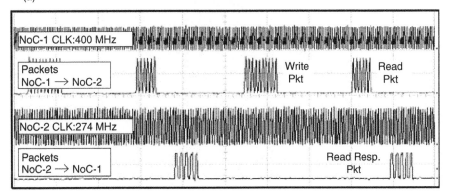

(b)

■ FIGURE 9.12

(a) Demonstration system: board, snapshots and (b) NiP signals between two NoC dies.

chip is fabricated with 0.18 μm CMOS process and successfully measured and demonstrated on an NiP system evaluation board running real-time multimedia applications.

### 9.1.3   BONE-3

The uniqueness of BONE-3 shown in Fig. 9.13 comes from the adaptive control schemes for high-speed flexible OCN design. Basically, it utilizes the *wave-front-train* (WAFT) scheme for high-speed serialization (see Section 3.3.3). In order to stabilize the WAFT operation against the power-supply-voltage variation, an adaptive reference voltage generation according to the supply-voltage variation is realized. The programmable delay synchronizers are used for run-time calibration of phase difference in a mesochronous communication links. Adaptive bandwidth control schemes are also adopted for effective energy reduction in global link wires.

**Supply-voltage-dependent reference voltage**

The supply-voltage difference between a sender and a receiver can cause unit delay time difference so that jitter at the receiver can occur. To resolve this problem, a delay element with current-starving inverter chain and a

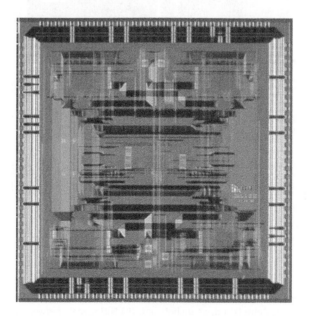

■ **FIGURE 9.13**

Die photograph of BONE-3 [8].

reference voltage generator of Fig. 9.14 are adopted for supply-voltage-independent delay time.

For supply-voltage-independent delay, the bias current of the current-starving inverter is controlled adaptively according the values of the supply voltage. Using this scheme, the receiver jitter is reduced from 30% to 11% for 10% supply-voltage variation.

### Self-calibrating phase difference

For mesochronous communication, many works used FIFO synchronizers or single-pipeline synchronizers. However, all of the solutions suffer from considerable additional power consumption and/or area. Otherwise, the success or failure of synchronization is random.

Instead of using such passive schemes, the BONE-3 performs self-calibrating for the skew between two clock domains so that it can actively adjust phase of input signal. As shown in Section 3.2.3, a *variable delay*

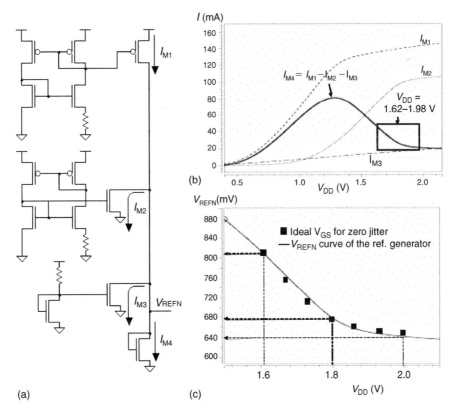

(a)  (b)  (c)

■ **FIGURE 9.14**

(a) Reference voltage generator, (b) current and (c) voltage profile.

(VD) is connected with a simple pipeline synchronizer, and the VD is controlled according to the network condition. The appropriate VD setting for a certain circumstance is obtained through calibration process. When the NoC turns on, it starts initialization which includes the self-calibration routine. When the NoC changes its configuration and the skew between the clock domains changes, the NoC also changes the VD setting according to the NoC configuration or environment.

**Adaptive link bandwidth control**

Bandwidth of a link is a function of the value of the supply voltage. When a network transfers packet signals through a link, the output bandwidth of a transmitter ($B_{OUT}$) must be lower than the maximum bandwidth of the link ($B_{MAX}$). In other words, the $B_{MAX}$ has to be slightly higher than the $B_{OUT}$. Since the $B_{OUT}$ varies according to traffic demands of PUs, the $B_{MAX}$ is set to the highest $B_{OUT}$ in conventional NoCs. The BONE-3 NoC, however, adaptively controls supply voltage of a link to lower the $B_{MAX}$ equal to current requirement of $B_{OUT}$ as shown in Fig. 9.15. Therefore, the power consumption in links is effectively reduced. The NoC link uses two kinds of supply voltages: 1.2 and 1.8 V. At normal mode, a serializer and a deserializer use 1.2 V while supporting the maximum bandwidth of 1.1 Gb/s. At this mode, the data rate is bounded to 0.8 Gb/s. When the traffic is congested, the link changes the supply voltage to 1.8 V so that the link supports the maximum bandwidth of 1.9 Gb/s. At this mode, the data rate is bounded to 1.6 Gb/s. At the normal mode, the power consumption of reduced by 57% compared to the enhanced bandwidth mode.

## 9.1.4    FONE: Flexible On-Chip Network

### NoC evaluation platform

NoC involves a complex design process such as the selection of suitable topology, switch parameters and communication protocol. Evaluation and optimization of such design parameters are essential for the efficient design of an application-specific SoC. An efficient NoC emulation platform is required to verify, evaluate and optimize a variety of application-specific NoC solutions. Moreover, an FPGA based emulation platform gives opportunities to offer and test a sufficient range of choices of NoC design parameters as well as IPs for various applications within a very fast execution time.

NoC evaluation board is implemented on three Altera Stratix EP1S60 series FPGAs to explore and evaluate a wide range of NoC solutions as shown in Fig. 9.16. The implemented system has various IPs: two masters (RISC CPU and LCD controller) and four slaves (*3D graphics processor* (3D-GP), SRAM, Flash and UART). The integrated NoC uses OCS technique to reduce the network area significantly and also supports

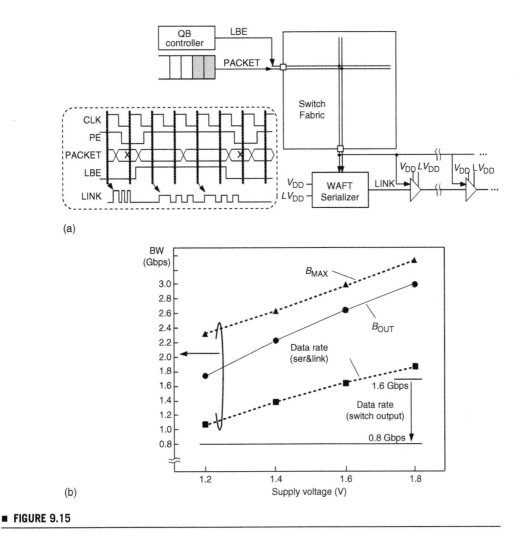

(a)

(b)

**■ FIGURE 9.15**

(a) Adaptive bandwidth control and (b) bandwidth variation according to the supply voltage.

plesiochronous communications among IPs for *globally asynchronous locally synchronous* (GALS) system. All implemented modules are designed in RTL level and six different clocks are used for the GALS operation.

The operation of this system is as follows. The 3D-GP is initialized by the RISC and then its instructions are fetched from the SRAM. After the rendering calculation of the 3D-GP, 3D scenes are stored in the SRAM. All of the transactions between the 3D-GP and the SRAM are conducted by the RISC because the 3D-GP is designed as a slave IP. The LCD controller continuously reads the SRAM frame memory and displays the 3D scenes on an LCD screen. As a result, 3D graphics applications are

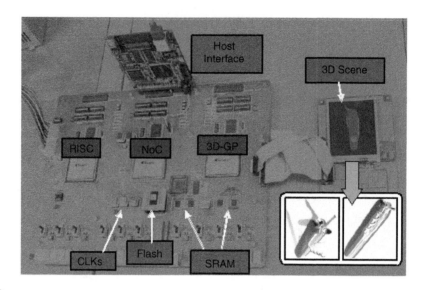

■ **FIGURE 9.16**

Implemented NoC evaluation board.

demonstrated on the NoC evaluation board, which shows feasibility of NoC in the implementation of a real system.

### NoC run-time traffic monitoring system

The NoC evaluation board contains a run-time traffic monitoring system for the accurate evaluation and optimization of an application-specific NoC. Three traffic parameters – (1) an end-to-end latency of each transaction, (2) a backlog in each queuing buffer and (3) output-conflict rate at each switch output port – are traced at run-time as the network performance indicators (see Fig. 9.17(a)). In addition, a wide range of dynamic statistics such as a communication bandwidth between integrated IPs, an evaluation time of the application and link/switch/buffer utilization can be also measured.

The main requirements of the NoC traffic monitoring are non-intrusiveness, scalability, real-time tracing and low cost. Figure 9.17(b) shows an example of the NoC traffic monitoring system to meet the requirements. It consists of three subsystems: host interface, central controller and traffic probes. The host interface, a bridge between the central controller and a host PC, transfers traffic monitoring results to the host PC via Ethernet line. The central controller enables/disables each traffic probe based on the requested monitoring regional scope and time interval. A traffic probe is connected to a switch or an NI in order to trace the real-time traffic parameters such as an end-to-end latency, a queuing buffer usage and an output-conflict rate on a switch. Then, the traffic traces are stored

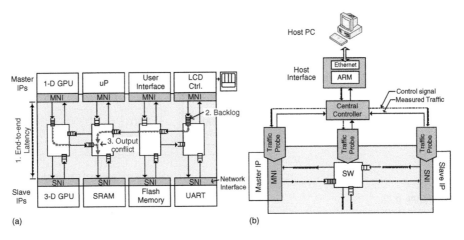

**■ FIGURE 9.17**

(a) Measurable traffic parameters and (b) NoC traffic monitoring system.

in its local trace memory which is accessible by host interface's ARM via the central controller. Since all the monitoring processes do not have any influence on the NoC packet flow, non-intrusiveness probing is achieved.

The traffic monitoring system has modular architecture. Thus, a traffic probe can be attached to any NoC components, which is a design-time choice. Since the internal traffic behavior of NoC core can be measured and analyzed through the monitoring system, the potential bottleneck of the NoC can be examined in the early stage of NoC design and also its design can be optimized by the suitable selection of various design parameters such as queuing buffer size, packet priority assignment and IP mapping. It will provide energy-efficient and performance-optimized communication structure for a large range of applications.

### Case study: portable multimedia system

*Target system description*
In this section, a case study in NoC evaluation and optimization is described to demonstrate the effectiveness of the NoC evaluation framework. A portable multimedia system is implemented on the NoC evaluation board with an H.264 decoder (HD) traffic generator as shown in Fig. 9.18. The traffic generator produces the real traces assuming that it decodes CIF (352*288) H.264 baseline profile at level 2 with 30 frames/s. The SRAM is used as a display frame buffer for HD and the 3D-GP. The LCD controller directly reads the frame data from the SRAM with burst operation (burst length = 8) continuously. This transaction has a hard real-time requirement to keep the display frame rate. User interface generates burst packets with burst length of 16b, once per 10,000 cycles.

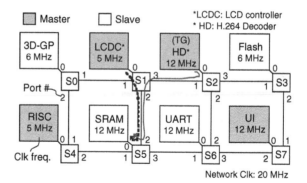

■ **FIGURE 9.18**

Portable multimedia system in mesh topology.

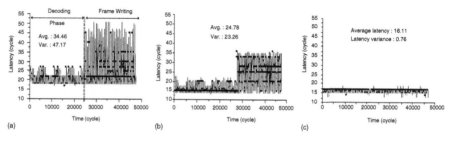

■ **FIGURE 9.19**

Latency distribution, average latency and its variation. (a) H.264 decoder → SRAM, (b) LCD controller → SRAM and (c) 3D-GP → RISC.

HD decodes encoded video stream which is already downloaded in the Flash memory. During the decoding process, the HD accesses the frame data in the SRAM with the bandwidth of 10 Mps. After the decoding process is completed, the HD writes the decoded 2D scenes into the SRAM frame memory with 145 Mps bandwidth.

*NoC evaluation*
Figure 9.19 shows latency distribution of three selected flows. The first two traffic flows from the HD and LCD controller to the SRAM experience relatively long latency and also large variations. After the decoding process of the HD, the traffic from the HD to the SRAM increases abruptly. Thus, the flow from the LCD controller to the SRAM is also affected seriously because the two flows are sharing a link between SW1 and SW5. Meanwhile, the third traffic flow from the 3D-GP to the RISC shows smaller and more constant latency since the flow does not share the network resource with others.

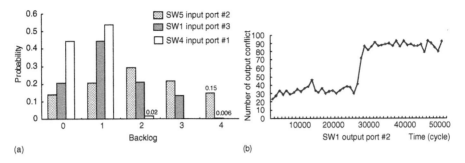

(a)                                             (b)

■ **FIGURE 9.20**

(a) Backlog distribution and (b) output-conflict status.

Figure 9.20(a) shows the backlog distribution on the three flows. Bursty traffics from the HD and the LCD controller to the SRAM cause the SW5 input port #2 queue to be in the full state. On the contrary, the backlog of SW4 input port #1 queue is almost 0 or 1.

Figure 9.20(b) presents the output-conflict counts/1000 cycles on the congested link between SW1 and SW5. After the decoding process is completed in the HD, the output conflict on the shared link increases rapidly (from the 25,000th cycle) and remains as highly congested around 90 conflicts/1000 cycles.

*NoC optimization*

In this section, the target system is optimized in three ways: buffer size optimization, packet priority assignment and topology remapping.

The input queuing buffers in a switch take a significant portion of the chip area of the NoC, thus, their size should be minimized without significant performance degradation. The initial NoC design has all input buffers of four-packet capacity uniformly. Although the uniform choice of the input buffer size is straightforward and widely used in current NoC designs, it may lead to excessive use of silicon area or poor performance. Based on the backlog monitoring results, the optimum buffer capacity of each input queue can be obtained. After the buffer size optimization, 36% total buffer size is reduced (see Fig. 9.21(a) and (b)). In addition, 10% latency reduction and 17% latency-variance reduction are also obtained on the congested flow.

Each packet has a priority field in its header. When more than two packets are destined to the same output port in a switch, that is an occurrence of an output conflict, a packet with higher priority gets a grant to the output port. Therefore, higher packet priority can be assigned to the latency-critical flow for the NoC optimization. To reduce the latency of the most critical flow (from the HD to the SRAM) in our application, a high priority is given to the packets generated by the HD right after the

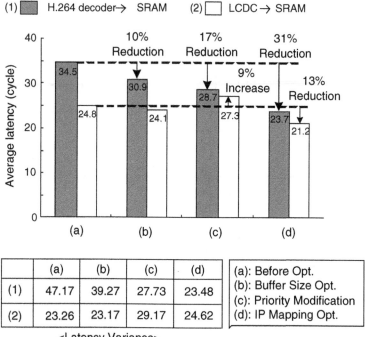

**■ FIGURE 9.21**

Average latency and its variance after the NoC optimization.

decoding process. Figure 9.21(c) shows that 17% average latency reduction is obtained and its variance is also diminished significantly.

If the position of the HD is interchanged with the UART in a given mesh topology, the HD and the LCD controller do not share a link any more. Moreover, the hop count between the HD and the SRAM is also minimized. As a result, the average latency of the HD and the LCD controller flows are reduced by 31% and 13%, respectively, as shown in Fig. 9.21(d).

### FONE platform summary

NoC emulation board is implemented on FPGAs to evaluate and optimize a variety of application-specific NoC designs. It provides dynamic network status such as a backlog, output conflict on a switch and an end-to-end communication latency for each packet flow using a traffic monitoring system. A portable multimedia system is implemented to demonstrate the effectiveness of the NoC evaluation framework. The target system is evaluated and optimized in three ways: buffer size optimization, packet priority assignment and topology remapping. As a result, buffering cost and latency reduction is obtained up to 36% and 31%, respectively.

## 9.2 NoC-BASED EXPERIMENTAL SYSTEMS

We describe here some experimental systems that use NoCs as the backbone for communication.

### 9.2.1 Pleiades: Heterogeneous Reconfigurable Processor for Baseband Wireless Application

The Pleiades processor approach combines an on-chip microprocessor (ARM8 core) with an array of heterogeneous programmable computational units of different granularities (called satellite processors) connected by a reconfigurable interconnect network [12]. It used a reconfigurable switched network. Its configuration is programmable according to the computation algorithm at run-time. It has a two-level hierarchy with mesh structure. Local interconnection is a local mesh with two busses per channel, and a universal switchbox at every intersection point (Fig. 9.22(b)). Global interconnections are supported by a second-level larger-granularity mesh (implemented on the higher metal layers) with two busses per channel and hierarchical switchboxes located at the key connection points. This hierarchical network architecture requires only a limited number of busses to achieve sufficient connection flexibility for target applications, and cuts the interconnect energy cost by a factor of seven compared to a straightforward crossbar network implementation. Communication energy is further reduced by employing a low-swing (0.4 V) pseudo-differential signaling [13]. The wire capacitance loads are also reduced by simplifying the switch network with NMOS-only switches.

### 9.2.2 Raw Machine: Scalable Multiprocessor Architecture

The search for processors that can exploit the increase in *instruction-level parallelism* (ILP) has lead to several creative choices. MIT' Raw processor has replaced the traditional bypass network with a more general interconnect for operand transport. Raw is probably the most significant example of a multiprocessors using a *scalar operand network* (SON), which can be defined as the set of mechanisms that joins the dynamic operands and operations of a program in space to enact the computation specified by a program graph [11]. Recent SONs incorporate point-to-point mesh interconnects, namely use NoCs for internal data and instruction transfers.

Multiprocessing has addressed the issue of scalability by distributing resources and by pipelining paths between distant components. Such approaches do not solve completely the bandwidth scalability problem, which occurs when the amount of information that needs to be transmitted and processed grows rapidly with the size of the system.

The MIT Raw processor [10] of Fig. 9.23 addresses bandwidth scalability using an NoC SON with a mesh topology. We report here on two

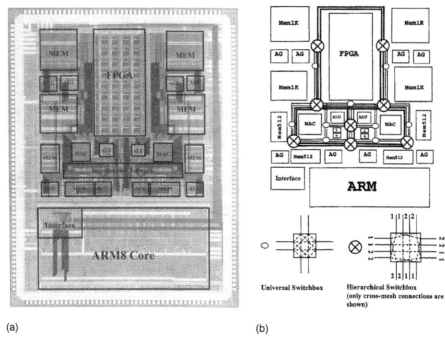

(a)                                                    (b)

■ **FIGURE 9.22**

Heterogeneous reconfigurable processor. (a) Chip micrograph and (b) Chip floorplan and switches.

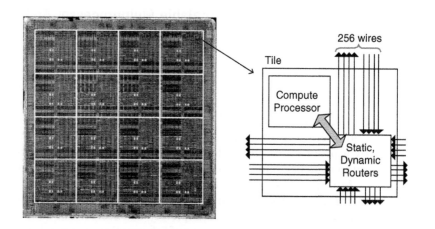

■ **FIGURE 9.23**

Raw processor chip micrograph and a tile organization.

variants of Raw: the former with static and the latter with dynamic transport. Both versions are based on the Raw processor, which consists of a 2D mesh of identical, programmable tiles, connected by two types of transport networks. Each tile is sized so that signals can travel through the tile in a clock cycle. The system can be scaled up by using more tiles, with frequency of operation being constant. Each Raw tile contains a single-issue in-order processor and a number of routers. The switching portion of the tile contains two dynamic routers (for two dynamic transport networks) and one static router (for a static transport network). The static version or Raw uses the static router, whereas the dynamic version uses one of the dynamic routers. A credit system is used to prevent FIFO overflow. The additional dynamic router is used for cache-miss traffic. Indeed, misses are turned into messages that are routed of the side of the mesh and eventually to the off-chip distributed DRAMs.

A mesh interconnect is overlaid on the tiles. Since all links can be programmed to route operands only to those tiles that need them, the bandwidth that is required is much smaller as compared to a shared bus approach. Each of the two SON relies upon a compiler to assign operations to tiles and to program the network transports to route operands between the corresponding instructions. Thus, assignment in Raw can be defined as static. Furthermore, Raw uses in-order execution. Both versions of Raw (static and dynamic) support exceptions. Branch conditions and jump pointers are transmitted over the NoC-like data. Raw's interrupt model allows each tile to take and process interrupts individually. Cache misses stall the processor that process the miss only.

The dynamic version of Raw uses a dynamic dimension-ordered, wormhole-based routing to route operands between tiles. The raw dynamic router is significantly more complex, because it handles deeper logic levels as compared to the static router to determine the path of the incoming operands.

An experimental version or Raw was realized with 16 tiles. Raw was implemented in a 180 nm technology with six levels of metallization. The area is 330 mm$^2$ and the pinout is 1657. The chip operates at 425 MHz at the nominal voltage of 1.8 V. Each of the 16 tiles contains an 8-stage, in-order, single-issue MIPS-like processor, a stage pipelined FPU, a 32-bit data cache, the three routers and 96 kb of instruction cache.

Tiles are connected to the nearest neighbors, using for separate networks, two static and two dynamic. The links consists of more than 1034 wires/tile.

Software code is generated by Rawcc, the Raw parallelizing compiler. Rawcc takes sequential C or Fortran source programs and schedules the code across the Raw tiles. The mapping of code to Raw tiles includes the following tasks: assigning instruction to tiles, scheduling the instructions on each tile and managing the delivery of the operands. Rawcc models the target network accurately, and minimizes the latency of operand transport

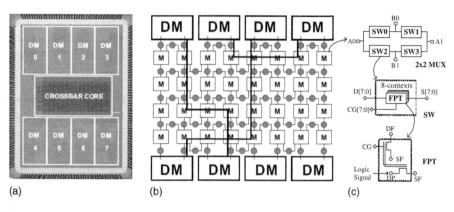

(a)        (b)        (c)

■ **FIGURE 9.24**

On-chip network using Flash-EEPROM switches. (a) Chip micrograph, (b) switch configuration and (c) circuits.

on critical paths of computation. We refer the interest reader to [11] for further details and experimental comparisons of Raw with other processors.

### 9.2.3 Multi-Context Network Using Flash-EEPROM Switches and Elastic Interconnects

The multi-stage reconfigurable crossbar architecture [1] has been implemented by using a combination of multi-context FPT (Flash-Programmable Pass Transistor) switches. Figure 9.24 shows its architecture and circuits; eight programmable finite state machines (DM), a 64b wide $2 \times 2$ multiplexer (M). Circles are symbolic representation of tri-state buffers. The eight finite state machines are programmed either as transmitter or receiver and are connected through the crossbar. The non-volatile multiplexer structures are reconfigured at run-time with eight different contexts changing connection patterns between blocks. The non-volatile switches enables reconfiguration settling time to be shorter than 50 ns. A communication throughput of 6.4 Gb/s for each channel is achieved.

## 9.3 SUMMARY

In this chapter, several SoC implementations based on NoC technology have been proposed so that the NoC can be used as a platform for SoC design. In academia, KAIST's BONE series are explained as the typical example to make the NoC practical. They cover various NoC implementation issues such as low power, synchronization and high-speed signaling techniques. Three NoC-based experimental systems, Berkeley's

Pleiades processor, MIT's Raw machine and STMicroelectronics' multistage reconfigurable crossbar architecture are shown as other trials of NoC implementation.

## REFERENCES

[1] M. Borgatti, et al., "A Multi-Context 6.4 Gb/s/Channel On-Chip Communication Network Using 0.18 μm Flash-EEPROM Switches and Elastic Interconnects," *IEEE International Solid-State Circuits Conference*, February 2003, pp. 466–467.

[2] D. Chung, et al., "A Chip-Package Hybrid DLL Loop and Clock Distribution Network for Low-Jitter Clock Delivery," *IEEE International Solid-State Circuits Conference*, February 2005, pp. 514–515.

[3] K. Lee, et al., "A 51 mW 1.6 GHz On-Chip Network for Low-Power Heterogeneous SoC Platform," *IEEE International Solid-State Circuits Conference*, February 2004, pp. 152–153.

[4] K. Lee, et al., "SILENT: Serialized Low-Energy Transmission Coding for On-Chip Interconnection Networks," *IEEE International Conference on Computer Aided Design*, November 2004, pp. 448–451.

[5] K. Lee, et al., "An Exploration of Hierarchical Topologies for Networks-on-Chip," submitted to *Design, Automation and Test in Europe Conference*, 2006.

[6] S.-J. Lee, et al., "An 800 MHz Star-Connected On-Chip Network for Application to Systems on a Chip," *IEEE International Solid-State Circuits Conference*, February 2003, pp. 468–469.

[7] S.-J. Lee, *Cost-Optimized System-on-Chip Implementation with On-Chlip Network*, Ph.D. thesis, Korea Advanced Institute of Science and Technology, 2005.

[8] S.-J. Lee, et al., "Adaptive Network-on-Chip with Wave-Front Train Serialization Scheme," *IEEE Symposium on VLSI Circuits*, Digest Technical Papers, June 2005, pp. 104–107.

[9] J.-H. Sohn, et al., "A 50 Mvertices/s Graphics Processor with Fixed-Point Programmable Vertex Shader for Mobile Applications," *IEEE International Solid-State Circuits Conference*, February 2005, pp. 192–193.

[10] M.B. Talyor, et al., "A 16-Issue Multiple-Program-Counter Microprocessor with Point-to-Point Scalar Operand Network," *IEEE International Solid-State Circuits Conference*, February 2003, pp. 170–171.

[11] M. Taylor, W. Lee, S. Amarasinghe and A. Agrawal, "Scalar Operand Networks," *IEEE Transactions on Parallel and Distributed Systems*, Vol. 16, No. 2, February 2005, pp. 1–18.

[12] H. Zhang, et al., "A 1 V Heterogeneous Reconfigurable Processor IC for Baseband Wireless Applications," *IEEE International Solid-State Circuits Conference*, February 2000, pp. 68–69.

[13] H. Zhang, et al., "Low-Swing On-Chip Signaling Techniques: Effectiveness and Robustness," *IEEE Transactions on VLSI systems*, Vol. 8, June 2000, pp. 264–272.

# Index

*Note*: Page numbers in italics refer to figures and tables.

Printed and bound by CPI Group (UK) Ltd, Croydon, CR0 4YY

03/10/2024

01040301-0010